FORMAL ASSESSMENT

Norman B. Patterson

International Standard Book Number: 0-395-87206-5

1 2 3 4 5 6 7 8 9 10 BEI 01 00 99 98 97

Evanston, Illinois • Boston • Dallas

Contents

Short Quizzes
The Short Quizzes cover two lessons and are of average difficulty.

Mid-Chapter Tests
For each chapter in the Student Text, there are two Mid-Chapter Tests, Forms A and B, which are of average difficulty.

Chapter Tests
There are three Chapter Tests for each chapter in the Student Text. Forms A and B are of average difficulty with Form B in multiple choice format. Form C is more challenging.

Cumulative Tests
Cumulative Tests are of average difficulty and are provided after every third chapter in the Student Text.

1.2 Short Quiz

Name ______________________

Date ______________________

In Problems 1–3, sort the figures below according to the indicated attribute.

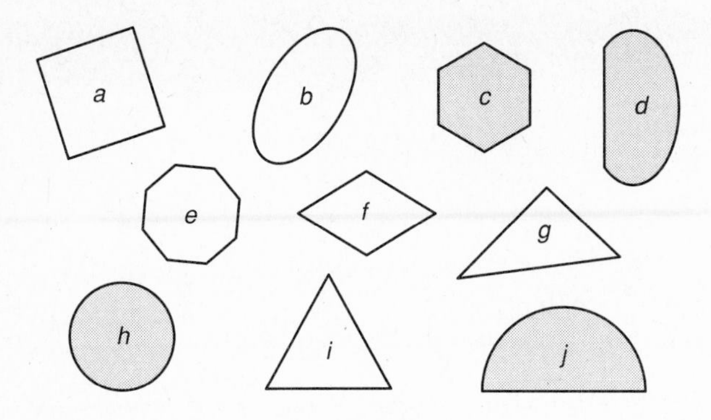

1. Each figure has 4 or 6 sides. **1.** ______________

2. Shaded figures. **2.** ______________

3. Figures having no curved sides. **3.** ______________

4. Add the x and y coordinates of each of the points plotted in the graph. Write the subset of points for which the sum is an even positive number. **4.** ______________

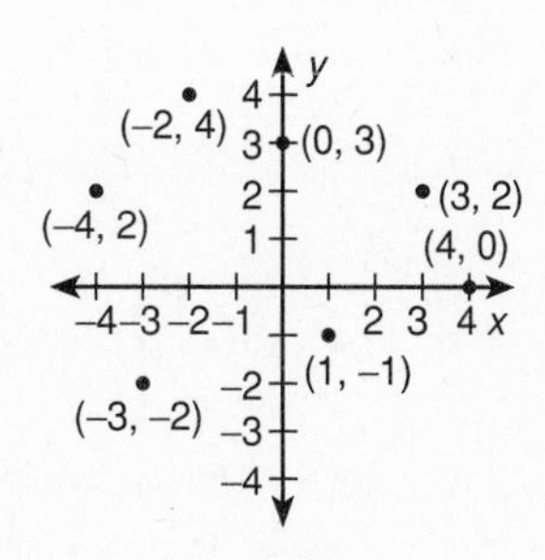

In Problems 5 and 6, classify the angles below according to the indicated attribute.

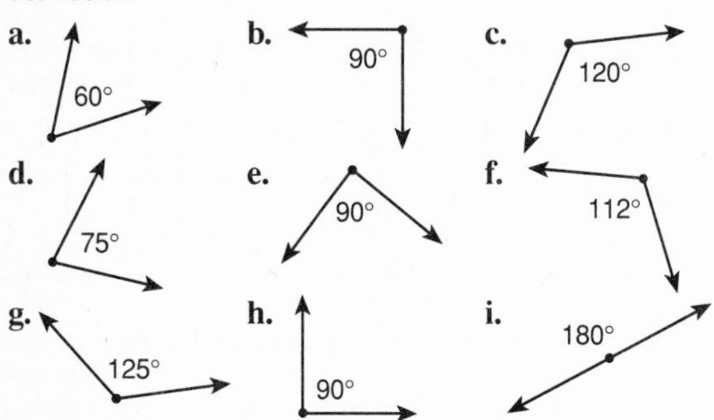

5. Right angles. **5.** ______________

6. Acute angles. **6.** ______________

1.4 Short Quiz

Name ______________________

Date ______________________

1. Divide the region into two congruent parts. Is (are) there one, two, or many solutions?

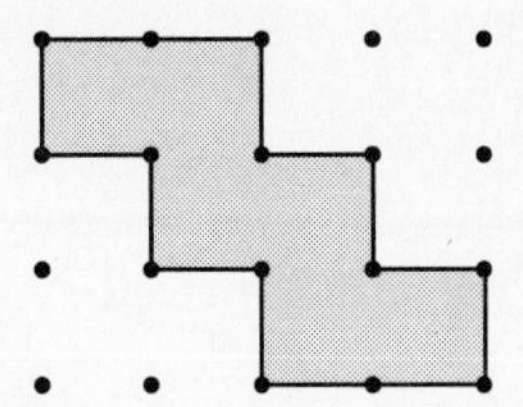

1. ______________ *Use figure at left.*

2. What is the geometric relationship between the image on a photographic transparency and its properly projected image on a screen?

2. ______________

3. Explain why no quadrilateral can be similar to a pentagon.

3. ______________

4. Two of these figures are congruent. Which are they?

4. ______________

a. 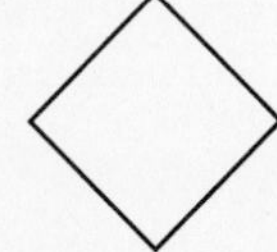

b. 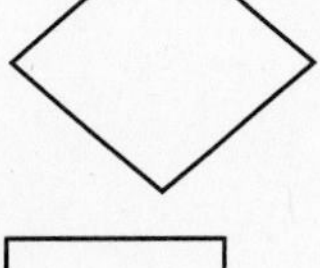

c.

d. 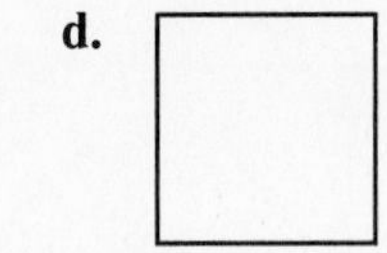

5. Divide the figure into four congruent parts.

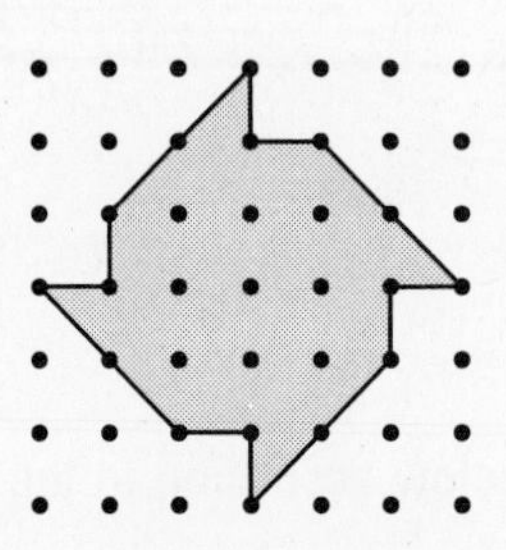

5. *Use figure at left.*

6. Refer to the figure in Problem 5. What type of symmetry does it have, if any?

6. ______________

7. Graph and label the points $A(-3, 4)$ and $B(2, -3)$. Find the coordinates of the midpoint of segment $\overline{AB}$.

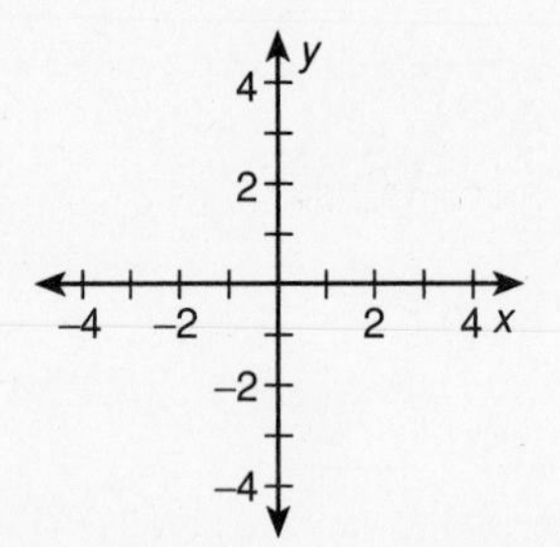

7. ______________ *Use graph at left.*

Mid-Chapter 1 Test Form A

(Use after Lesson 1.4)

Name ______________________

Date ______________________

1. How would you define the difference between an obtuse and an acute angle? Draw figures to illustrate your definitions.

1. 

2. Divide the shaded region into four congruent figures.

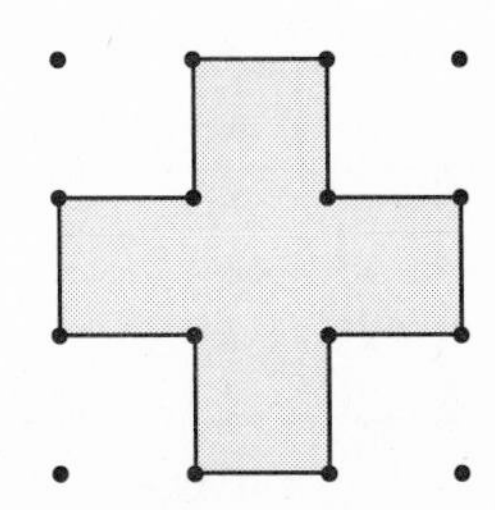

2. *Use figure at left.*

In Problems 3 and 4, shade only one square in each grid to produce the indicated symmetry.

3. Rotational symmetry

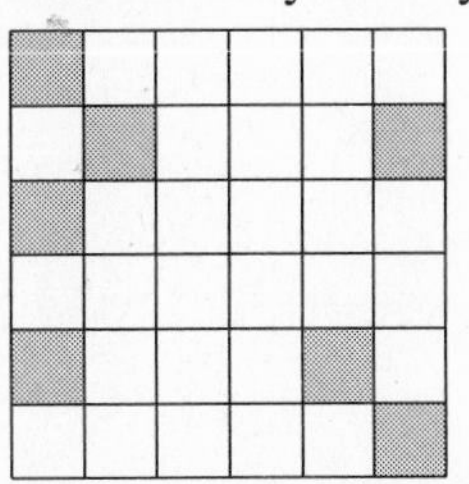

4. Line symmetry

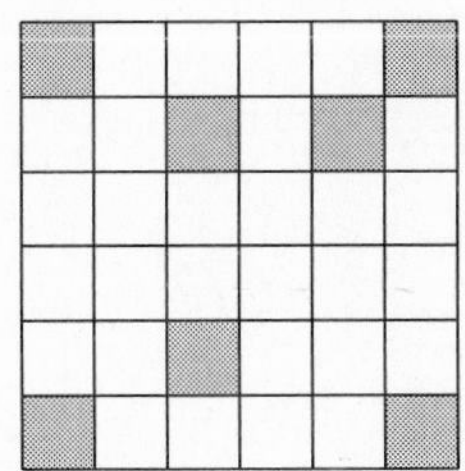

3. *Use grids at left.*

5. Find the midpoint of $\overline{PQ}$, where $P = (-3, 7)$ and $Q = (8, -3)$.

5. 

6. Two of these figures are similar. Two of these figures are congruent. Which pairs meet those descriptions?

a.

b. 

c.

d. 

6. 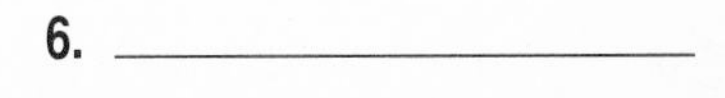

7. Which of the figures shown has line symmetry?

a. 

b. 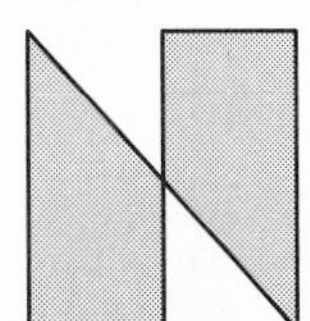

c.

7. ______________________

Mid-Chapter 1 Test Form B

(Use after Lesson 1.4)

Name ______________________

Date ______________________

1. How would you define the difference between a right angle and a straight angle? Draw figures to illustrate your definitions.

1. ______________________

2. Divide the shaded region into two congruent figures.

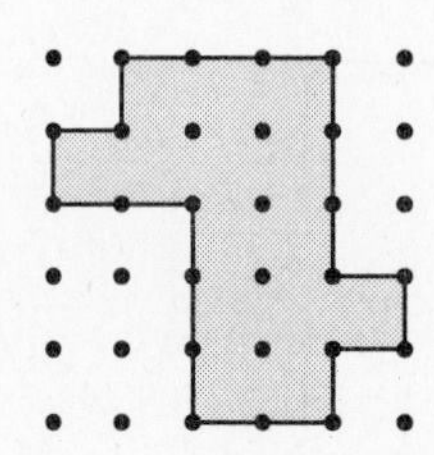

2. Use figure at left.

In Problems 3 and 4, shade only one square in each grid to produce the indicated symmetry.

3. Line symmetry

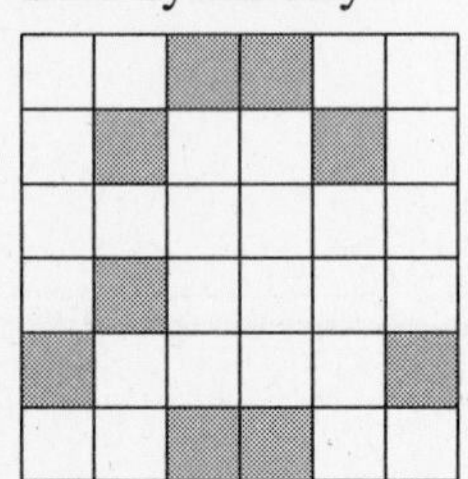

4. Rotational symmetry

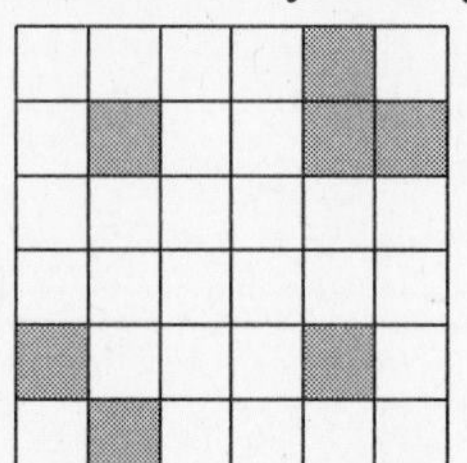

3. Use grids at left.

5. Find the midpoint of $\overline{TS}$, where $T = (5, -4)$ and $S = (8, 0)$.

5. ______________________

6. Two of these figures are similar. Two of these figures are congruent. Which pairs meet those descriptions?

a. 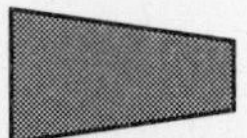

b. 

c. 

d.

6. ______________________

7. Which of the figures shown has rotational symmetry?

a. 

b. 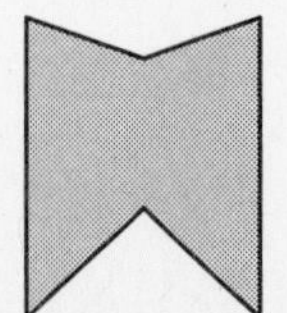

c. 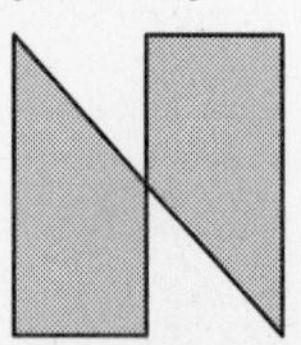

7. ______________________

1.6 Short Quiz

Name ______________________

Date ______________________

1. Find the slope of the line that passes through points $A(-2, 3)$ and $B(4, -1)$.

1. ______________

2. What is the slope of the line $y = -3x + \frac{1}{2}$?

2. ______________

3. Which pair of equations matches the graph?

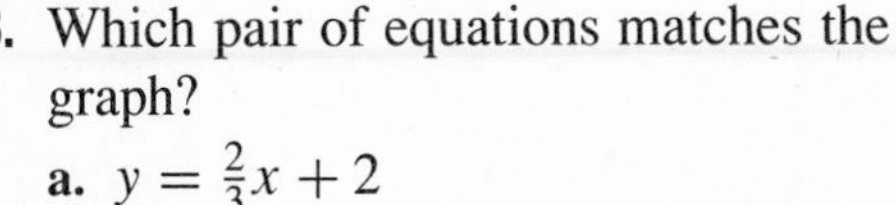

   a. $y = \frac{2}{3}x + 2$
      $y = -\frac{2}{3}x + 2$
   b. $y = \frac{2}{3}x + 2$
      $y = -\frac{3}{2}x + 2$
   c. $y = \frac{3}{2}x + 2$
      $y = -\frac{2}{3}x + 2$

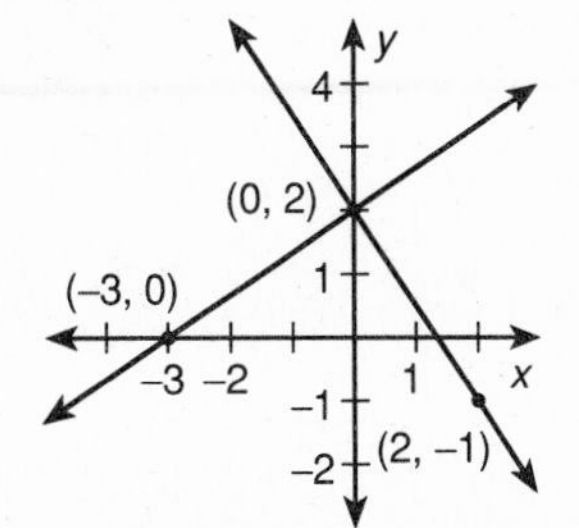

3. ______________

4. Refer to the graph in Problem 3. Are the lines parallel, perpendicular, or neither? Explain.

4. ______________

5. Graph the points $A(-1, -3)$, $B(2, 0)$, and $C(4, 3)$ and decide whether they all lie on the same line. Explain.

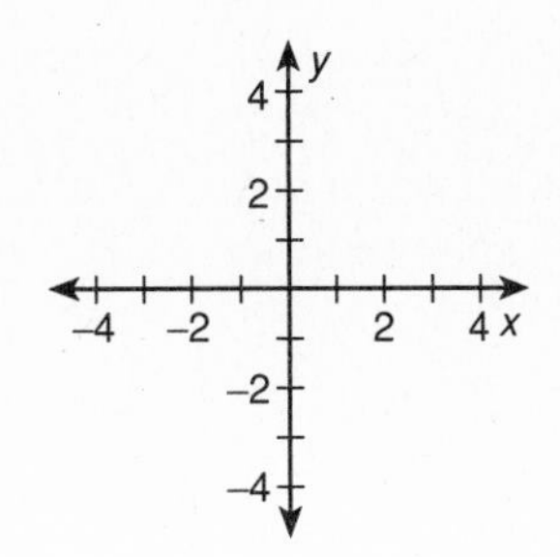

5. ______________
Use graph at left.

6. The dots are one unit apart horizontally and vertically and $\sqrt{2}$ units apart diagonally. Find the perimeter and area of the shaded figure.

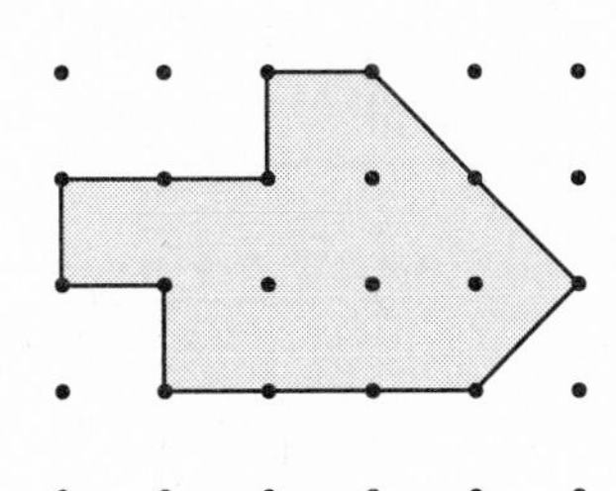

6. ______________

Chapter 1 Test

Form A

(Page 1 of 3 pages)

Name ____________________

Date ____________________

1. Which two figures are exactly alike?

a. 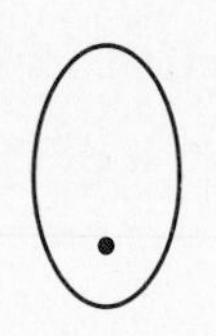

b.

c. 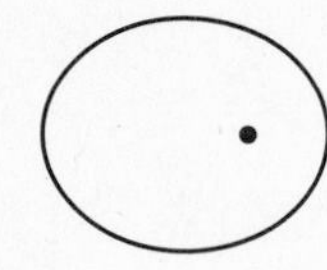

d.

1. 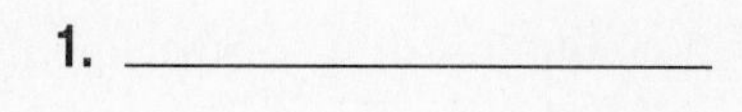

In Problems 2–4, use the figures below.

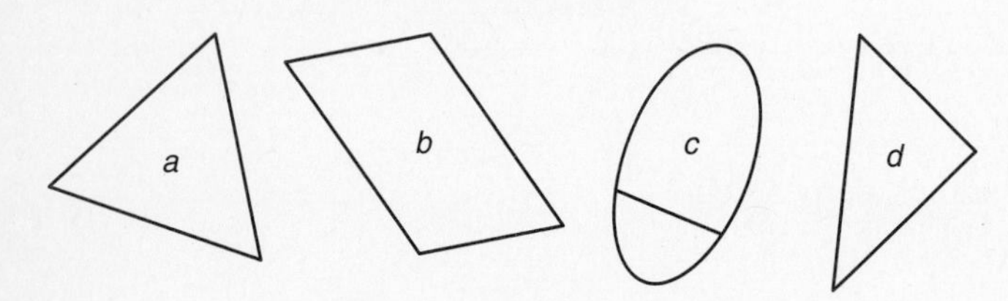

2. Two of the figures shown have line symmetry. Which are they? **2.** ____________________

3. Two of the figures shown have rotational symmetry. Which are they? **3.** ____________________

4. Which, if any, of the figures shown has neither rotational nor line symmetry? **4.** ____________________

5. How many different triangles are in the figure? Are there any similar triangles? Are there any congruent triangles?

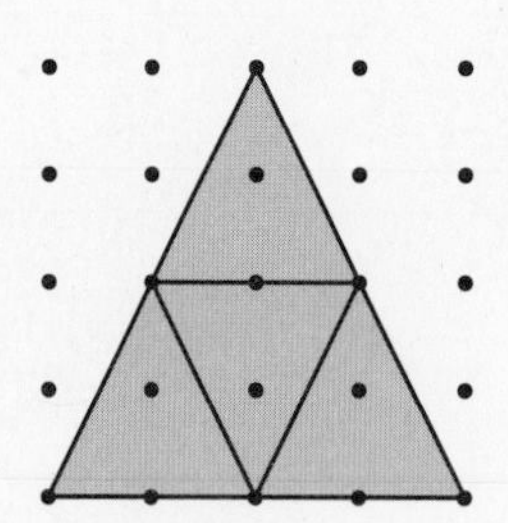

5. ____________________

6. What is the relationship between a full-sized race car and a scale model of it? **6.** ____________________

Name ______________________

Date ______________________

7. A floor plan for a conference room is shown. The scale of $\frac{1}{16}$ inch in the drawing represents 1 foot in the actual room. What are the actual dimensions of the room?

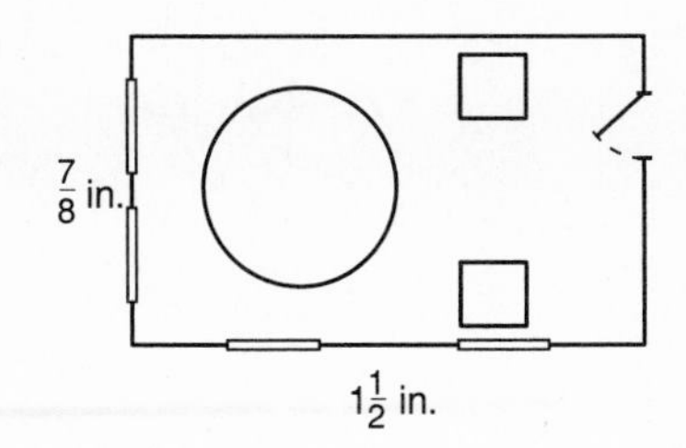

7. ______________________

8. Divide the shaded region into three congruent parts.

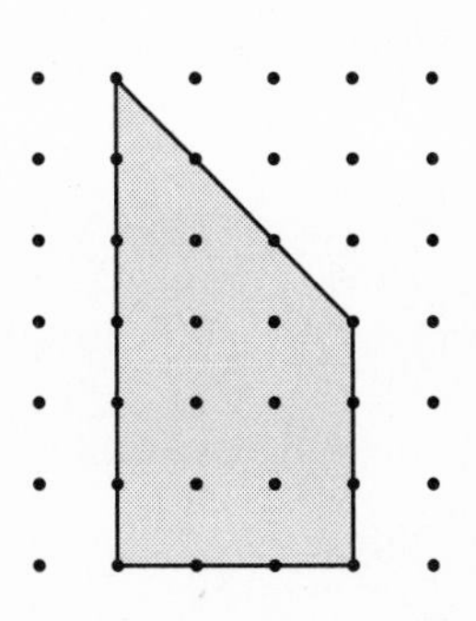

8. *Use figure at left.*

9. Two of these figures are congruent. Which two?

a. 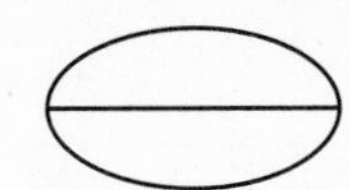

b. 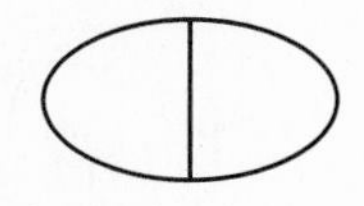

c. 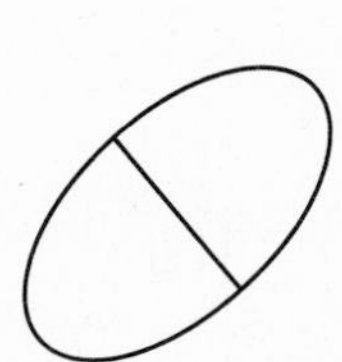

d. 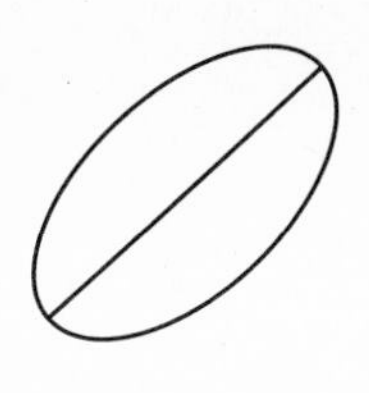

e. 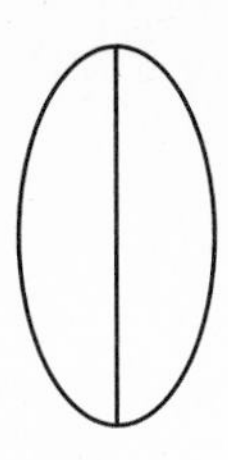

9. ______________________

10. Identify the types of symmetry which the propeller has.

10. ______________________

11. Given the points $A(-7, -3)$ and $B(5, 5)$, find the coordinates of the midpoint of $\overline{AB}$.

11. ______________________

Name ______________________

Date ______________________

12. Does the figure have any lines of symmetry? If so, how many?

12. ______________________

13. Identify any symmetry that the figure appears to have.

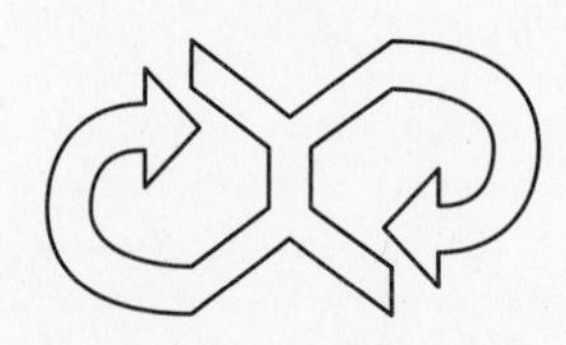

13. ______________________

14. Identify any symmetry in the figure.

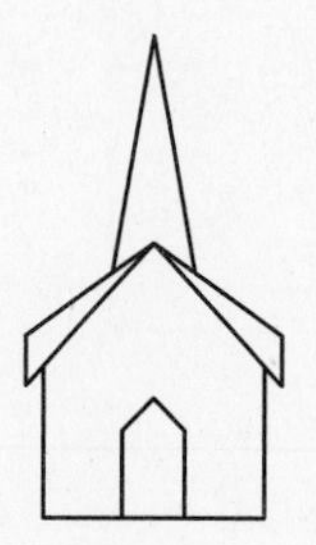

14. ______________________

15. Find the slope of the line that passes through points A and B.

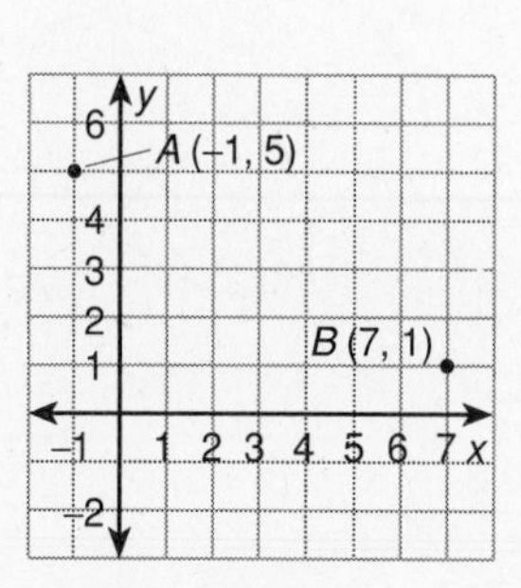

15. ______________________

16. State a relationship between the given lines. Explain.
$y = -\frac{1}{3}x + 2$ and $y = -\frac{1}{3}x - 2$.

16. ______________________

17. A can of paint will cover 70 square feet. How many cans are needed to paint a wall 5 feet high and 98 feet long?

17. ______________________

18. Use construction tools to bisect the angle. Show any construction lines.

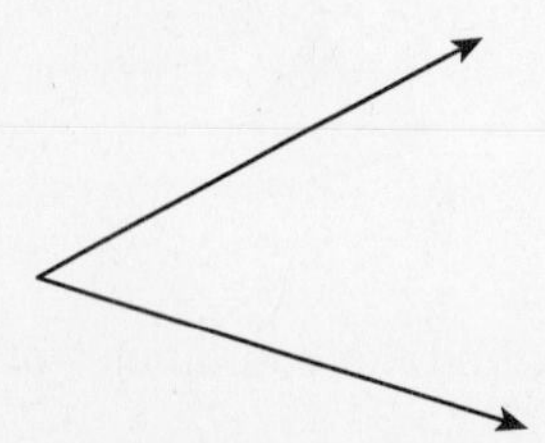

18. *Use figure at left.*

Name ____________________

Date ____________________

1. Which two figures are exactly alike? 1. ______

   a. A and B b. B and C c. A and C

   A. 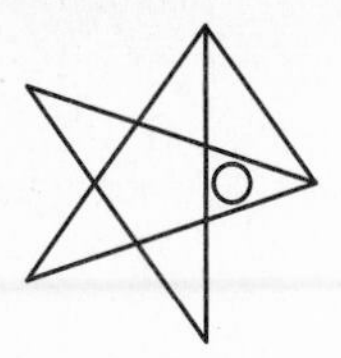 B.  C. 

2. The figure shown [?]. 2. ______

   a. has rotational symmetry b. has line symmetry
   c. is a pentagon d. is a quadrilateral

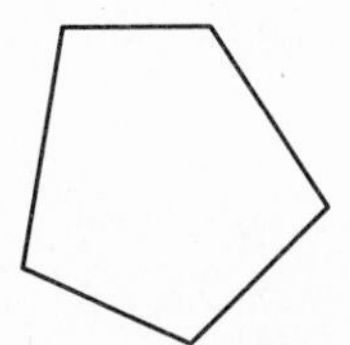

3. The figure shown is [?]. 3. ______

   a. an obtuse angle
   b. a straight angle
   c. a right angle
   d. an acute angle

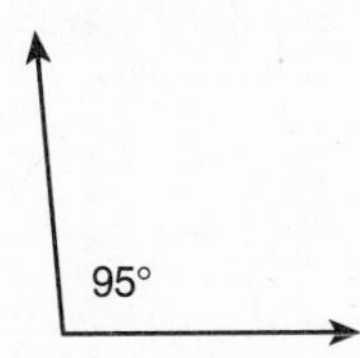

4. A checkerboard has [?]. 4. ______

   a. rotational symmetry only
   b. no symmetry
   c. line symmetry only
   d. both line symmetry and rotational symmetry

5. How many congruent triangles are there in this figure? 5. ______

   a. 5 b. 3 c. 4 d. 6

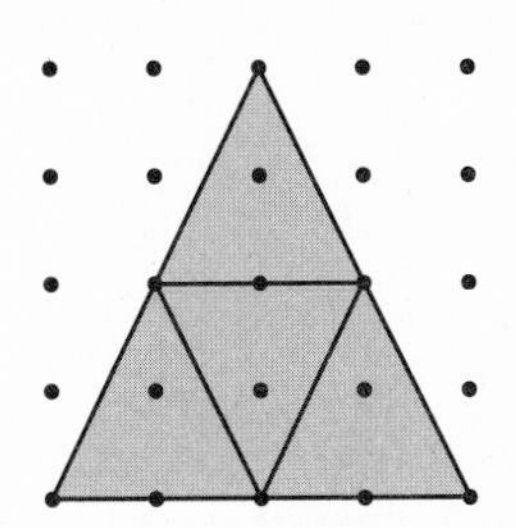

6. A floor plan for a room is shown. The scale of $\frac{1}{16}$ inch in the drawing represents 1 foot in the actual room. The dimensions of the room are [?]. 6. ______

   a. 12 feet by 21 feet
   b. 16 feet by 12 feet
   c. 28 feet by 16 feet
   d. 24 feet by 16 feet

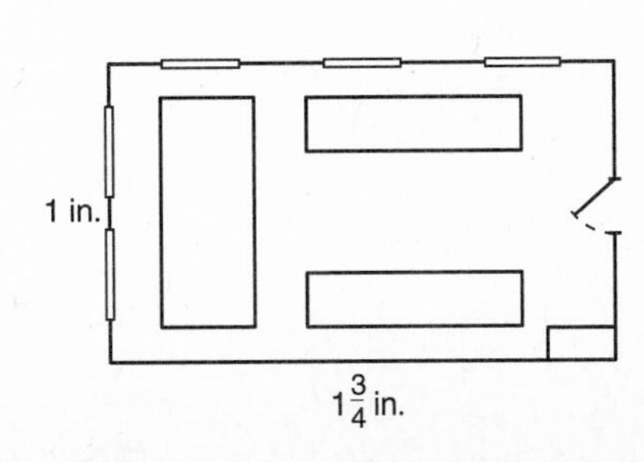

Name ______________________

Date ______________________

7. A full-sized Victorian house and a scale model of it are [?].

a. congruent b. the same size

c. not similar d. similar

7. ________

8. Which two figures shown are congruent?

a. A and D b. B and C

c. B and D d. A and B

8. ________

A.

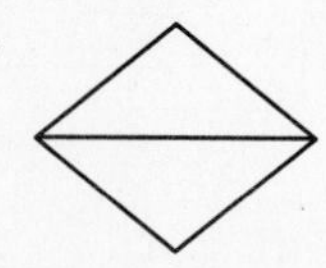

B.

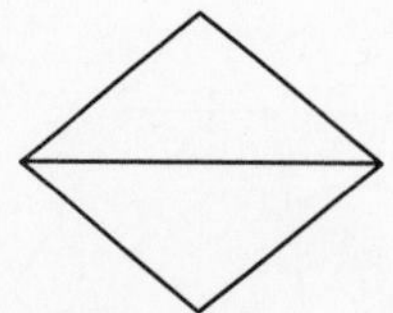

C.

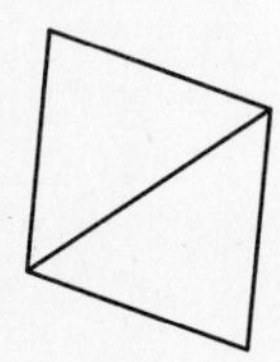

D.

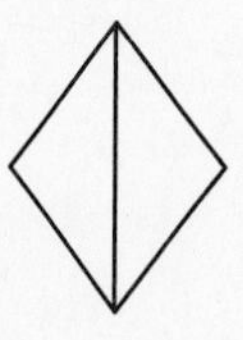

9. One way to divide the shaded region into two congruent parts is with line [?].

a. ℓ_4 b. ℓ_1

c. ℓ_2 d. ℓ_3

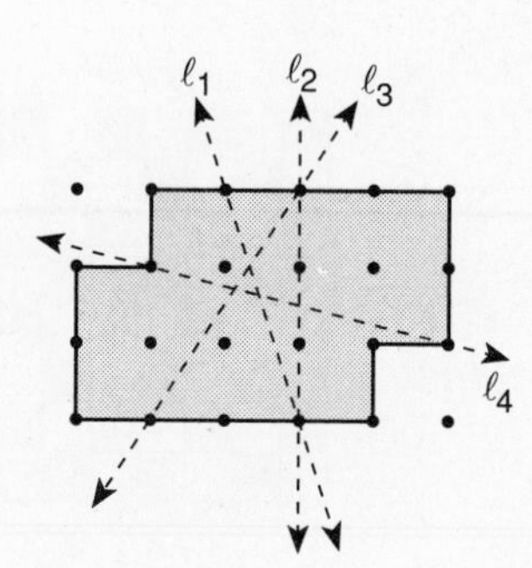

9. ________

10. How many lines of symmetry does the figure appear to have?

a. 0 b. 1 c. 2 d. 3

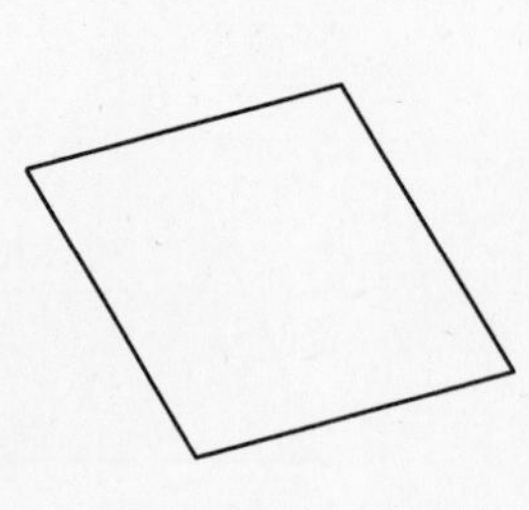

10. ________

11. Given the points $A(-3, 7)$ and $B(5, -1)$, the coordinates of the midpoint of segment $\overline{AB}$ are [?].

a. $(1, 3)$ b. $(4, 4)$ c. $(-1, -3)$ d. $(2, 6)$

11. ________

Name ______________________

Date ______________________

In Problems 12 and 13, use the following figures.

A. 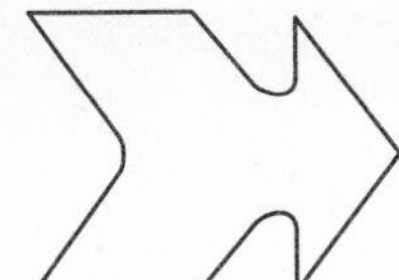 B. 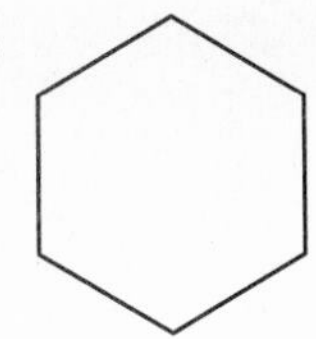 C. 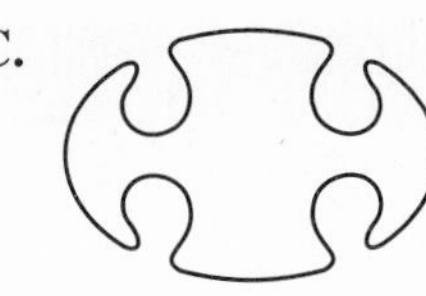 D. 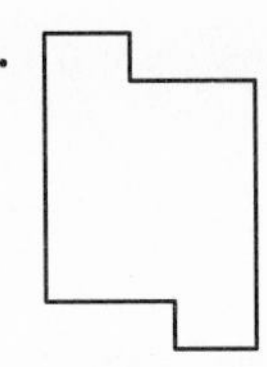

12. Which of the above figures has only *rotational* symmetry?
a. A **b.** B **c.** C **d.** D

12. ________

13. Which of the above figures has only *line* symmetry?
a. A **b.** B **c.** C **d.** D

13. ________

14. This figure has [?].
a. two lines of symmetry
b. horizontal line symmetry
c. rotational symmetry
d. vertical line symmetry

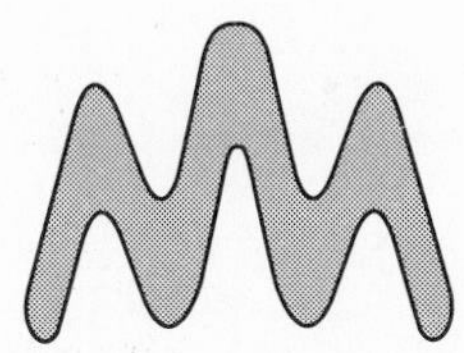

14. ________

15. The line $y = -\frac{1}{2}x + 3$ is perpendicular to the line [?].
a. $y = -2x$ **b.** $y = 2x - 3$
c. $y = \frac{1}{2}x + 1$ **d.** $y = -\frac{1}{2}x + 6$

15. ________

16. The slope of the line that passes through points A and B is [?].
a. $\frac{7}{6}$ **b.** 0 **c.** $\frac{6}{7}$ **d.** $-\frac{6}{7}$

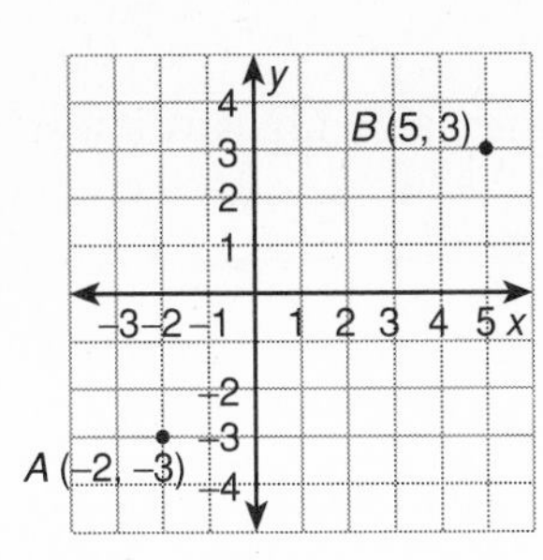

16. ________

17. The construction shown can be used to [?].
a. bisect an angle
b. copy the segment AB
c. copy an angle
d. bisect the segment AB

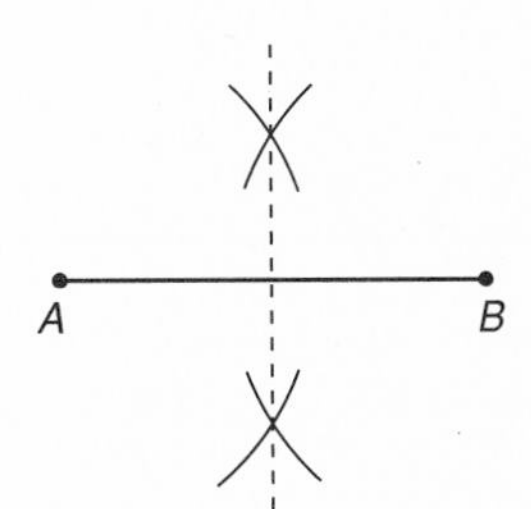

17. ________

Chapter 1 Test **Form C**

(Page 1 of 3 pages)

Name ______________________

Date ______________________

1. Two of the figures are exactly alike. Which two?

a. 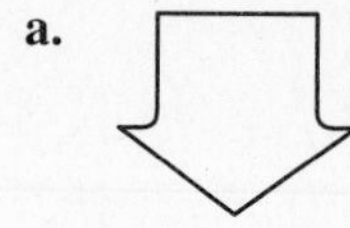

b. 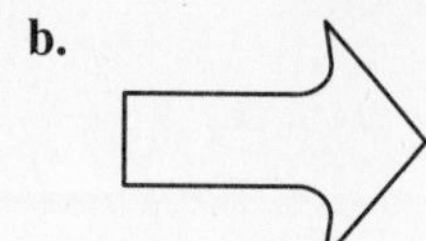

c. 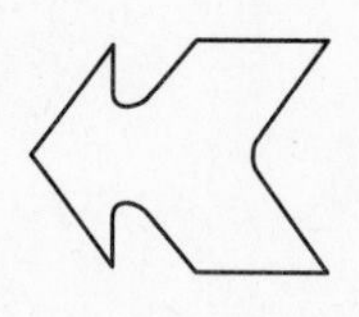

d. 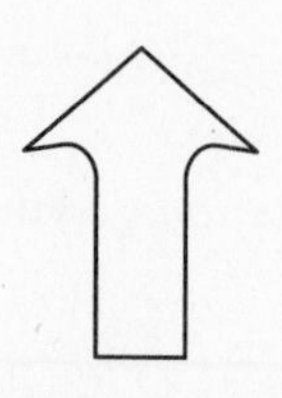

1. ______________

2. Draw a triangle with rotational symmetry. Explain your answer.

2.

3. List the subset of points for which the y-coordinate is the square of the x-coordinate.

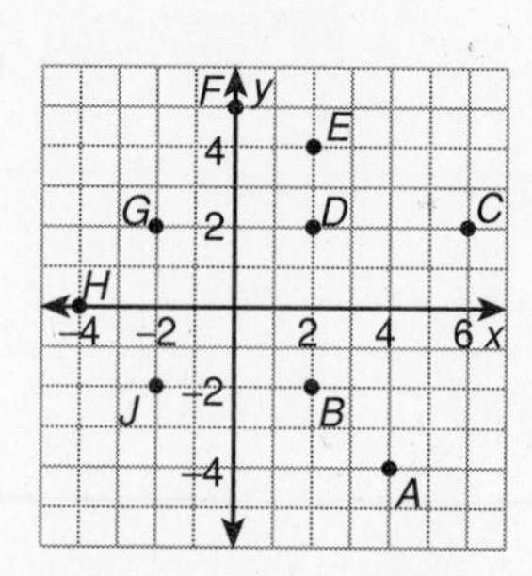

3. ______________

4. Draw a straight angle. State any criteria that makes it a straight angle and show its degree measure.

4.

5. 3×3 squares are to be obtained by cutting up a checkerboard. At most, how many can be obtained?

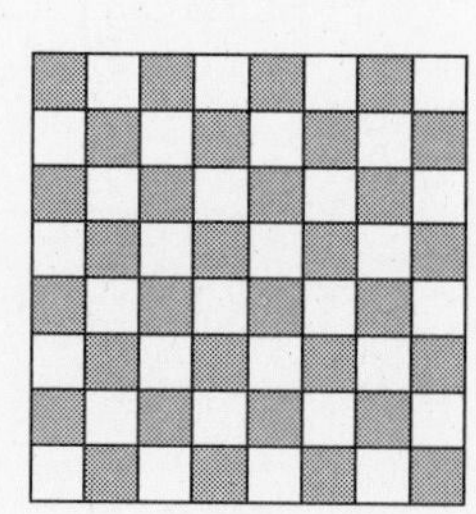

5. ______________

6. What is the relationship between a business letter and a 1 to 1 scale photocopy of it?

6. ______________

Name ______________________

Date ______________________

7. The floor plan for a meeting room is shown. The scale of $\frac{1}{32}$ inch in the drawing represents 1 foot in the actual room. Find the dimensions of the room.

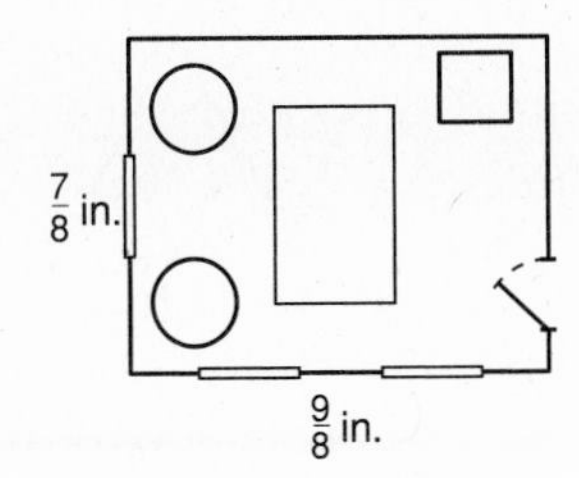

7. ______________________

8. Divide the shaded region into four congruent parts. (Draw appropriate additional lines.)

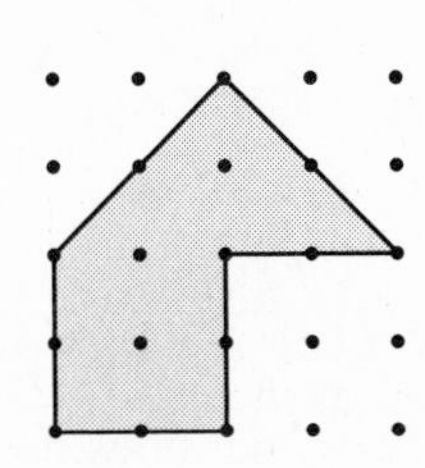

8. *Use figure at left.*

9. Two of the figures are similar. Which two?

a.

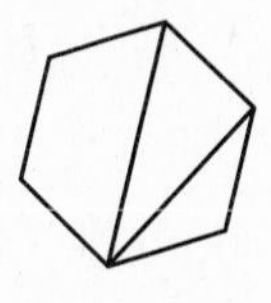

b.

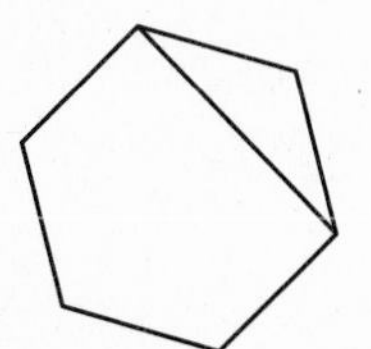

c.

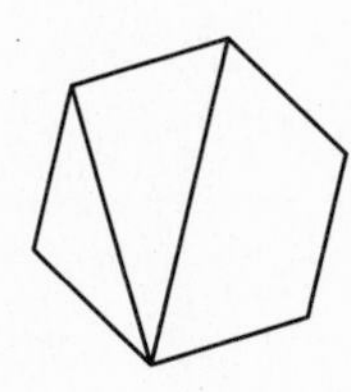

d.

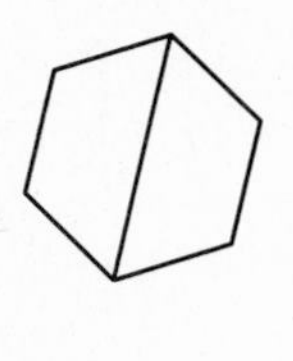

9. ______________________

10. For the badge shown, state the type(s) of symmetry it appears to have. If it has line symmetry, how many lines of symmetry are there?

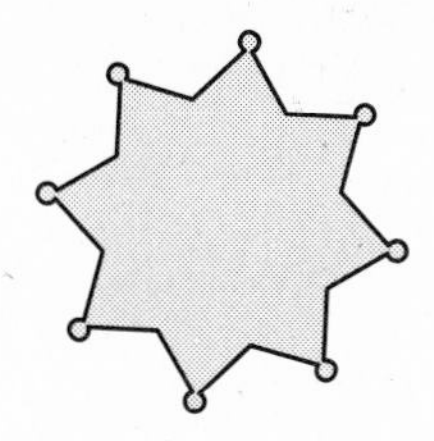

10. ______________________

11. Given the points $A(-2, 7)$ and $B(6, 3)$; if C is the midpoint of the line segment $\overline{AB}$, find the midpoint, D, of the line segment $\overline{CB}$.

11. ______________________

12. For the figure shown, describe any symmetry it appears to have.

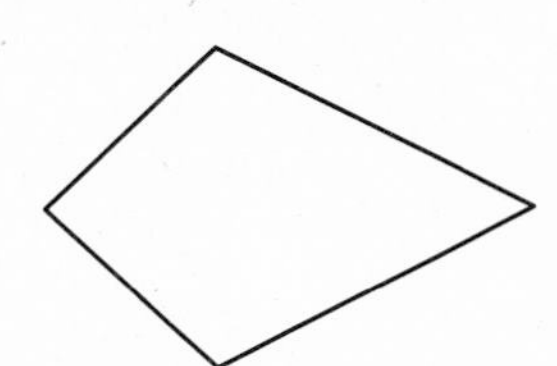

12. ______________________

Name ______________________

Date ______________________

13. In the grid, each small triangle has sides of 1 unit. How many congruent triangles with sides of 2 units are in the grid?

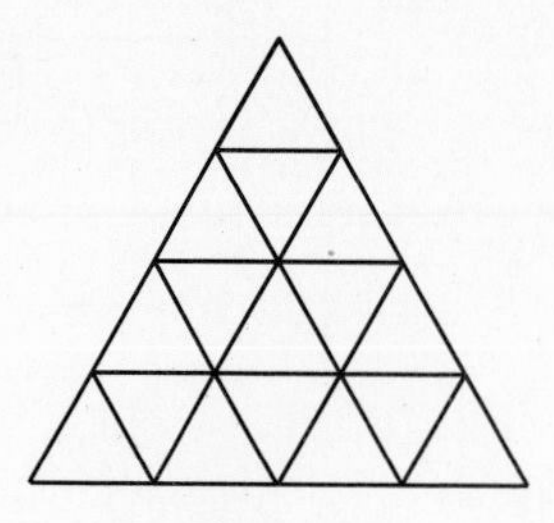

13. ______________

14. Describe any types of symmetry that appear to apply to the figure.

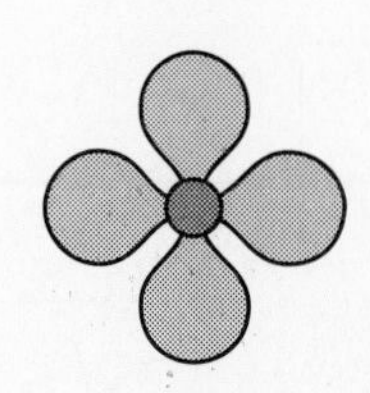

14. ______________

15. Draw a figure which has one line of symmetry, but does not have rotational symmetry.

15.

16. Find the slope of the line passing through points A and B.

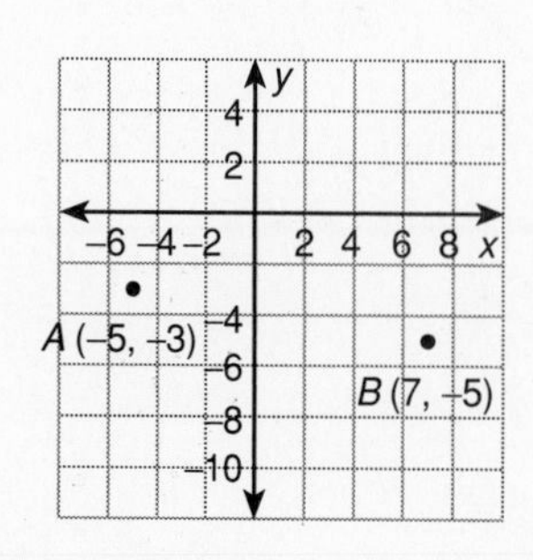

16. ______________

17. State a relationship between the lines $y = -\frac{2}{3}x + 2$ and $y = \frac{3}{2}x - 2$. Explain your answer.

17. ______________

18. A can of paint will cover 108 square feet. How many cans of paint are needed to paint a wall 6 feet high and 75 feet long?

18. ______________

19. Using construction tools, bisect the line segment $\overline{AB}$. Show construction lines.

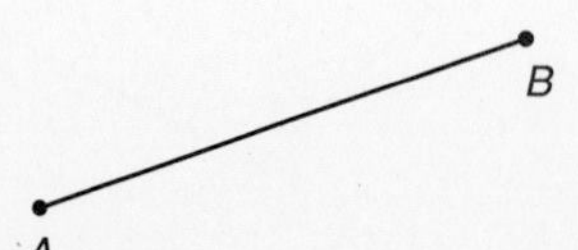

19. *Use figure at left.*

2.2 Short Quiz

Name ______________________

Date ______________________

1. Write a definition for $\overleftrightarrow{TR}$. Illustrate your definition with a sketch. — 1.

2. L, M, and N are collinear. $\overrightarrow{ML}$ and $\overrightarrow{MN}$ are opposite rays. Which statement is true: L is between M and N; M is between L and N; or N is between L and M? — 2. ______________

3. a. Name all of the angles whose vertex is O.
   b. Point B is in the interior of which angle? — 3.

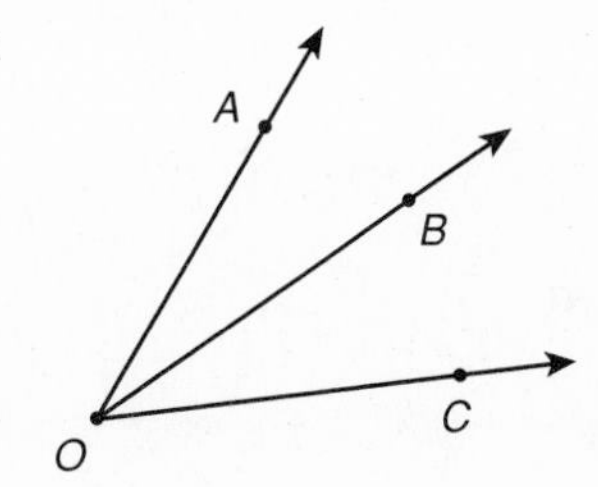

4. L is between J and M. K is between J and L. $JM = 18$, $KL = 2$, and $JK = LM$. Make a sketch and find LM. — 4.

5. B is in the interior of $\angle AOC$. C is in the interior of $\angle BOD$. $m\angle COD = m\angle AOB$, $m\angle AOD = 85°$, and $m\angle BOC = 27°$. Find $m\angle BOD$. — 5.

6. Sketch a point exterior to an acute angle. — 6.

7. Complete the table. — 7. *Use table at left.*

$n = 1$ $n = 2$ $n = 3$ $n = 4$

n	1	2	3	4	5
nth square number	1	4			

Mid-Chapter 2 Test Form A

(Use after Lesson 2.3)

Name ____________________

Date ____________________

1. R, S, and T are three distinct collinear points. R is on $\overrightarrow{ST}$ and R is on $\overrightarrow{TS}$. Draw a sketch to show the position of R with respect to S and T.

1.

2. V is interior to $\angle SOT$. T is interior to $\angle VOW$. W is interior to $\angle TOE$. $m\angle SOE = 160°$; $m\angle TOE = 86°$; and $m\angle SOV = m\angle TOW = m\angle WOE$. Draw a sketch to show the relationship between the specified angles. Find $m\angle VOT$.

2.

3. Find the distance from the midpoint of $\overline{AB}$ to point C.

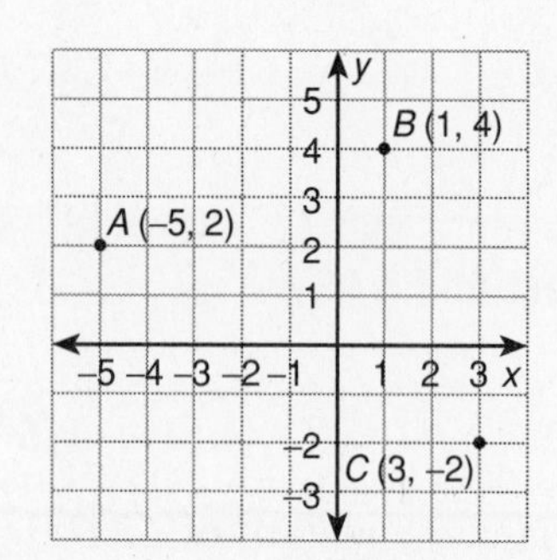

3. ____________________

In Problems 4–6, use the sketch.

Given: $\overleftrightarrow{RP}$ is a line of symmetry of $\triangle RMS$.
$\overrightarrow{SQ}$ bisects $\angle RSM$.
$\overrightarrow{MT}$ bisects $\angle RMS$.

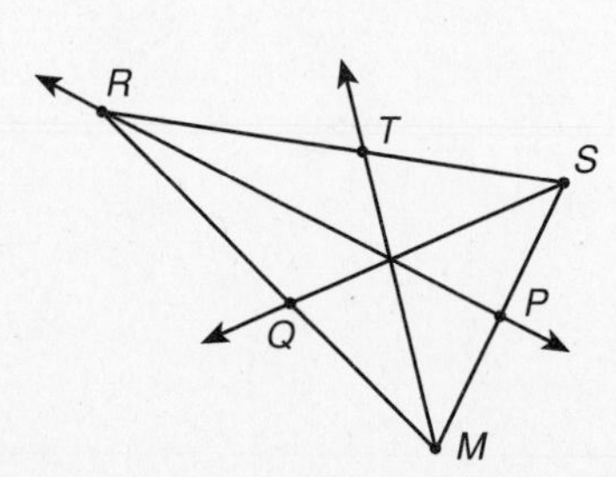

4. Find a segment that is congruent to $\overline{RS}$.

4. ____________________

5. Find an angle that is congruent to $\angle MTS$.

5. ____________________

6. Find a point that is interior to $\angle RSM$.

6. ____________________

Mid-Chapter 2 Test Form B

(Use after Lesson 2.3)

Name ______________________

Date ______________________

1. R, S, and T are three distinct colinear points. S is on $\overrightarrow{RT}$ but S is not on $\overrightarrow{TR}$. Draw a sketch to show the position of R, S and T.

1.

2. W is interior to $\angle SOT$. T is interior to $\angle WOV$. V is interior to $\angle TOE$. $m\angle SOE = 135°$; $m\angle SOT = 66°$; and $m\angle VOE = m\angle SOW = m\angle WOT$. Draw a sketch to show the relationship between the specified angles. Find $m\angle TOV$.

2.

3. Find the distance from the midpoint of $\overline{AC}$ to point B.

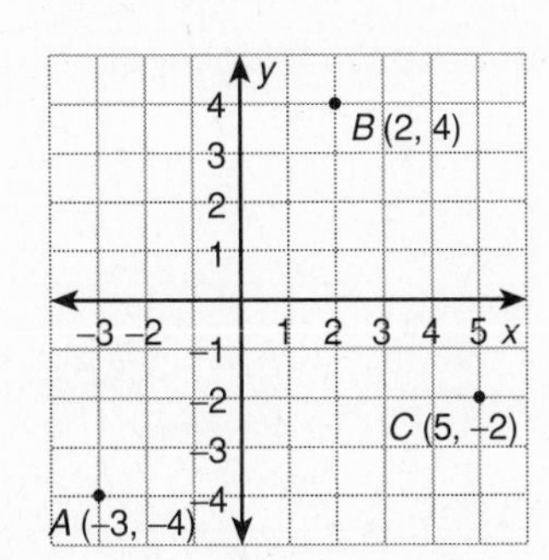

3. ______________________

In Problems 4–6, use the sketch.

Given: $\overleftrightarrow{RP}$ is a line of symmetry of $\triangle RMS$.
$\overrightarrow{SQ}$ bisects $\angle RSM$.
$\overrightarrow{MT}$ bisects $\angle RMS$.

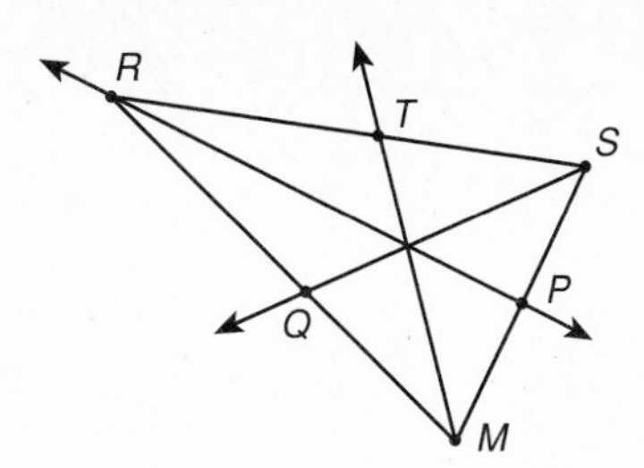

4. Find a segment that is congruent to $\overline{MR}$.

4. ______________________

5. Find an angle that is congruent to $\angle SQM$.

5. ______________________

6. Find a point that is interior to $\angle RMS$.

6. ______________________

2.4 Short Quiz

Name ______________________

Date ______________________

1. Use a ruler and protractor to draw a bisector of the angle shown.

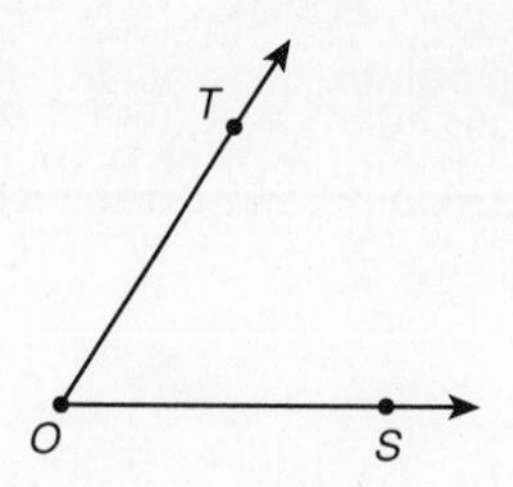

1. *Use figure at left.*

2. Use a ruler and protractor to draw a triangle ($\triangle PQR$) which is congruent to $\triangle ABC$, as shown.

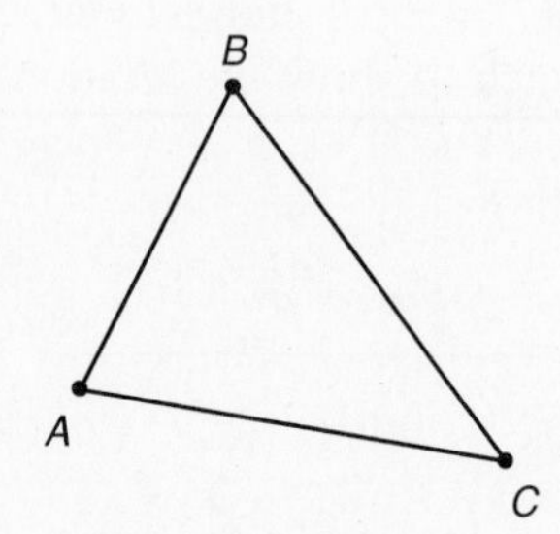

2. ______________________

3. Find the distance between points A and B.

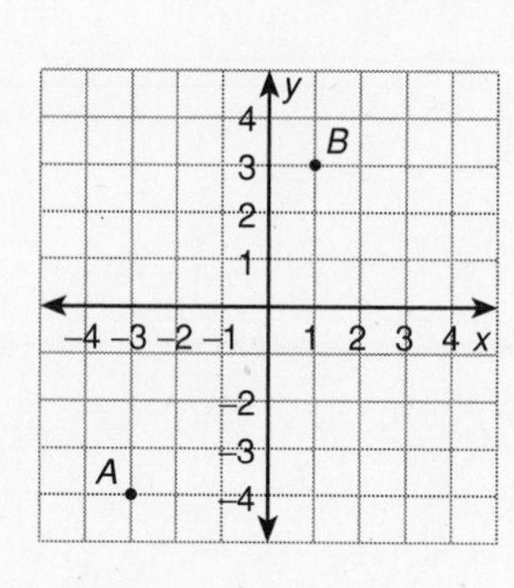

3. ______________________

4. Bisecting a *straight angle* results in what type of angles?

4. ______________________

5. Given the conditional statement, "If at least two distinct points exist, then a line exists." Write the converse statement. Is the converse true? Why or why not?

5.

6. "If x is equal to three, then x^2 is equal to nine."

a. Is the conditional statement true?

b. Write the converse statement.

c. Decide whether the converse statement is true or false. Explain your reason.

6.

2.6 Short Quiz

Name ______________________

Date ______________________

In Problems 1–5, match the statement with one of the properties below.

a. Transitive Property of Congruence
b. Reflexive Property of Congruence
c. Symmetric Property of Equality
d. Transitive Property of Equality
e. Symmetric Property of Congruence

1. If $\angle POQ \cong \angle WOV$, then $\angle WOV \cong \angle POQ$. **1.** ______________

2. If $AB = CD$, then $CD = AB$. **2.** ______________

3. If $TS = PR$ and $PR = UV$, then $TS = UV$. **3.** ______________

4. If $\overline{PQ} \cong \overline{TR}$ and $\overline{TR} \cong \overline{AB}$, then $\overline{PQ} \cong \overline{AB}$. **4.** ______________

5. $\angle RST \cong \angle RST$ **5.** ______________

6. Complete the reasons for each step in the proof. **6.**

P Q R S

Given: $PQ = RS$

$QR = 4,\ PS = 10$

Prove: $PQ = 3$

Statements	*Reasons*
1. $PS = 10,\ QR = 4,\ PQ = RS$	1. Given
2. $PQ + QR + RS = 10$	2. Segment Addition Post.
3. $2PQ + QR = 10$	3. [?]
4. $2PQ + 4 = 10$	4. Substution Prop. of Equality
5. $2PQ = 6$	5. Subtraction Prop. of Equality
6. $PQ = 3$	6. [?]

7. Sketch a pair of angles to fit each description. **7.**

a. Obtuse vertical angles

b. A linear pair of angles

8. Assume that $\angle 1$ and $\angle 2$ are complementary. Complete the table. **8.** *Use table at left.*

$m\angle 1$	15°	30°	47°	65°
$m\angle 2$				

9. Assume that $\angle 1$ and $\angle 2$ are supplementary. Complete the table. **9.** *Use table at left.*

$m\angle 1$	35°	72°	123°	169°
$m\angle 2$				

Chapter 2 Test Form A

(Page 1 of 3 pages)

Name ____________________

Date ____________________

1. Write the correct notation for a ray from Q through P. 1. ____________

2. Y is a point interior to $\angle AOB$. Draw a sketch. Name two adjacent angles. 2.

3. Complete the table. 3. ____________ *Use table at left.*

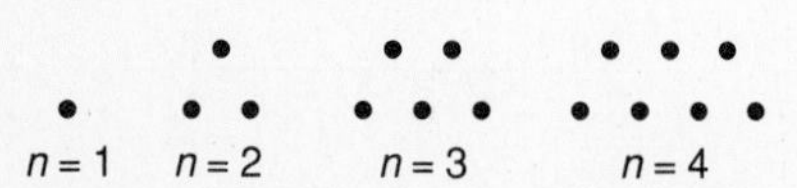

n	1	2	3	4	5	6
nth number	1	3	5	?	?	?

4. Draw four points, A, B, C, and D, on a line so that $\overrightarrow{AC}$ and $\overrightarrow{AB}$ are opposite rays and $\overrightarrow{AC}$ and $\overrightarrow{AD}$ are the same ray. 4.

5. Point S is between points R and T. P is the midpoint of $\overline{RS}$. $RT = 20$ and $PS = 4$. Draw a sketch to show the relationship between the specified segments. Find ST. 5.

6. B is in the interior of $\angle AOC$. C is in the interior of $\angle BOD$. D is in the interior of $\angle COE$. $m\angle AOE = 162°$, $m\angle COE = 68°$, and $m\angle AOB = m\angle COD = m\angle DOE$. Draw a sketch to show the relationship between the specified angles. Find $m\angle DOA$. 6.

7. Find the length of the segment AB. 7. ____________

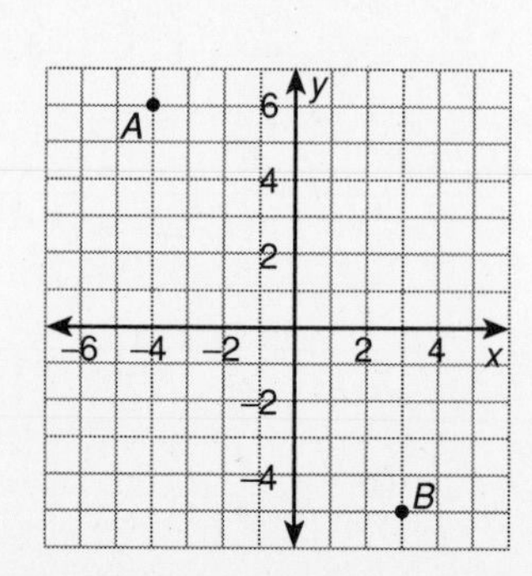

Name ______________________

Date ______________________

8. Use a ruler and protractor to draw a 40° angle and its bisector. **8.**

9. $\overleftrightarrow{EF}$ is a line of symmetry of quadrilateral $ABCD$. Find a segment congruent to $\overline{CB}$. **9.** ______________

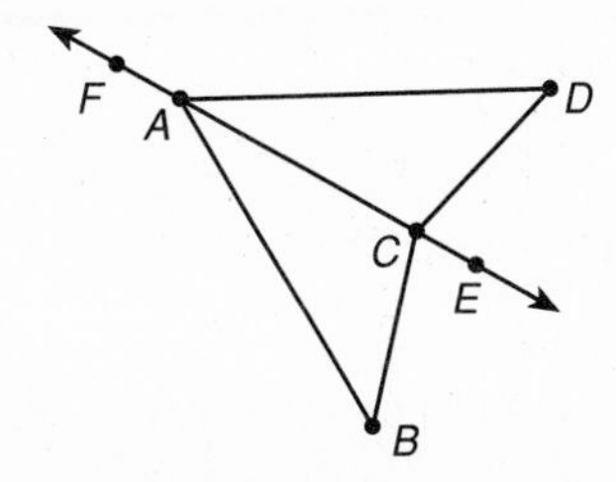

10. Find the length of $\overline{PR}$ if $\overrightarrow{MN}$ bisects $\overline{PR}$ at Q and $QR = 19$. Draw a sketch that shows the given information. **10.**

11. "If an obtuse angle is bisected, then two acute angles are obtained." Write the converse of this conditional statement. Is the converse true? **11.**

12. "If I am invited, then <u>I will go</u>." What is the underlined portion called in this conditional statement? **12.** ______________

13. Construct a Venn diagram for the statement. Then translate the statement to if-then form.
"I must be a member of the club to use the tennis courts." **13.**

14. According to Postulate 8, how many points determine a plane? What must be true about these points? **14.**

Name ____________________

Date ____________________

15. *Given:* $RQ = 5$ and $2(PQ) + 3(RQ) = 27$.
Use the Substitution Property of Equality to find the value of PQ.

15.

16. Given a conditional statement such as *"If it is a woggle, then it is a boggle"*, what is sufficient to disprove it, that is, how can you demonstrate that it is false?

16.

17. For which of the following properties does the relationship "is darker than" hold true?

a. reflexive **b.** symmetric **c.** transitive

17. ____________________

18. $\angle 1$ and $\angle 2$ are a linear pair. $m\angle 1 = 73°$. Find $m\angle 2$.

18. ____________________

19. Write a definition for supplementary angles.

19.

20. Give a reason for each step in the proof.
Given: $\angle 1$ and $\angle 2$ are a linear pair; $m\angle 2 = 100°$.
Prove: $m\angle 1 = 80°$

20.

Statements	*Reasons*
1. $m\angle 2 = 100°$	**1.** Given
2. $\angle 1$ and $\angle 2$ are a linear pair.	**2.** Given
3. $m\angle 1 + m\angle 2 = 180°$	**3.** ?
4. $m\angle 1 + 100° = 180°$	**4.** Substitution Prop. of Equality
5. $m\angle 1 = 80°$	**5.** ?

Chapter 2 Test

Form B

Name ______________________

Date ______________________

1. The notation for the length of the segment between P and Q is $\boxed{?}$.

   a. $\overline{PQ}$ b. $\overrightarrow{QP}$ c. $\overleftrightarrow{PQ}$ d. PQ

1. ______

2. C is an interior point of $\boxed{?}$.

   a. $\angle AOB$ b. $\angle AOD$

   c. $\angle COD$ d. $\angle BOC$

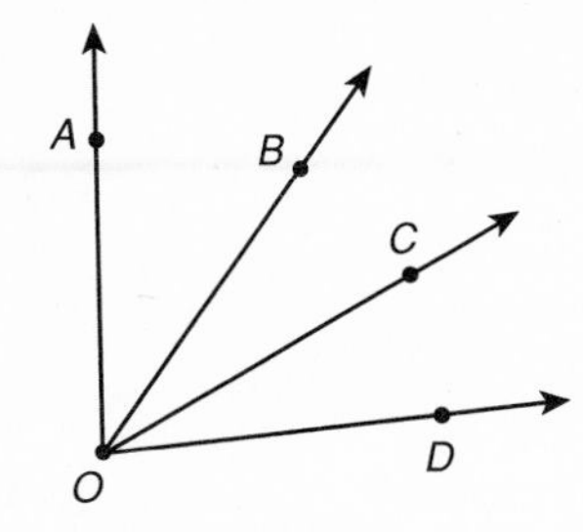

2. ______

3. Ray PR is represented by which sketch?

   a. 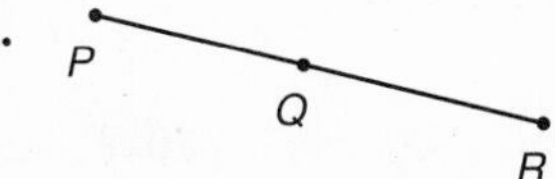

   b. 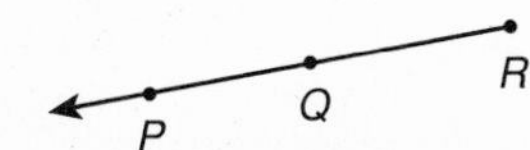

   c. 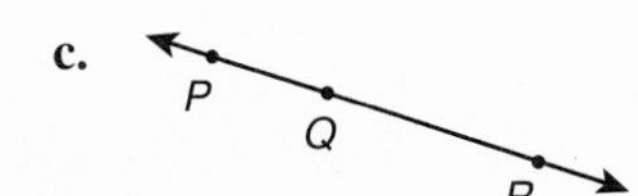

   d. 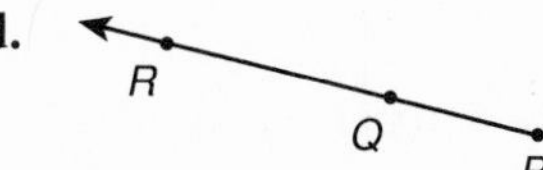

3. ______

4. The first three members of a sequence are shown. How many dots are in the fourth member of the sequence?

   a. 7 b. 16 c. 14 d. 30

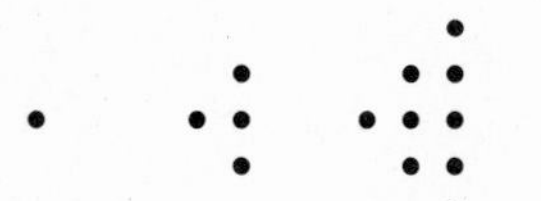

4. ______

5. $A = (1, 3)$, $B = (-1, 1)$, $C = (0, -2)$. A point interior to $\angle ABC$ is $\boxed{?}$.

   a. $(-1, 3)$ b. $(1, 5)$

   c. $(0, 0)$ d. $(0, 2)$

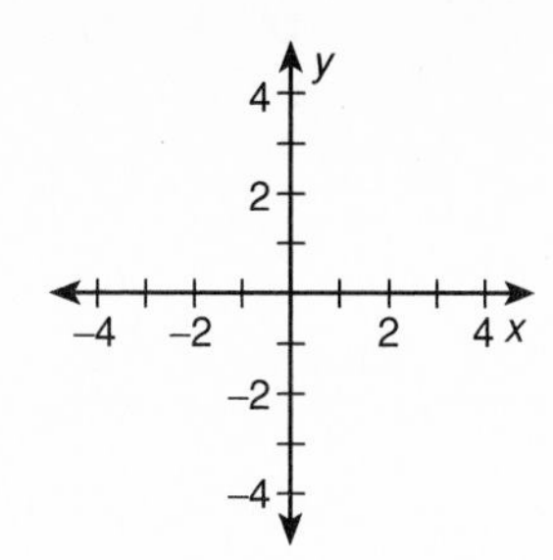

5. ______

6. C is in the interior of $\angle BOD$. B is in the interior of $\angle AOC$. Then, B is also in the interior of angle $\boxed{?}$.

   a. $\angle AOB$ b. $\angle AOD$ c. $\angle DOB$ d. $\angle DOC$

6. ______

7. R, S, and T are collinear. S is between R and T. $RS = 2w + 1$, $ST = w - 1$, and $RT = 18$. Use the Segment Addition Postulate to solve for w. Then determine the length of segment RS.

   a. 6 b. 5 c. 13 d. 16

7. ______

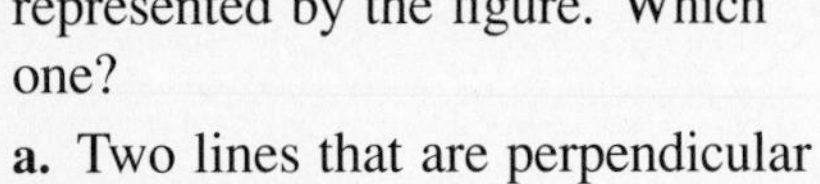

Name ______________________

Date ______________________

8. One of the following statements is represented by the figure. Which one? **8.** ______

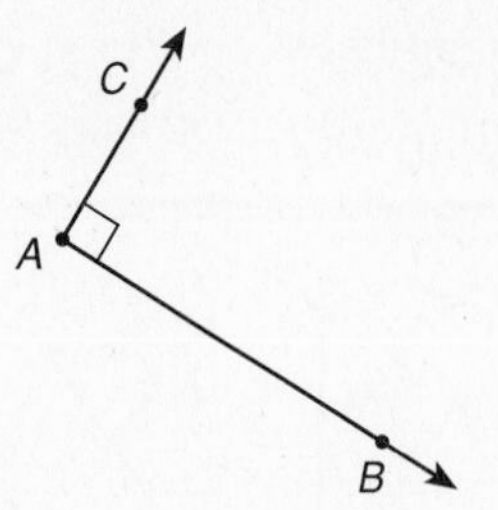

a. Two lines that are perpendicular
b. $AB = AC$
c. A straight angle
d. Two rays that are perpendicular

9. T is the midpoint of $\overline{PQ}$. Which one of the following is not an appropriate statement? **9.** ______

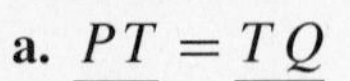

a. $PT = TQ$ **b.** $\overline{PT} \cong \overline{TQ}$
c. $\overline{PT} = \overline{TQ}$ **d.** $PT + TQ = PQ$

10. The distance between points A and B is [?]. **10.** ______

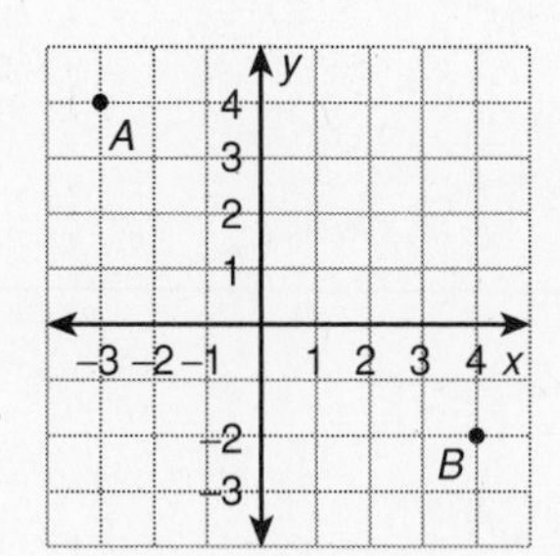

a. $\sqrt{85}$ **b.** $\sqrt{13}$
c. $\sqrt{11}$ **d.** 85

11. If an obtuse angle is bisected, the resulting angles are [?]. **11.** ______

a. never congruent **b.** always acute
c. always obtuse **d.** right angles

12. "If <u>I get a chance</u>, I will succeed." For this conditional statement, the underlined portion is [?]. **12.** ______

a. the conclusion **b.** the converse
c. the hypothesis **d.** the argument

13. Decide which one of the following statements is false. **13.** ______

a. Any three points lie on a distinct line.
b. A line contains at least two points.
c. Through any two distinct points there exists exactly one line.
d. Three noncollinear points determine a plane.

Name ______________________

Date ______________________

14. Consider the conditional statement, "If $x^2 = 25$, then $x = -5$." All of the following are true statements except [?].

a. $(-5)^2 = 25$ **b.** the statement is false
c. the converse is true **d.** the converse is false

14. ________

15. If $PQ = 3$ and $PQ + RS = 5$, then $3 + RS = 5$ is an example of the [?].

a. Multiplication Property of Equality
b. Substitution Property of Equality
c. Reflexive Property of Equality
d. Transitive Property of Equality

15. ________

16. Given the Venn diagram as shown.
Relationship: "is inside of."
Example: Circle G is inside of circle A, since no points of circle G lie on circle A or outside of circle A.
The relationship "is inside of" is [?].

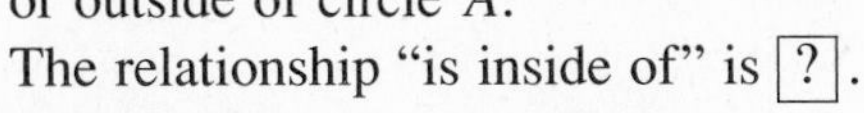

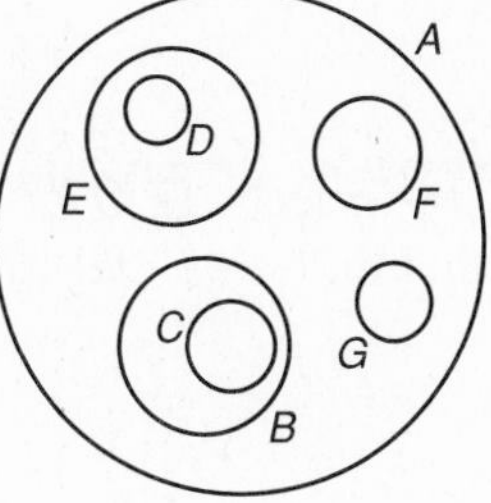

a. reflexive **b.** symmetric
c. transitive **d.** none of these

16. ________

17. The relationship "is a pen pal of" is [?].

a. reflexive **b.** symmetric
c. transitive **d.** none of these

17. ________

18. The sides of two angles form two pairs of opposite rays. The angles are [?].

a. vertical angles **b.** a linear pair
c. complementary **d.** supplementary

18. ________

19. $\angle 1$ and $\angle 2$ are a linear pair. $m\angle 2 = 67°$. $m\angle 1 =$ [?].

a. 67° **b.** 23° **c.** 33° **d.** 113°

19. ________

20. $\angle 1$ and $\angle 2$ are supplementary angles. $\angle 1$ and $\angle 3$ are vertical angles. $m\angle 2 = 72°$. $m\angle 3 =$ [?].

a. 108° **b.** 72° **c.** 18° **d.** 28°

20. ________

Chapter 2 Test

Form C

(Page 1 of 3 pages)

Name ______________________

Date ______________________

1. Describe what the notation $\overrightarrow{RS}$ stands for. Illustrate with a sketch. **1.**

2. What do $\overrightarrow{PQ}$ and $\overrightarrow{QP}$ have in common? **2.**

3. Draw four points, A, B, C, and D, on a line so that $\overrightarrow{CB}$ and $\overrightarrow{CA}$ are opposite rays and $\overrightarrow{CD}$ and $\overrightarrow{CA}$ are the same ray. **3.**

4. $\overrightarrow{OR}$ and $\overrightarrow{OP}$ are opposite rays. $\overrightarrow{OQ}$ bisects $\angle TOR$. $m\angle TOQ = 41°$. Draw a sketch and find $m\angle TOP$. **4.**

5. A pilot, P, is flying from City B toward City A. Use a centimeter ruler and a protractor to measure the course correction (from the $\overleftrightarrow{AB}$ line) at P to divert to City C and the mileage from P to C. **5.** ______________

1 cm = 100 km

6. Find the length of the segment from point C to the midpoint of AB. **6.** ______________

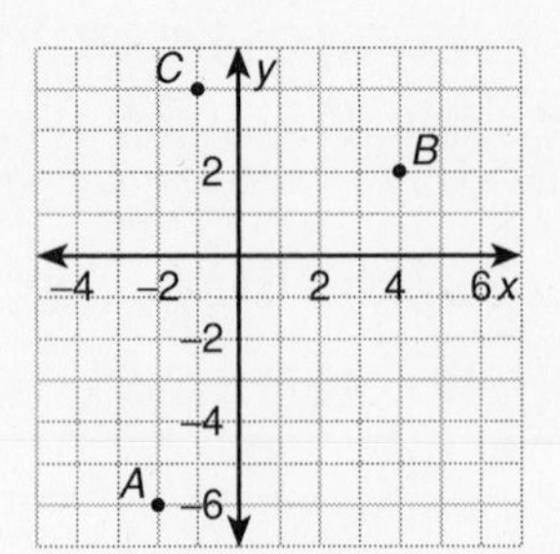

Name ______________________

Date ______________________

7. Use a straightedge and protractor to draw two rays that are perpendicular. **7.**

8. $\overleftrightarrow{PQ}$ is a line of symmetry of quadrilateral $ABCD$. Find an angle congruent to $\angle DCB$. **8.** ______________

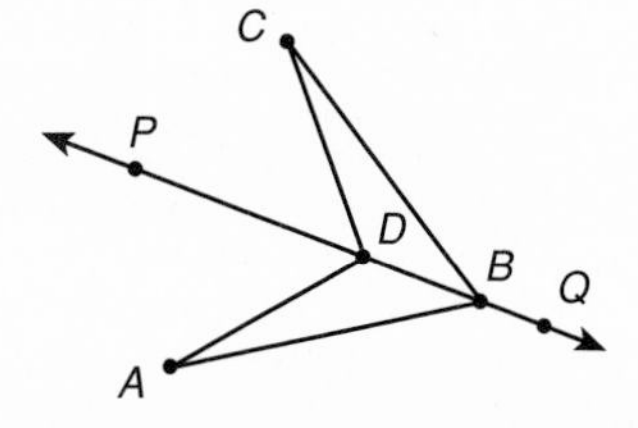

9. $\overline{AB}$ bisects $\overline{LM}$ at T, and $\overline{AB} \cong \overline{TM}$. If $AB = 17$, find LM. **9.** ______________

10. "If an obtuse angle is bisected, then two acute angles are obtained." Decide whether the statement and its converse are true. If false, explain. **10.** ______________

11. "If it doesn't rain, then I will go to the game." What is the underlined portion called in this conditional statement? **11.** ______________

12. Construct a Venn diagram for the statement and then translate the statement to if-then form. "I can't be in the starting line-up unless I'm on the team." **12.**

13. Two distinct planes intersect. Describe their intersection. Draw a sketch to support your answer. **13.**

Name ______________________

Date ______________________

14. Explain what inductive reasoning is and give an example. **14.**

15. Explain what is required to disprove a conditional (if-then) statement. **15.**

16. Consider the whole number relationship "is divisible by." Is this relationship [?]. **16.** ____________

a. reflexive **b.** symmetric **c.** transitive

17. $\angle 1$ and $\angle 2$ are supplementary angles. $\angle 1$ and $\angle 3$ are vertical angles. $m\angle 2 = 67°$. Find $m\angle 3$. **17.** ____________

18. Define complementary angles. **18.**

19. *Given*: $\angle 1$ and $\angle 2$ are vertical angles; $\angle 1$ and $\angle 3$ are a linear pair.
Prove: $\angle 2$ and $\angle 3$ are supplementary angles. Justify each step. **19.** *Use space at left.*

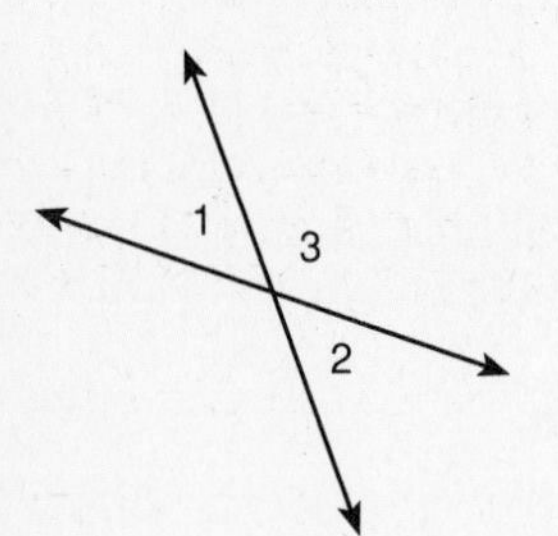

3.2 Short Quiz

Name ______________________

Date ______________________

1. Given three coplanar lines: ℓ_1, ℓ_2, and ℓ_3. If ℓ_1 is parallel to ℓ_3 and if ℓ_2 is parallel to ℓ_3, what conclusion can you draw? What theorems did you use?

1.

2. Define skew lines.

2.

3. Sketch a pair of intersecting planes (an appropriate portion).

3.

4. Name and sketch the three possible cases for two coplanar lines.

4.

5. Solve the system.

$$\begin{cases} 2x - y = 7 \\ 3x + y = 3 \end{cases}$$

Check your answer algebraically.

5.

6. Write an equation of the line that is parallel to $y = \frac{1}{3}x + 1$ and passes through the point $(0, -1)$. Then graph the line.

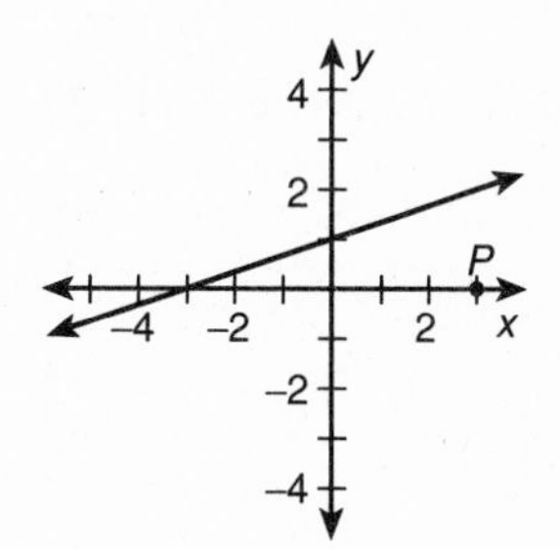

6. ______________________

Use graph at left.

3.4 Short Quiz

Name ______________________

Date ______________________

In Problems 1 and 2, use the statement:

"If no one filed a lien, then the property is mine."

1. Write the contrapositive of the statement. **1.**

2. Write the converse of the statement. **2.**

3. Assume these statements are true: **3.**

a. If A, B, and C are distinct points on a circle, then they are not collinear.

b. If A, B, and C are not collinear, then the slope of $\overline{AB}$ and the slope of $\overline{BC}$ are not equal.

Write the conclusion of the syllogism.

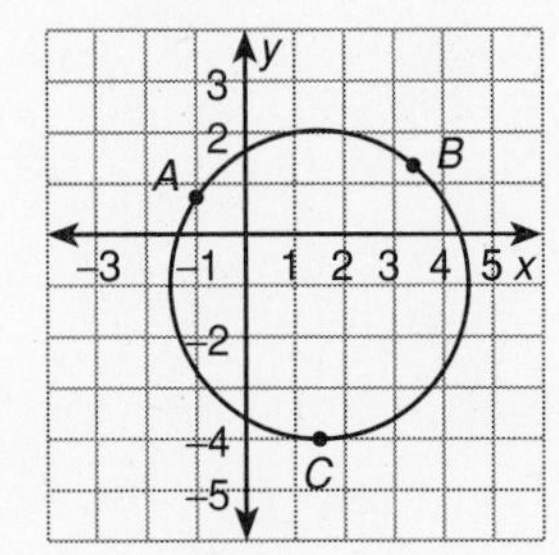

4. Write the statements and reasons of the proof in a logical order. **4.**

Given: $\angle 2$ and $\angle 6$ are supplementary.

Prove: $\ell \parallel m$

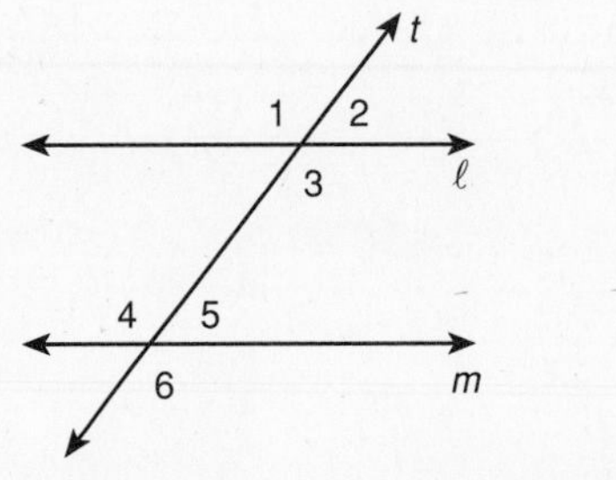

a. Linear Pair Postulate
b. Congruent Supplements Theorem
c. $\angle 2 \cong \angle 5$
d. $\angle 2$ and $\angle 6$ are supplementary.
e. $\ell \parallel m$
f. Given
g. Corresponding Angles Converse
h. $\angle 6$ and $\angle 5$ are supplementary.

Mid-Chapter 3 Test Form A

(Use after Lesson 3.4)

Name ______________________

Date ______________________

1. Define parallel lines. 1.

2. Two coplanar lines are each perpendicular to a third line. What is their relationship to each other? 2. ____________

3. Draw a pair of oblique lines. Explain what makes them oblique. 3.

4. Write the equation of a line that is perpendicular to $y = -3x + 4$ and passes through $(-3, 1)$. 4. ____________

In Problems 5 and 6, use the following statement.

"If I get an 85 or better on the final test, then I get a B in the course."

5. Write the contrapositive statement. 5.

6. Write the converse statement. 6.

7. Use the angles labeled in the diagram at the right to write a two-column proof for the Vertical Angles Theorem. 7.

   Given: $\angle 1$ and $\angle 2$ are supplementary.
   $\angle 2$ and $\angle 3$ are supplementary.

   Prove: $\angle 1 \cong \angle 3$

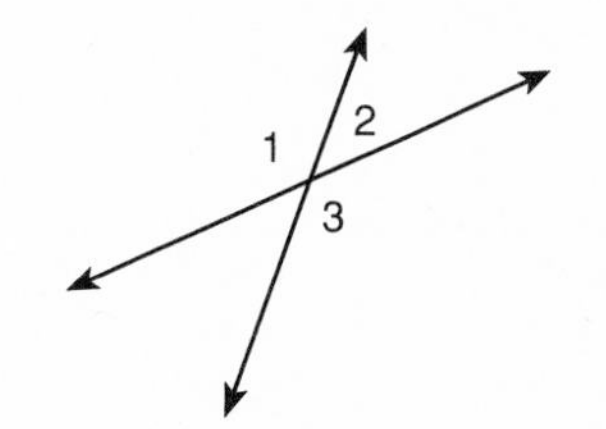

Mid-Chapter 3 Test Form B

(Use after Lesson 3.4)

Name ______________________

Date ______________________

1. Define perpendicular lines.

1.

2. Two lines are each parallel to a third line. What is their relationship to each other?

2. ______________________

3. ℓ_1 and ℓ_2 are parallel lines. ℓ_3 is also parallel to ℓ_1. Must ℓ_2 also be parallel to ℓ_3? Must all three lines be coplanar?

3.

4. Write the equation of a line that is parallel to $y = -3x + 4$ and passes through $(-3, 1)$.

4. ______________________

In Problems 5 and 6, use the following statement.

"If my grade point average is at least 3.00, then I can get a scholarship."

5. Write the contrapositive statement.

5.

6. Write the converse statement.

6.

7. Mark the diagram with the given information. Write a paragraph proof justifying each step.

Given: $\angle 2$ and $\angle 3$ are complementary.
$m\angle 4 = 50°$
Prove: $m\angle 1 = 40°$

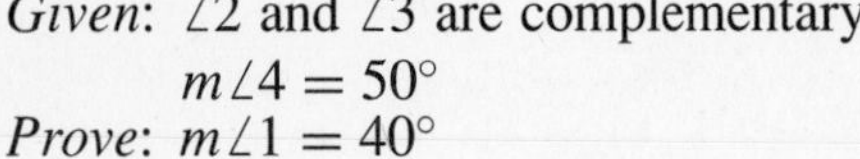

7. *Use diagram at left.*

3.6 Short Quiz

Name ______________________

Date ______________________

In Problems 1 and 2, use the figure below.

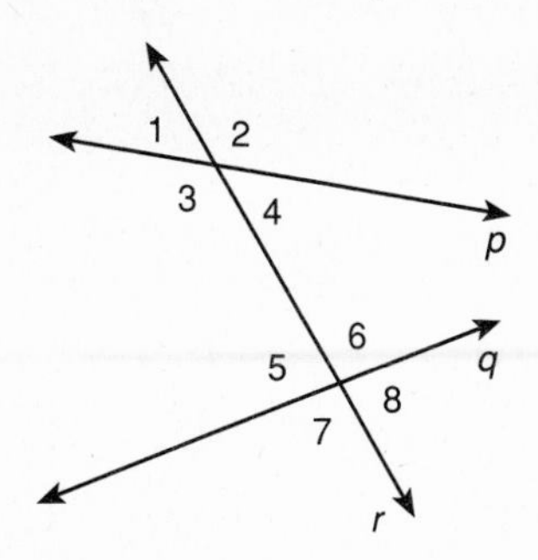

1. Name a pair of alternate exterior angles. **1.** ______________

2. Name a pair of consecutive interior angles. **2.** ______________

3. Use a ruler and a protractor to draw a pair of alternate interior angles, each with a measure of 45°. **3.**

4. Find the measures of ∠1 and ∠2. Explain your reasoning. **4.**

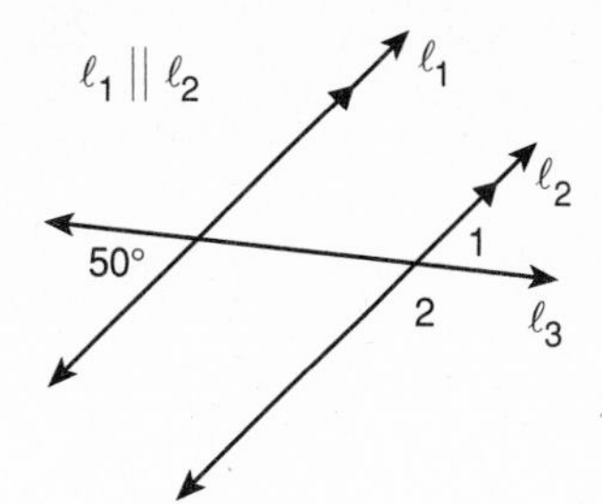

5. Theorem 3.11 states, "If two lines are cut by a transversal so that consecutive interior angles are supplementary, then the lines are parallel." Write the converse of the theorem. Is the converse true? **5.**

Chapter 3 Test Form A

Name ______________________

Date ______________________

1. Draw the three types of pairs of distinct coplanar lines: perpendicular, oblique, and parallel.

1. 

2. $\ell_1 \parallel \ell_2$ and $\ell_1 \parallel \ell_3$. Is $\ell_2 \parallel \ell_3$? Must the three lines be coplanar?

2. 

3. List two pairs of skew lines in this cube.

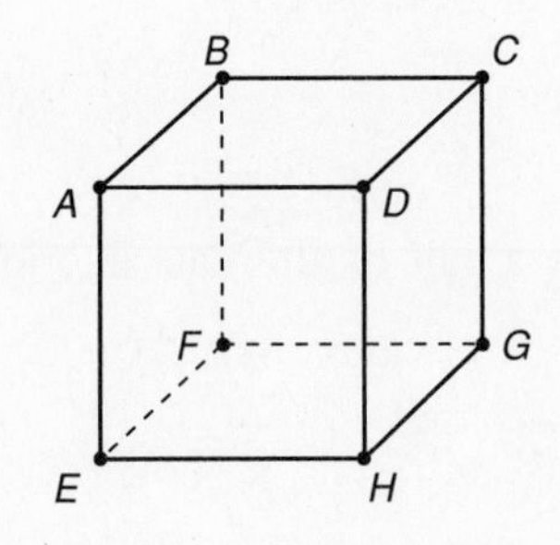

3. ______________________

4. Draw a pair of intersecting planes.

4.

5. Solve the system.

$$\begin{cases} 2x - 3y = 8 \\ x - 2y = 2 \end{cases}$$

Check your solution algebraically.

5. ______________________

6. Write the equation of the line that is parallel to $y = \frac{1}{3}x - 3$ and passes through point (6, 2).

6. ______________________

Name ____________________

Date ____________________

7. According to the Parallel Postulate, if there is a line and a point not on the line, then how many parallels to the given line can be drawn through the point?

7. ____________________

For Problems 8 and 9, use the following statement.

"If we won the division title, then we are in the playoffs."

8. Write the converse of the statement.

8.

9. Write the contrapositive of the statement.

9.

10. Write the conclusion for the syllogism:
"If A, B, and C are distinct points on a circle, then they are not collinear. If three distinct points do not lie on the same line, then they always form a triangle."

10.

11. Write the conclusion for the syllogism:
"If the three points, A, B, and C are not collinear, then $\triangle ABC$ is a right triangle with the right angle at B.
If $\angle ABC$ is a right angle, then the slope of $\overleftrightarrow{AB}$ is the negative reciprocal of the slope of $\overleftrightarrow{BC}$.

11.

12. Write the reason for each step of the proof.
Given: $\overline{AC} \cong \overline{BD}$
Prove: $\overline{AB} \cong \overline{CD}$

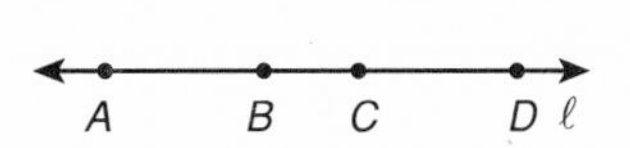

12.

Chapter 3 Test Form A

(Page 3 of 3 pages)

Name ______________________

Date ______________________

13. Sketch an example of consecutive interior angles.

13.

14. Draw an example of the Alternate Interior Angles Theorem.

14.

15. *Given:* $\angle 1 \cong \angle 3$

Use the figure and the given information to determine which lines must be parallel.

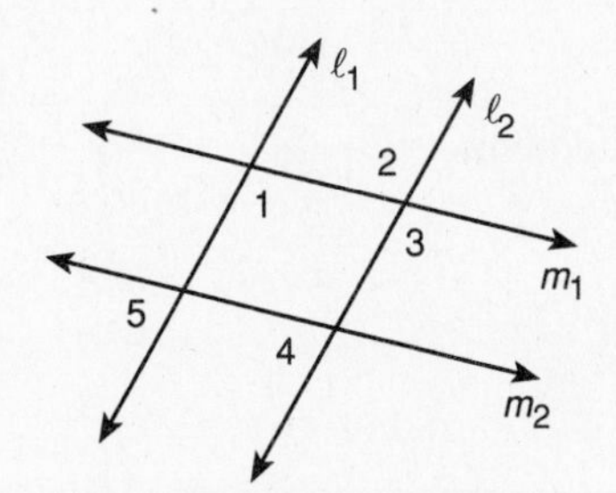

15. ______________________

16. Use construction tools to construct a line through point P perpendicular to line m. Show construction marks.

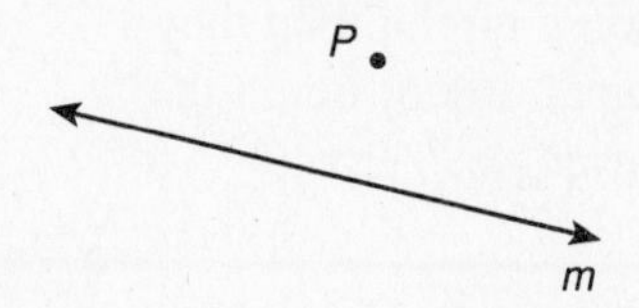

16. *Use figure at left.*

17. Write the ordered pair representation of $\overrightarrow{u}$ and find its length.

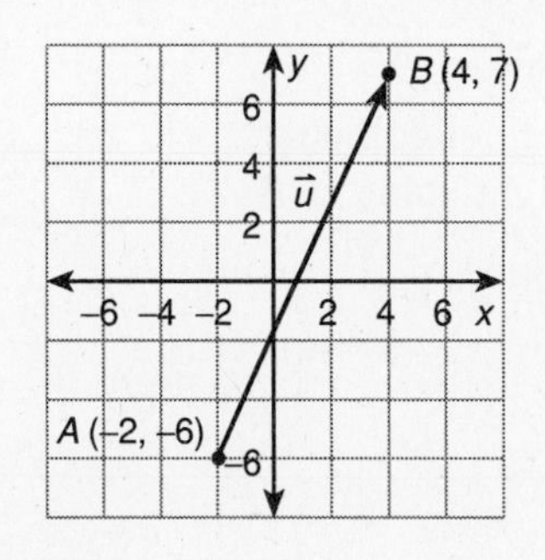

17. ______________________

18. $\overrightarrow{u} = \langle -2, 7 \rangle$; $\overrightarrow{v} = \langle 3, 4 \rangle$; $\overrightarrow{w} = \langle -1, 5 \rangle$

a. Find $\overrightarrow{u} + \overrightarrow{w}$.

b. Find $\overrightarrow{v} \cdot \overrightarrow{w}$.

18.

Name ____________________

Date ____________________

1. $\ell_1 \parallel \ell_2$ and $\ell_2 \parallel \ell_3$. Then [?]. 1. ______
   a. $\ell_1 \perp \ell_3$ b. ℓ_1 intersects ℓ_3
   c. $\ell_1 \parallel \ell_3$ d. ℓ_1 and ℓ_3 are coincident

2. Two lines that are not coplanar and do not intersect are called [?]. 2. ______
   a. skew lines b. perpendicular c. parallel d. oblique

3. If two lines are perpendicular, the product of their slopes is [?]. 3. ______
   a. 1. b. −1. c. 0. d. m^2.

In Problems 4 and 5, use the figure below.

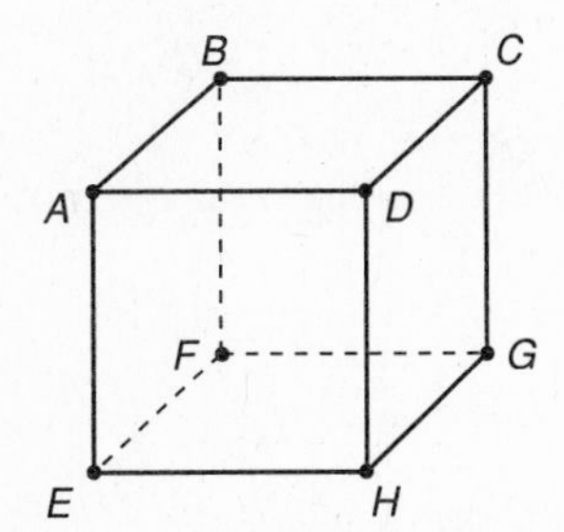

4. For the cube shown, $\overleftrightarrow{AD}$ and $\overleftrightarrow{HG}$ are [?]. 4. ______
   a. parallel lines b. skew lines
   c. oblique lines d. perpendicular lines

5. For the cube shown, $\overleftrightarrow{AE}$ and $\overleftrightarrow{AB}$ are [?]. 5. ______
   a. parallel lines b. skew lines
   c. oblique lines d. perpendicular lines

6. Four lines lie in a plane. The maximum number of intersections among the lines is [?]. 6. ______
   a. 4 b. 5 c. 8 d. 6

7. A line parallel to $y = \frac{2}{3}x - 7$ is [?]. 7. ______
   a. $y = \frac{2}{3}x + 1$ b. $y = -\frac{2}{3}x - 7$
   c. $y = \frac{3}{2}x + 2$ d. $y = -\frac{3}{2}x + 7$

8. A line perpendicular to $y = -4x + 2$ is [?]. 8. ______
   a. $y = -4x - 2$ b. $y = -\frac{1}{4}x + 1$
   c. $y = \frac{1}{4}x + 1$ d. $y = 4x + 2$

Name ____________________

Date ____________________

9. A line parallel to $y = \frac{1}{2}x + 3$ and passing through (0, 0) is $\boxed{?}$. **9.** ______

a. $y = 2x$ **b.** $y = \frac{1}{2}x$

c. $y = \frac{1}{2}x + 6$ **d.** $y = \frac{1}{2}x - 3$

10. For the conditional statement, "If <u>I build it</u>, they will come," the underlined portion is called the $\boxed{?}$. **10.** ______

a. conclusion **b.** proof **c.** hypothesis **d.** negation

11. If $p \rightarrow q$ is the original statement, then $\sim q \rightarrow \sim p$ is the $\boxed{?}$. **11.** ______

a. converse **b.** contrapositive **c.** negation **d.** conclusion

12. Assume the statements are true: "If it doesn't rain, we play the game. It didn't rain." The conclusion is $\boxed{?}$. **12.** ______

a. we may or may not play

b. we play the game

c. we do not play the game

d. we wait for rain

13. "If we get enough rain, we will have a good harvest. If we have a good harvest, we can pay off our debts." The conclusion of the syllogism is $\boxed{?}$. **13.** ______

a. It rained and we had a good harvest

b. We had a good harvest and paid off our debts

c. If we get enough rain, we can pay off our debts

d. If we paid off our debts, then it rained

14. In the figure, $\angle 1$ and $\angle 2$ are $\boxed{?}$. **14.** ______

a. corresponding angles **b.** alternate interior angles

c. alternate exterior angles **d.** consecutive interior angles

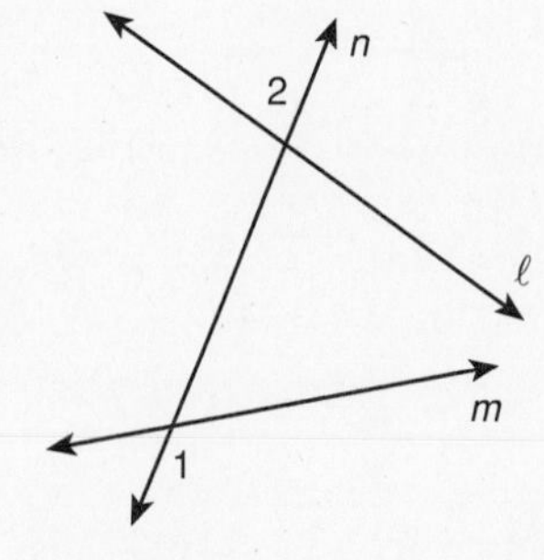

Name ____________________

Date ____________________

15. In the figure, $\ell \parallel n$ and r is a transversal. Which of the following is not true?

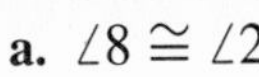

a. $\angle 8 \cong \angle 2$ **b.** $\angle 7 \cong \angle 4$
c. $\angle 2 \cong \angle 6$ **d.** $\angle 5 \cong \angle 3$

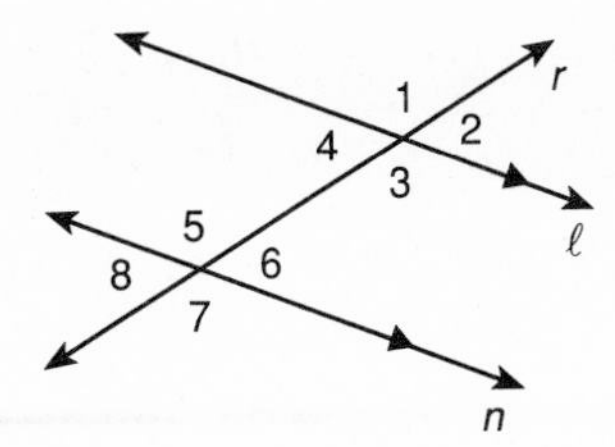

15. ________

16. In the sketch for Problem 15, $\angle 6$ and $\angle 3$ are ?.

16. ________

a. consecutive interior angles
b. alternate interior angles
c. alternate exterior angles
d. corresponding angles

17. In the sketch for Problem 15, $\angle 6$ and $\angle 2$ are ?.

17. ________

a. alternate interior angles **b.** alternate exterior angles
c. consecutive interior angles **d.** corresponding angles

18. Refer to the figure at right. ℓ_1 and ℓ_2 are guaranteed to be parallel because of which theorem?

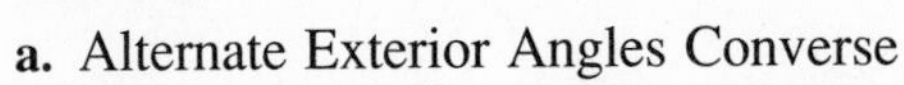

a. Alternate Exterior Angles Converse
b. Alternate Interior Angles Converse
c. Corresponding Angles Converse
d. Consecutive Interior Angles Converse

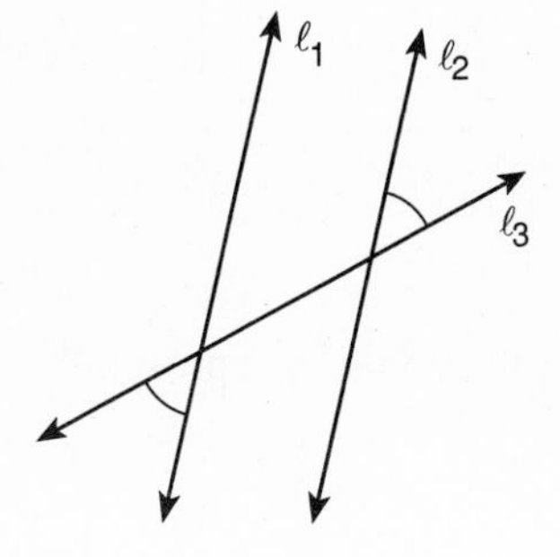

18. ________

19. The ordered pair representation for the vector $\vec{u}$ as shown in the figure is ?.

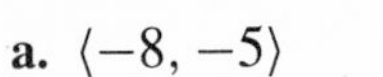

a. $\langle -8, -5\rangle$ **b.** $\langle 5, 8\rangle$
c. $\langle -5, -8\rangle$ **d.** $\langle 8, 5\rangle$

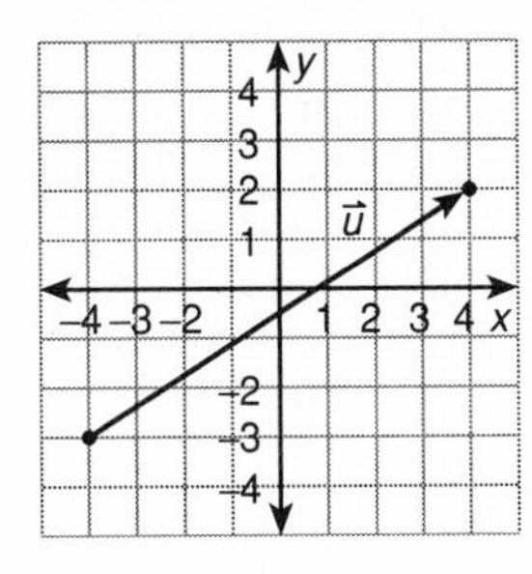

19. ________

20. The length of the vector $\vec{u} = \langle 6, -3\rangle$ is ?.

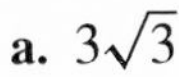

a. $3\sqrt{3}$ **b.** 45 **c.** $3\sqrt{5}$ **d.** $\sqrt{3}$

20. ________

Chapter 3 Test

Form C

(Page 1 of 3 pages)

Name ______________________

Date ______________________

1. Two lines are coplanar. Draw and name the three possible cases.

1.

2. ℓ_1, ℓ_2, and ℓ_3 are coplanar. $\ell_1 \perp \ell_2$ and $\ell_3 \perp \ell_2$. Draw a figure and state the relationship between ℓ_1 and ℓ_3.

2. ______________________

3. Define skew lines.

3. ______________________

4. Draw a pair of intersecting planes.

4.

5. Solve the system. What does the solution represent?

$$\begin{cases} 3x + y = 4 \\ x - 2y = 13 \end{cases}$$

5. ______________________

6. A line is perpendicular to $y = \frac{1}{3}x - 2$ and passes through point (6, 2). Write its equation.

6. ______________________

7. According to the Perpendicular Postulate, if there is a line and a point not on the line, then how many perpendiculars to the given line can be drawn through the point?

7. ______________________

Name ______

Date ______

In Problems 8 and 9, use the statement:

"If I wash the car, then I can drive to the mall."

8. Write the converse of the statement.

8. ______

9. Write the contrapositive of the original statement.

9. ______

10. Write the conclusion for the syllogism (see the figure): "If $\overline{AC}$ is a diameter of the circle, then $\angle ABC$ is a right angle. If $\angle ABC$ is a right angle, then the slope of $\overleftrightarrow{AB}$ is the negative reciprocal of the slope of $\overleftrightarrow{BC}$."

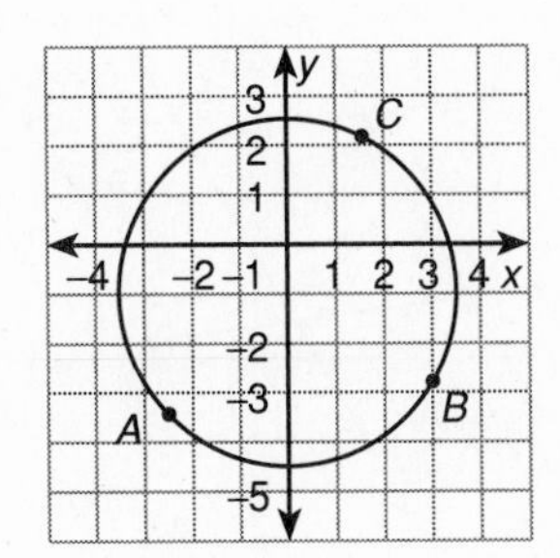

10. ______

11. Use the Law of Detachment to write the conclusion: "If Sean is lying, our case falls apart. If our case falls apart, then we don't get the money. If we do not get the money, we will have to sell the house. Nothing Sean has said is true."

11. ______

12. Mark the given information on the diagram. Write a proof justifying each statement.
Given: $\ell \perp n$, $m \perp n$
Prove: $\angle 3$ and $\angle 6$ are supplementary.

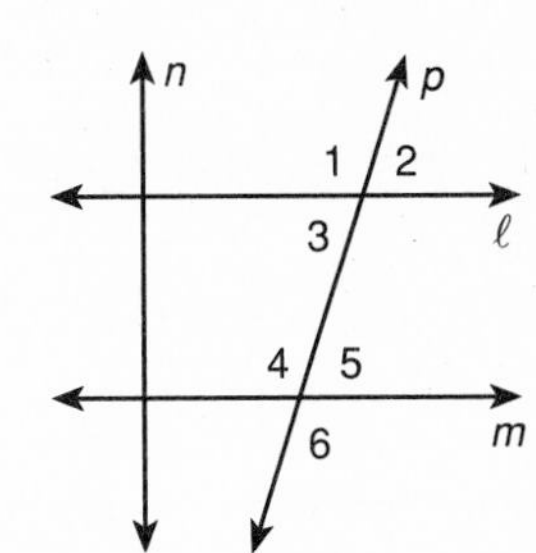

12. *Use figure at left.*

Name ______________________

Date ______________________

13. Draw a pair of corresponding angles. **13.**

14. Draw a sketch for the Alternate Exterior Angles Theorem. State the given information and what is to be proven. Label the diagram. **14.**

15. Use the figure and the given information to determine which lines must be parallel.
Given: $\angle 1 \cong \angle 3$

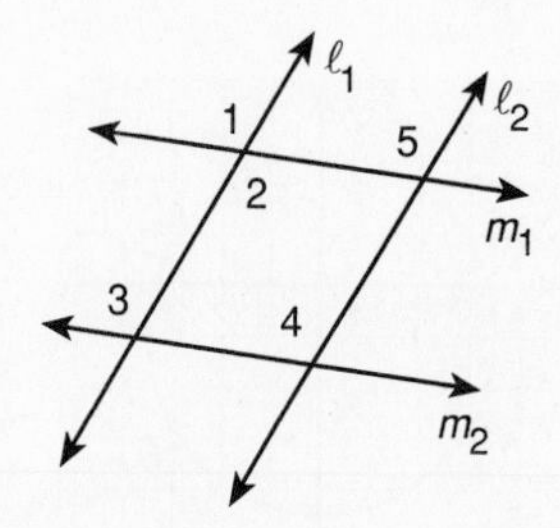

15. ______________________

16. Use construction tools to construct a line through P parallel to m. Show construction lines.

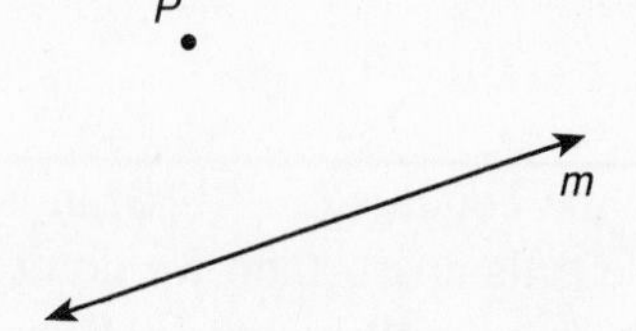

16. *Use figure at left.*

17. Write the ordered pair representation of $\vec{u}$ and find its length.

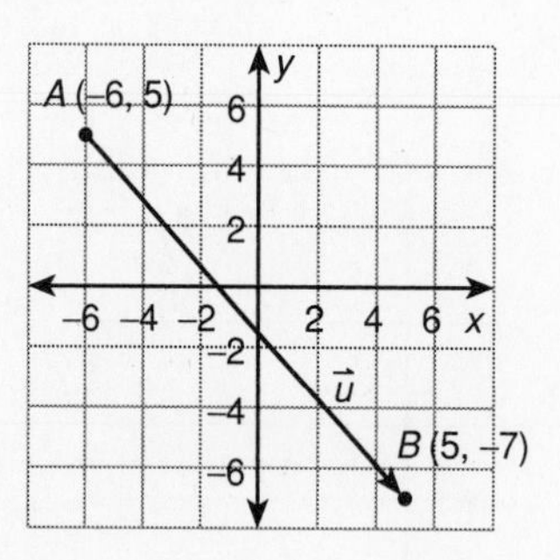

17. ______________________

18. $\vec{u} = \langle 3, -7 \rangle$; $\vec{v} = \langle -2, -5 \rangle$; $\vec{w} = \langle 4, -3 \rangle$

a. Find $\vec{u} + \vec{w}$.

b. Find $\vec{v} \cdot \vec{w}$.

18. a. ______________________

b. ______________________

Cumulative Test

Chapters 1–3

Name ______________________

Date ______________________

1. Describe any symmetry that each figure may have.

a. 

b. 

1. ______________________

2. Isolate the diamond and club symbols in the previous problem and describe the symmetry that each figure may have.

a. 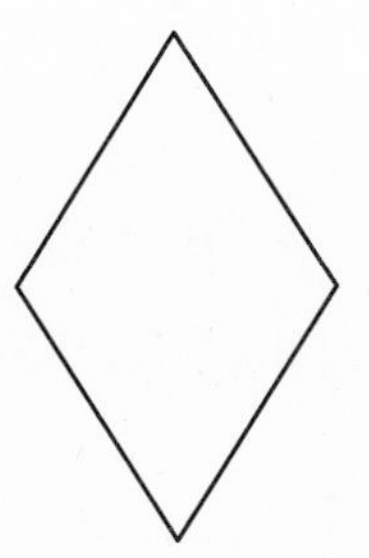

b. 

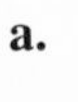

2. ______________________

3. Draw an oblique triangle. Explain what makes it oblique.

3.

4. Divide the shaded region into six congruent figures.

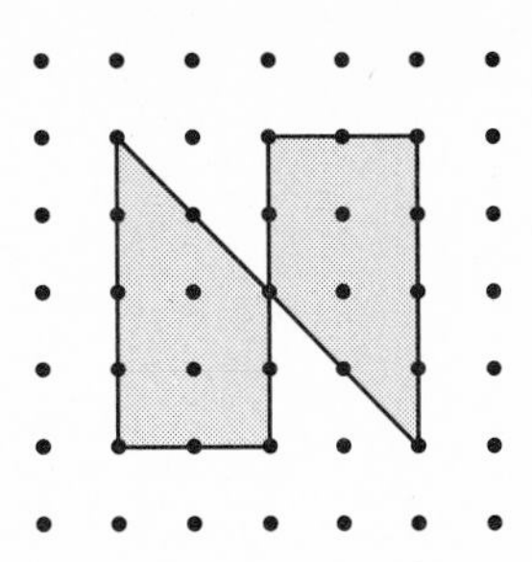

4. *Use figure at left.*

5. Write a definition for similar figures. Draw an example to illustrate your definition.

5.

6. Find the midpoint of $\overline{CD}$, where $C = \left(\frac{1}{2}, 7\right)$ and $D = \left(\frac{7}{2}, -2\right)$.

6. ______________________

7. A floor plan for a small classroom is shown. The scale of $\frac{1}{16}$ inch in the drawing represents one foot in the actual room. What are the actual dimensions of the room?

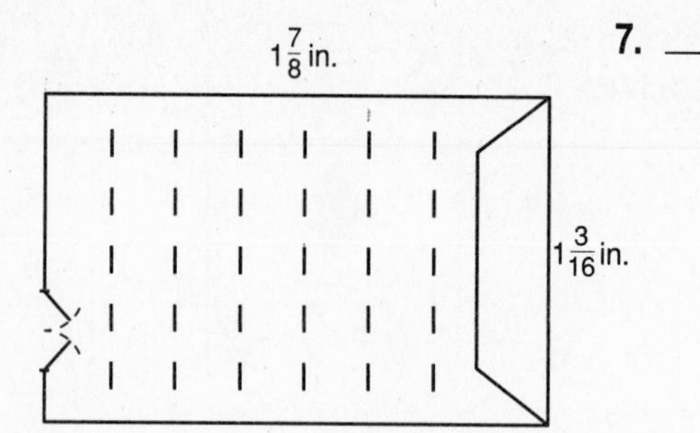

7. ______

8. Two of the figures are congruent. Which two?

a.

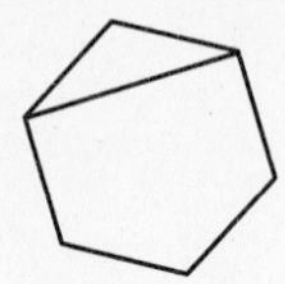

b.

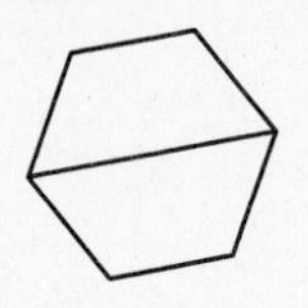

c.

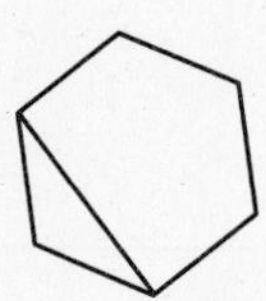

d.

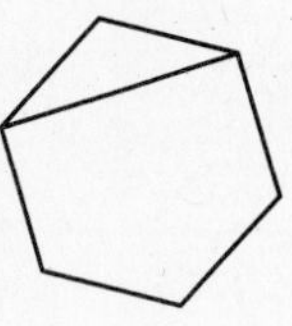

8. ______

9. How many lines of symmetry does the badge appear to have?

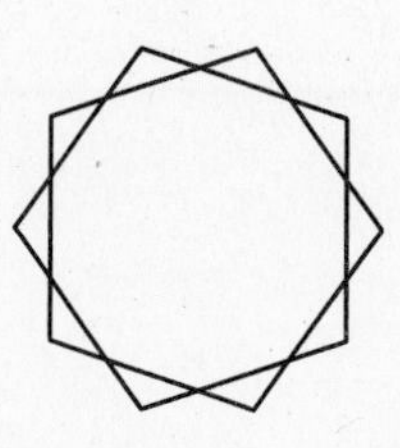

9. ______

10. Draw a figure that has rotational symmetry but does not have line symmetry.

10.

11. What does the slope of the lines tell you about the relationship between the lines $y = \frac{3}{2}x + 1$ and $y = -\frac{2}{3}x - 1$. Explain.

11.

12. A quart of liquid fertilizer will cover an area of 400 square feet. How many quarts of liquid fertilizer are needed to cover a garden that is 16 feet by 45 feet?

12. ______

Name ______________________

Date ______________________

13. Find the slope of the line that passes through points $A(-3, -2)$ and $B(5, 1)$.

13. ______________________

14. Use construction tools to construct a line that is perpendicular to line ℓ from point P. Show construction lines.

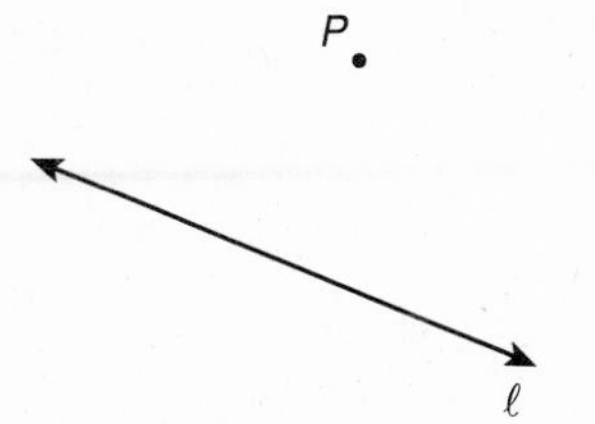

14. *Use figure at left.*

15. Write the notation for the line that passes through points R and S.

15. ______________________

16. A, B, and C are three distinct collinear points. C is on $\overrightarrow{AB}$ and C is on $\overrightarrow{BA}$. Make a sketch that illustrates these conditions.

16.

17. Draw four points on a line such that $\overrightarrow{AB}$ and $\overrightarrow{AD}$ are opposite rays and that $\overrightarrow{DB}$ and $\overrightarrow{DC}$ are opposite rays.

17.

18. Complete the sketch of a fifth object in the sequence and complete the table.

$n = 1$ $n = 2$ $n = 3$ $n = 4$ $n = 5$

n	1	2	3	4	5
Number of line segments	0	1	3	?	?

18. *Use table at left.*

19. B is interior to $\angle AOC$. D is interior to $\angle BOC$. $m\angle AOC = 125°$ and $m\angle BOD = 40°$. $m\angle AOB = m\angle BOD$. Sketch the figure and find $m\angle COD$.

19.

Name ____________

Date ____________

20. Find the distance from point A to the midpoint of $\overline{BC}$.

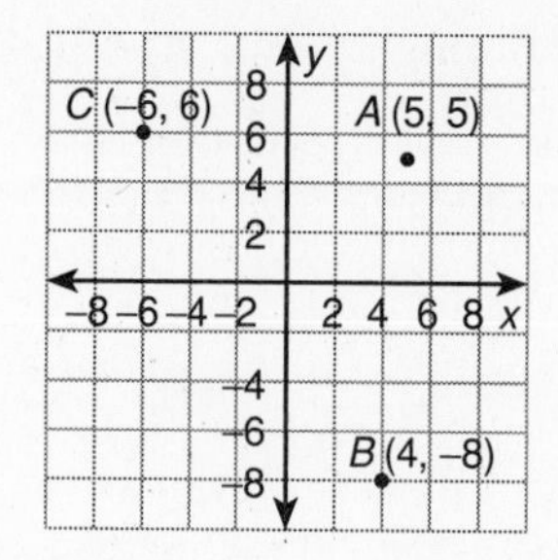

20. ____________

21. Use a compass and a straightedge to draw the bisector of a 70° angle.

21.

22. Find the length of $\overline{QR}$ if $\overleftrightarrow{MN}$ bisects $\overline{PR}$ at Q and $PR = 26$.

22. ____________

23. If an acute angle is bisected, two acute angles are obtained. Write the converse of this conditional statement. Is the converse true? Explain.

23.

24. $\overleftrightarrow{FG}$ is a line of symmetry of the figure shown. Which segment is congruent to $\overline{BA}$?

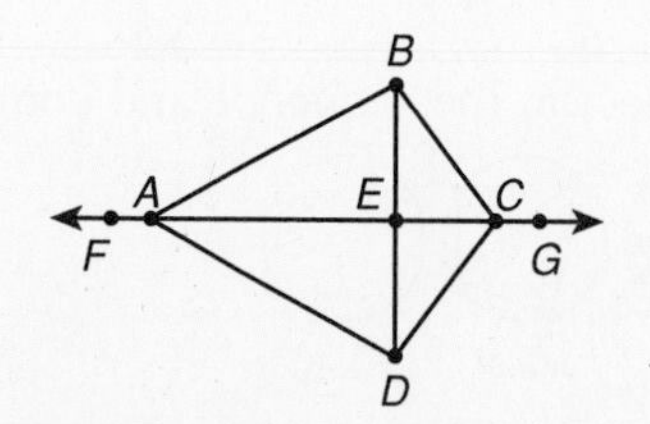

24. ____________

25. "If I pass this course, then it will be because I kept up with the homework." What is the underlined portion called in this conditional statement?

25. ____________

26. Construct a Venn diagram for the statement. Then translate the statement into if-then form.

"As a senior I am assured of getting each course I preregister for."

26.

Name ______________________

Date ______________________

27. According to Postulate 7, what do you know exists through any three noncollinear points,?

27. ______________________

28. *Given:* $BC = 10$ and $2(AB) + 3(BC) = 54$. Use the Substitution Property of Equality to find the value of AB.

28.

29. $\angle 3$ and $\angle 4$ are a linear pair. $m\angle 4 = 73°$. Find $m\angle 3$.

29. ______________________

30. If a statement such as "If A is true, then B is true" is false, what is sufficient to disprove it?

30.

31. Consider the relationship, "is a partner of." Which of the properties (reflexive, symmetric, and transitive) does the relationship have?

31. ______________________

32. What conditions must be true for two angles to be a linear pair?

32. ______________________

33. Give a reason for each step in the proof.

33.

Given: $\angle 1$ and $\angle 2$ are supplementary angles.
$m\angle 1 = 75°$

Prove: $m\angle 2 = 105°$

Statements	*Reasons*
1. $\angle 1$ and $\angle 2$ are supplementary.	**1.** ?
2. $m\angle 1 + m\angle 2 = 180°$	**2.** ?
3. $m\angle 1 = 75°$	**3.** ?
4. $75° + m\angle 2 = 180°$	**4.** ?
5. $m\angle 2 = 105°$	**5.** ?

Name ______________________

Date ______________________

34. Given coplanar lines ℓ_1, ℓ_2, and ℓ_3. If $\ell_1 \perp \ell_2$, $\ell_3 \perp \ell_2$, what is the relationship between ℓ_1 and ℓ_3?

34. ______________________

35. Define skew lines.

35.

36. Sketch a pair of parallel planes with a pair of skew lines, one in each of the planes.

36.

37. Draw a set of intersecting planes.

37.

38. List two pairs of perpendicular lines in this cube.

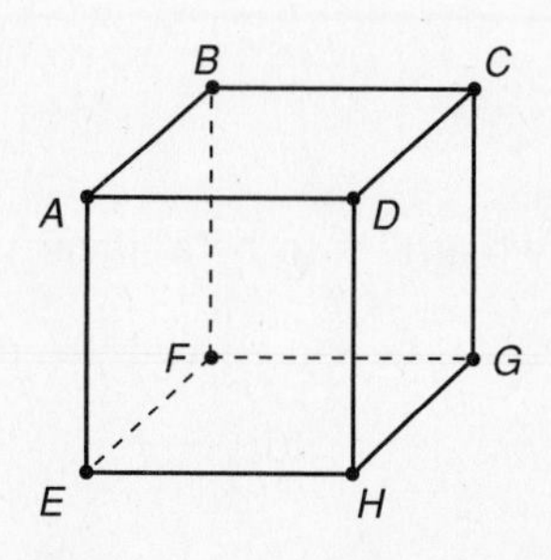

38. ______________________

39. Solve the system.

$$\begin{cases} 6x + 5y = 0 \\ 5x + 4y = 1 \end{cases}$$

Check your answer algebraically.

39. ______________________

Name ____________________

Date ____________________

40. Write an equation of the line that is parallel to $y = -4x + 1$ and passes through the point $(-4, 10)$.

40. ____________________

41. Write an equation of the line that is perpendicular to $y = -4x + 1$ and passes through the point $(-4, 10)$.

41. ____________________

In Problems 42 and 43, use the following statement.

"If I cannot raise the down payment, then I cannot buy the car."

42. Write the contrapositive statement.

42.

43. Write the converse statement.

43.

44. Complete the proof.

Given: $\angle AOD \cong \angle COB$

Prove: $\angle AOC \cong \angle DOB$

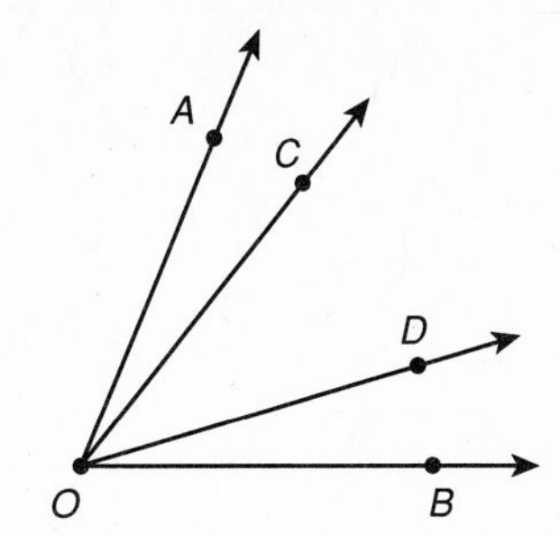

44.

Name ____________________

Date ____________________

45. Write the *conclusion* for the syllogism: If I get a 92 on the final test, then I get an A in the course. If I get an A in the course, then I will make the Dean's List. If I make the Dean's List, then I can get a scholarship. If I get a scholarship, then I can finish college.

45.

46. Draw an example of alternate exterior angles.

46.

47. According to the Alternate Interior Angles Converse, if a transversal of two lines (see figure) forms alternate interior angles which are congruent, then what do we know about the lines?

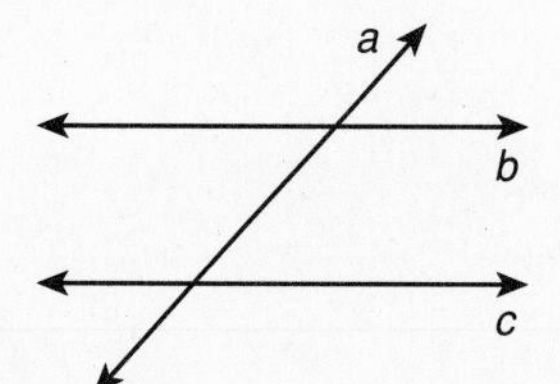

47. ____________________

48. Write the ordered pair representation of $\vec{u}$ and find its length.

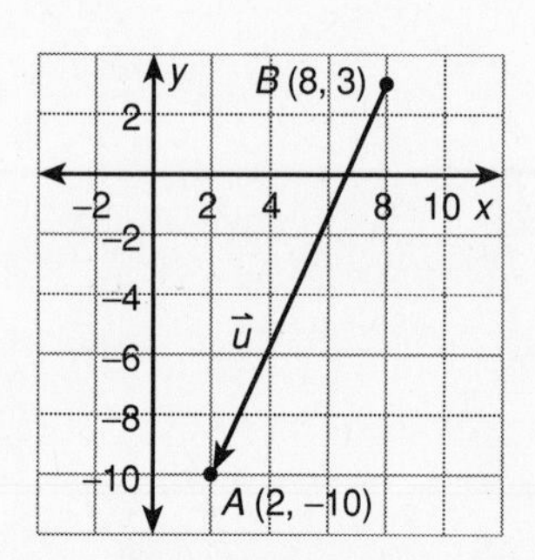

48.

49. Given the points, $A(-3, 2)$, $B(7, 1)$, and $C(-2, -4)$, let $\vec{u}_1$ be the vector from A to C; and, let $\vec{u}_2$ be the vector from A to B.

a. Find $\vec{u}_1 + \vec{u}_2$.

b. Find $\vec{u}_1 \cdot \vec{u}_2$.

49. a. ____________________

b. ____________________

4.2 Short Quiz

Name ______________________

Date ______________________

1. A triangle has side lengths 5 cm, 7 cm, and 5 cm. Name the type of triangle.

1. ______________

2. *Given:* $\triangle ABC \cong \triangle LMN$. Finish marking $\triangle LMN$ and complete the statements.

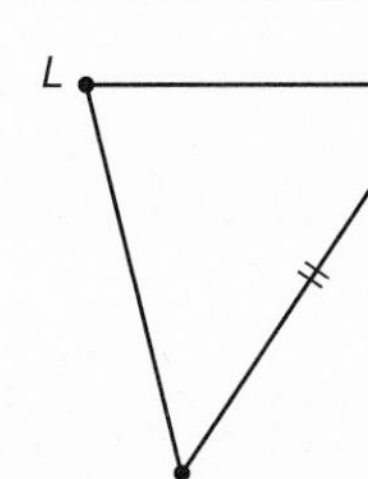

a. $\angle L \cong \boxed{?}$

b. $\overline{AC} \cong \boxed{?}$

2. a. ______________

b. ______________

3. Is it possible to have an isosceles triangle that is also scalene? Is it possible to have an isosceles triangle that is also equilateral? Explain.

3.

4. In the figure, $m\angle 2 = 65°$ and $m\angle 4 = 135°$.

Find $m\angle 3$.

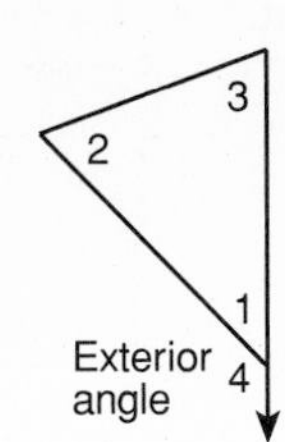

4. ______________

5. Use the figure to complete the following statements.

a. $m\angle 2 = \boxed{?}$

b. $m\angle 3 = \boxed{?}$

c. $m\angle 4 = \boxed{?}$

d. $m\angle 6 = \boxed{?}$

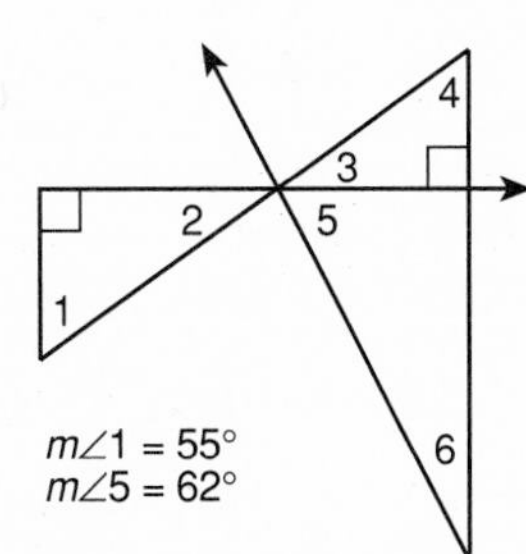

5. a. ______________

b. ______________

c. ______________

d. ______________

4.4 Short Quiz

Name ______________________

Date ______________________

1. Is there enough information in the diagram to conclude that $\triangle ABC \cong \triangle DEC$? Explain?

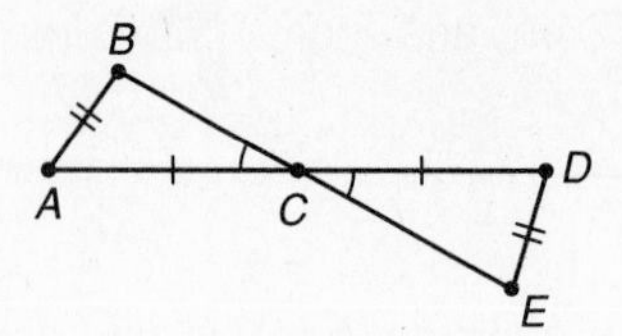

1. ______________

2. Is there enough information in the diagram to conclude that $\triangle ABC \cong \triangle ADC$? Explain?

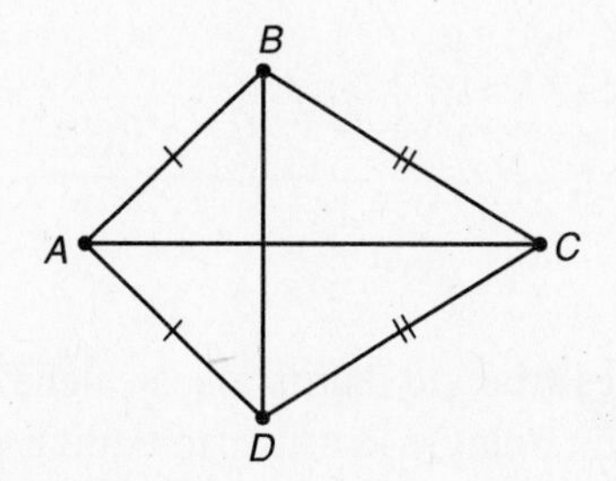

2. ______________

3. Are all equilateral triangles congruent? If not, why not? Are all equiangular triangles congruent? If not, why not?

3.

4. What theorem or postulate can be used to conclude that $\triangle ABC \cong \triangle EDC$?

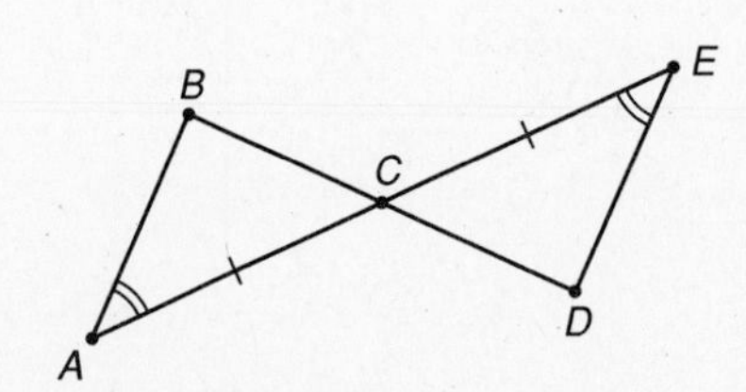

4. ______________

5. Which side or angle triples are insufficient to prove two triangles are congruent? Draw a diagram to support your answer.

5. ______________

Mid-Chapter 4 Test Form A

(Use after Lesson 4.4)

Name ______________________

Date ______________________

1. If $\triangle PQR \cong \triangle STR$, then $\overline{PR} \cong$ [?].

1. ______________

2. If $m\angle D = m\angle E$, then $\angle E \cong \angle D$. What does this statement mean?

2.

3. Can an isosceles triangle have two right angles? Explain.

3.

4. Sketch the following types of triangles. Mark each sketch appropriately.
   a. Right isosceles **b.** Equilateral **c.** Obtuse scalene

4.

5. *Given:* Isosceles $\triangle ABC$ with $\overline{AB} \cong \overline{BC}$.
   a. Solve for x.
   b. Is the triangle equilateral?

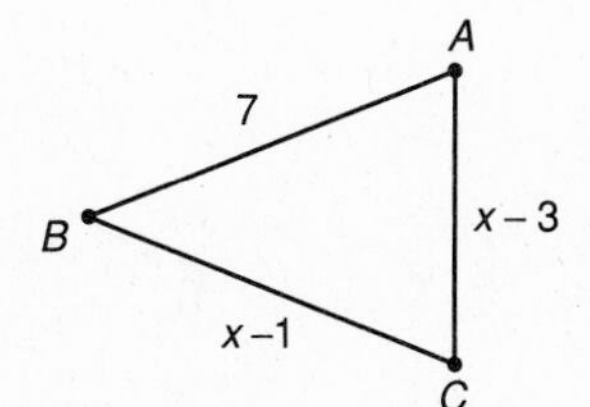

5. **a.** ______________
   b. ______________

6. Do postulates require proof? Explain.

6.

7. State the postulate or theorem that can be used to conclude that $\triangle ACB \cong \triangle DCE$.

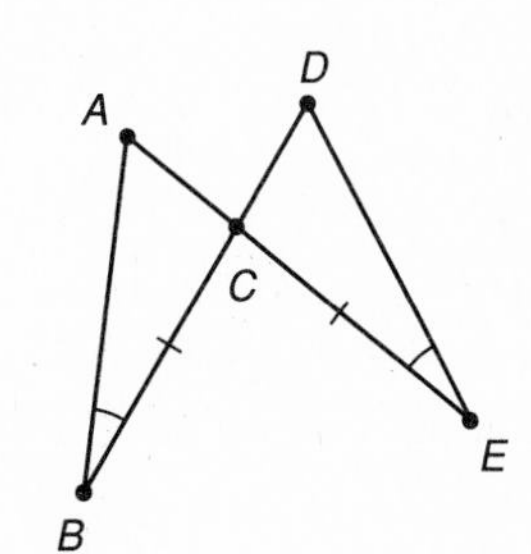

7.

Mid-Chapter 4 Test Form B

(Use after Lesson 4.4)

Name ______________________

Date ______________________

1. If $\triangle AMN \cong \triangle LKJ$, then $\overline{AN} \cong \boxed{?}$. 1. ______________

2. If $\angle A \cong \angle B$, then $m\angle B = m\angle A$. What does this statement mean? 2.

3. Can an acute triangle contain an obtuse angle? Explain. 3.

4. Sketch the following types of triangles. Mark each sketch appropriately. 4.

   a. Right isosceles **b.** Equiangular **c.** Obtuse scalene

5. *Given:* Isosceles $\triangle ABC$ with $\overline{AB} \cong \overline{BC}$.

   a. Solve for x.

   b. Is the triangle equilateral?

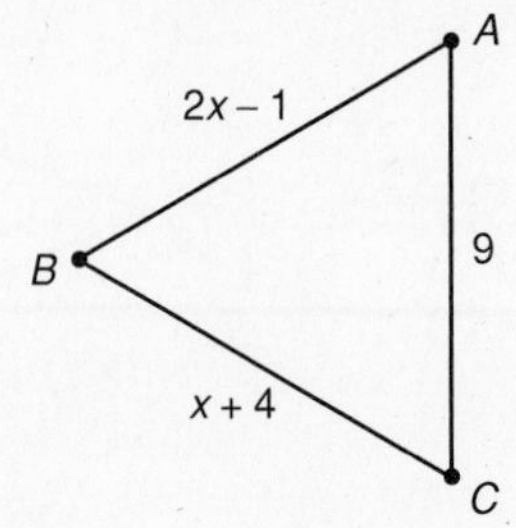

5. a. ______________

   b. ______________

6. Can a triangle be congruent to itself? Explain. 6.

7. State the postulate or theorem that can be used to conclude that $\triangle BAC \cong \triangle CDB$. 7.

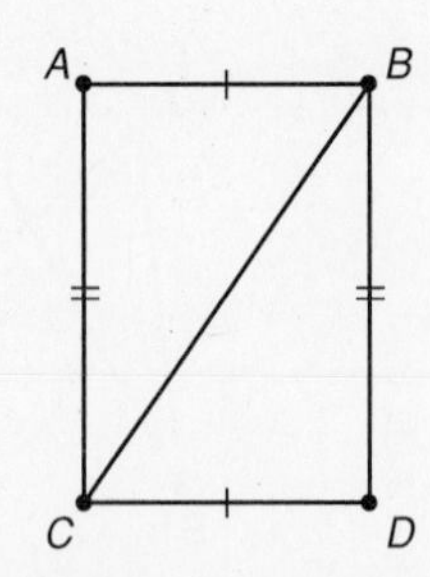

4.6 Short Quiz

Name ______________________

Date ______________________

1. State the reason why $\triangle ABC \cong \triangle PQR$.

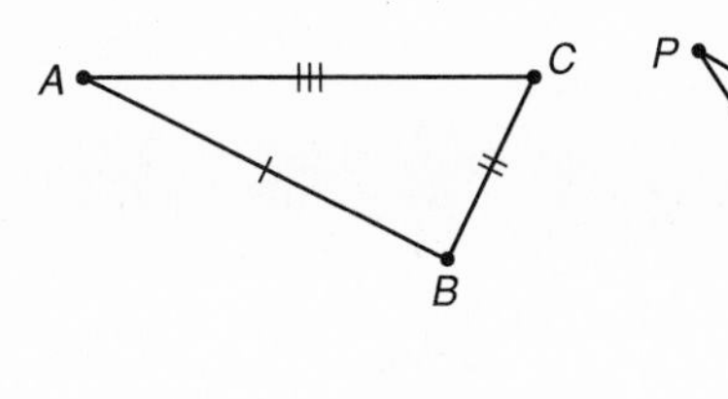

1. ______________________

2. Which sides are congruent in the given triangle? Explain.

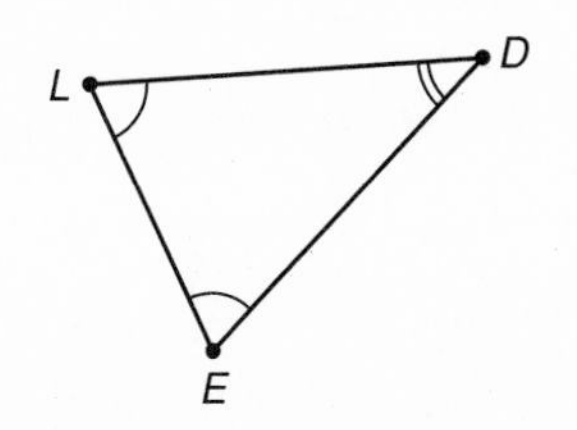

2.

3. What is the special name for a triangle with all sides congruent?

3. ______________________

4. Identify the base of $\triangle REM$.

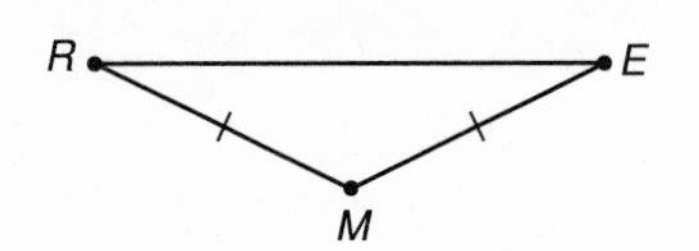

4. ______________________

5. State the theorem that verifies that $\triangle RST \cong \triangle WYX$.

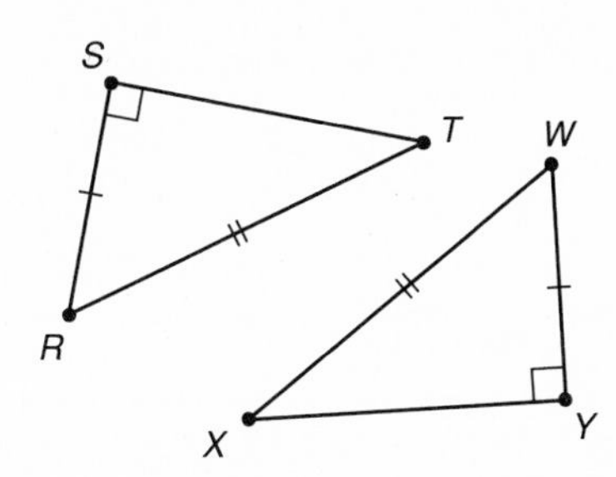

5. ______________________

Chapter 4 Test

Form A

(Page 1 of 3 pages)

Name ______________________

Date ______________________

1. If $\angle P \cong \angle Q$ and $m\angle Q = 67°$, then $m\angle P = \boxed{?}$.

1. ______________________

2. How many acute angles can a scalene triangle have? Explain your reasoning.

2.

3. *Given:* $\triangle LMN \cong \triangle UVW$. Complete the statements.

a. $\overline{UW} \cong \boxed{?}$

b. $\angle LMN \cong \boxed{?}$

3. a. ______________________

b. ______________________

4. *Given:* $\triangle ABC \cong \triangle GHJ$; $\triangle DEF \cong \triangle GHJ$. Then $\triangle ABC \cong \triangle DEF$. What property of congruence does this statement represent?

4. ______________________

5. Find the measure of exterior angle $\angle BCD$.

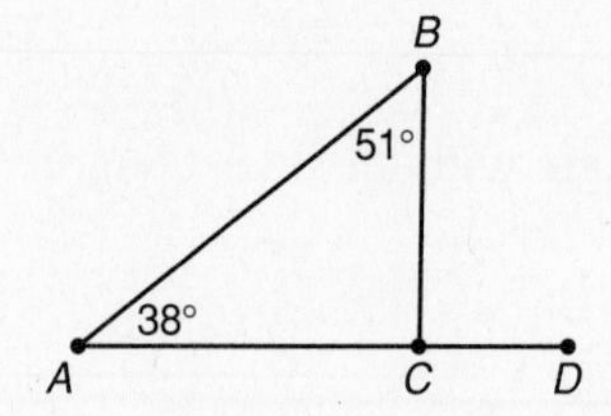

5. ______________________

6. *Given*: $\overline{AB} \cong \overline{BC}$

a. Solve for x.

b. Is the triangle equilateral?

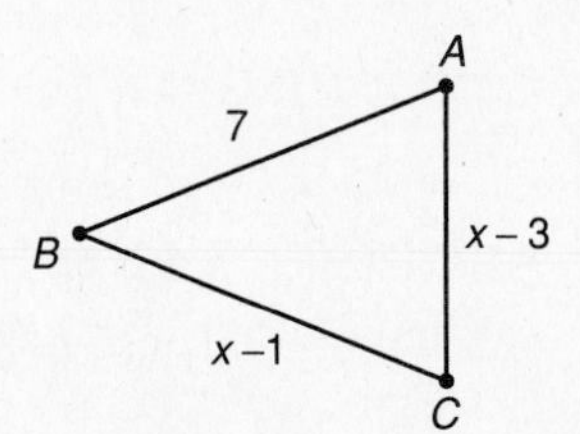

6. a. ______________________

b. ______________________

7. Find the measures of angles A, B, and C.

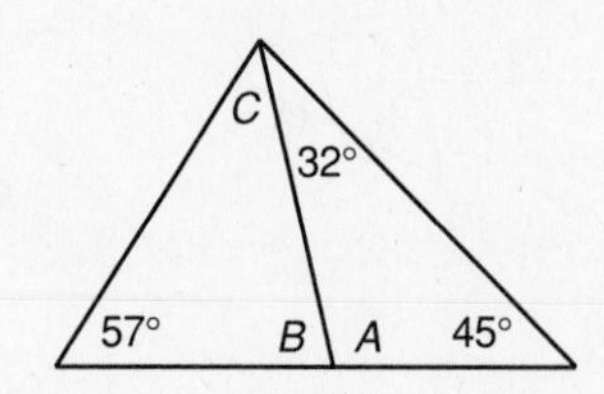

7. ______________________

Name ______________________

Date ______________________

8. Sketch the following, if possible. If not possible, state why.

a. Right scalene triangle

b. Two obtuse isosceles triangles that are not congruent.

c. A triangle with an exterior angle of 30°

d. A triangle with two acute angles and one obtuse angle

8.

9. Solve for x.

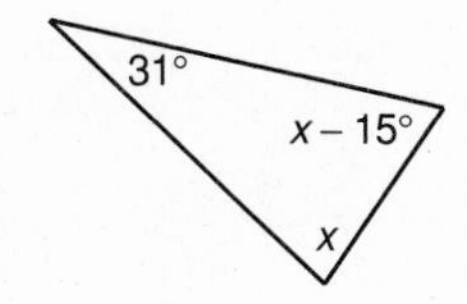

9. ______________________

10. How many acute angles can an isosceles triangle have? Explain.

10.

11. State the postulate or theorem that can be used to conclude that $\triangle OCD \cong \triangle OAB$.

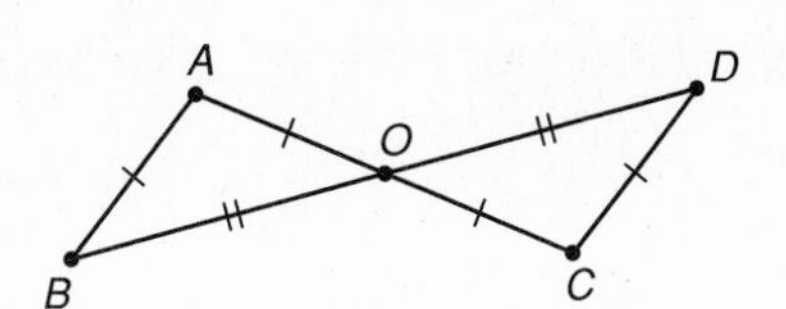

11. ______________________

12. Refer to the figure for Problem 11. Give a reason to justify that $\angle BOA \cong \angle DOC$.

12.

13. Refer to the figure at the right.

a. Is $\overline{QR} \cong \overline{SR}$? If so, why?

b. What type of triangle is $\triangle QRS$?

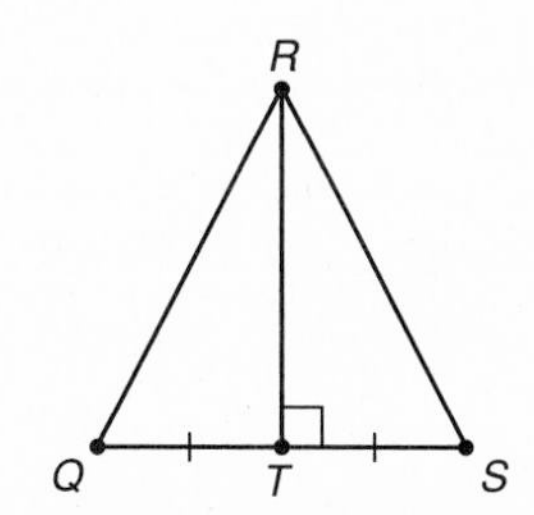

13.

Name ____________________

Date ____________________

14. Find the measure of exterior angle A.

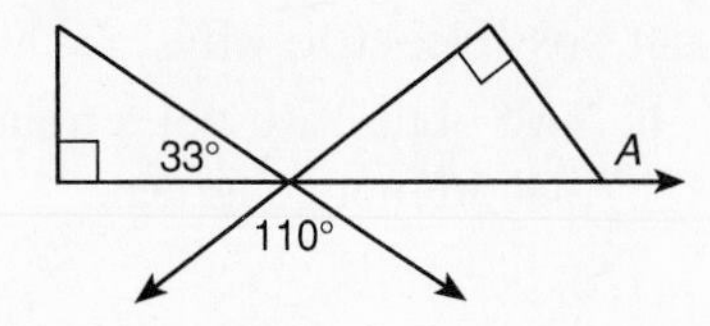

14. ____________________

15. If $\triangle ABC \cong \triangle DEF$, does it follow that $\triangle CBA \cong \triangle FED$? Explain.

15.

16. Find the measure of $\overline{LM}$. State the postulate or theorem you used.

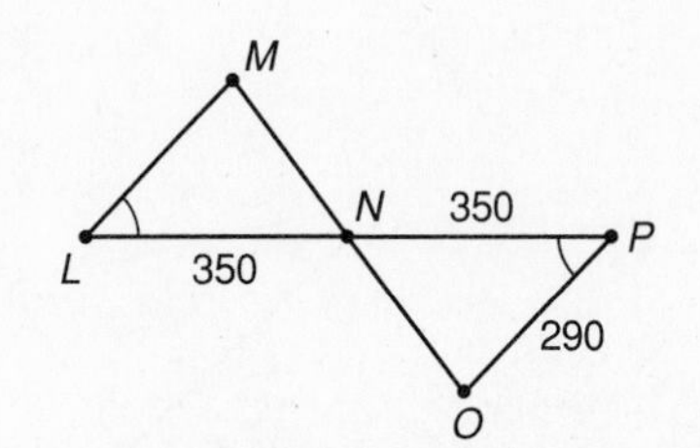

16.

17. Complete the missing step in the proof.

Given: $\overline{AB} \cong \overline{CD}$;
$\overline{AB} \parallel \overline{CD}$

Prove: $\triangle ABD \cong \triangle CDB$

17.

Statements	*Reasons*
1. $\overline{AB} \cong \overline{CD}$	1. Given
2. $\overline{BD} \cong \overline{BD}$	2. Reflexive Prop. of $\cong$
3. $\overline{AB} \parallel \overline{CD}$	3. Given
4. [?]	4. [?]
5. $\triangle ABD \cong \triangle CDB$	5. SAS Congruence Post.

18. *Given*: $\overline{BC} \cong \overline{BA}$,
$\overline{CD} \cong \overline{AD}$

Prove: $\triangle ABD \cong \triangle CBD$

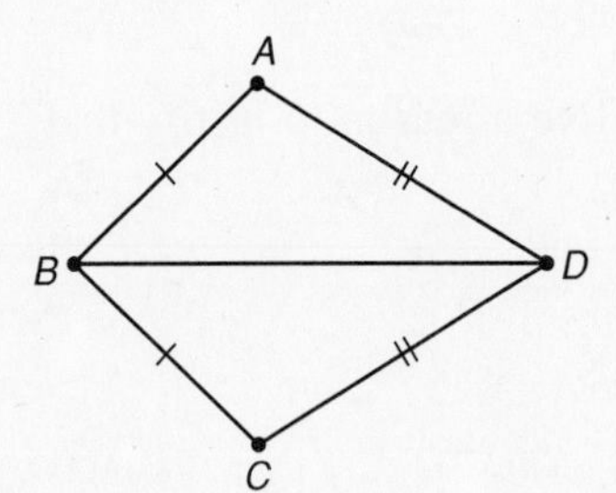

18.

19. How many lines of symmetry does an isosceles right triangle have? Draw a diagram to illustrate.

19.

Chapter 4 Test

Form B

Name ____________________

Date ____________________

1. In $\triangle ABC$, if $\overline{AB} \cong \overline{BC}$ and $m\angle A = 39°$, then $m\angle C =$ $\boxed{?}$. **1.** ______

a. 102° **b.** $m\angle B$ **c.** 141° **d.** 39°

2. How many obtuse angles can an isosceles triangle have? **2.** ______

a. 0 **b.** 1 **c.** 2 **d.** 3

3. Refer to the figure. $\triangle ABC \cong$ $\boxed{?}$. **3.** ______

a. $\triangle CDE$
b. $\triangle ACE$
c. $\triangle EDC$
d. $\triangle EDA$

4. Refer to the figure. $m\angle C =$ $\boxed{?}$. **4.** ______

a. 100°
b. 17°
c. 148°
d. 15°

5. Refer to the figure. $m\angle A =$ $\boxed{?}$. **5.** ______

a. 67°
b. 74°
c. 35°
d. 141°

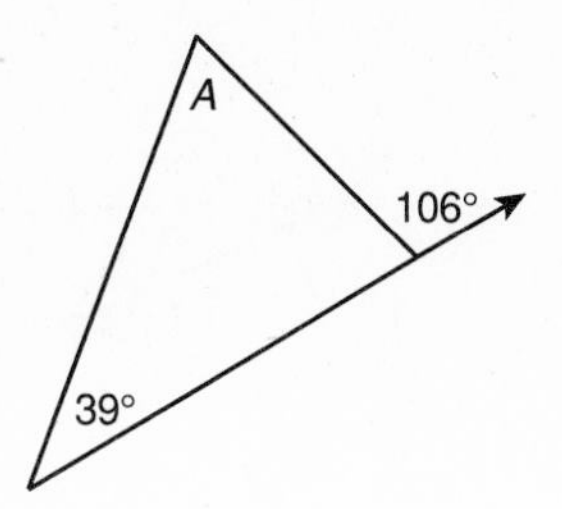

6. In $\triangle ABC$, $\overline{AB} \cong \overline{BC}$. Which term does *not* describe the triangle? **6.** ______

a. Obtuse
b. Isosceles
c. Equilateral
d. Acute

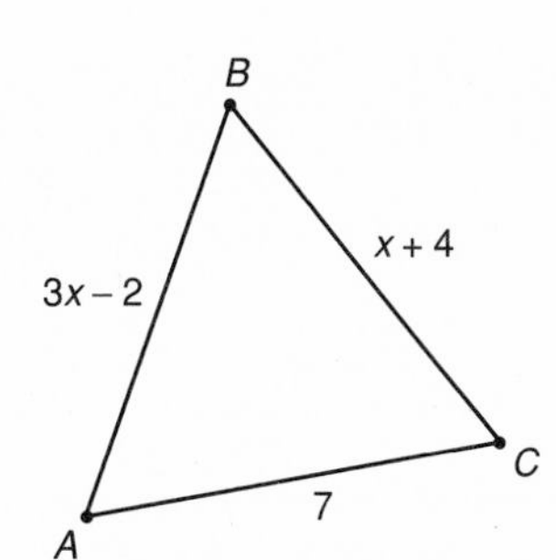

7. If $\triangle ABC \cong \triangle PQR$ and $\triangle PQR \cong \triangle LMK$, then $\triangle ABC \cong \triangle LMK$. This illustrates which property of congruence? **7.** ______

a. Reflexive **b.** Symmetric **c.** Transitive **d.** Commutative

Name ______________________

Date ______________________

8. A triangle has angle measures of 60°, 60°, and 60°. Choose the term that describes the triangle. **8.** ______

a. Right **b.** Acute **c.** Scalene **d.** Obtuse

9. Use the figure to solve for x. **9.** ______

a. 145°
b. 90°
c. 45°
d. 55°

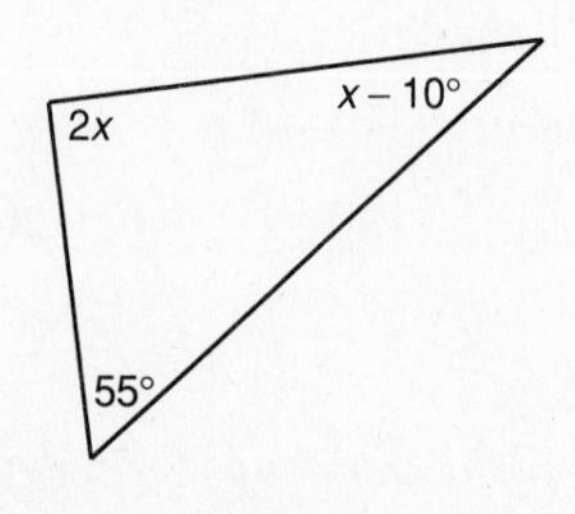

10. How many obtuse angles can an equilateral triangle have? **10.** ______

a. 0 **b.** 1 **c.** 2 **d.** 3

11. The postulate or theorem that can be used to conclude that $\triangle CAB \cong \triangle CED$ is [?]. **11.** ______

a. SSS Congruence Postulate
b. AAS Congruence Theorem
c. SAS Congruence Postulate
d. ASA Congruence Postulate

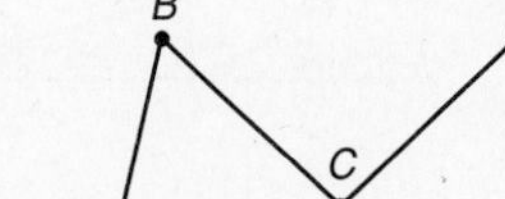
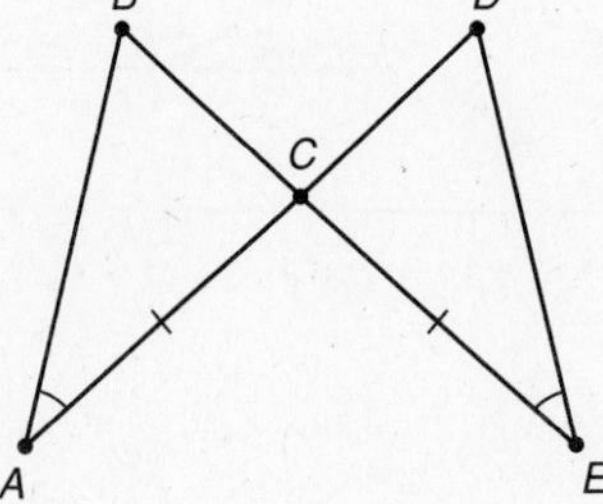

12. What must be true in order for $\triangle ABC \cong \triangle EDC$ by the SAS Congruence Postulate? **12.** ______

a. $\overline{AB} \cong \overline{DE}$
b. $\angle A \cong \angle E$
c. $\overline{AC} \cong \overline{CE}$
d. $\angle B \cong \angle D$

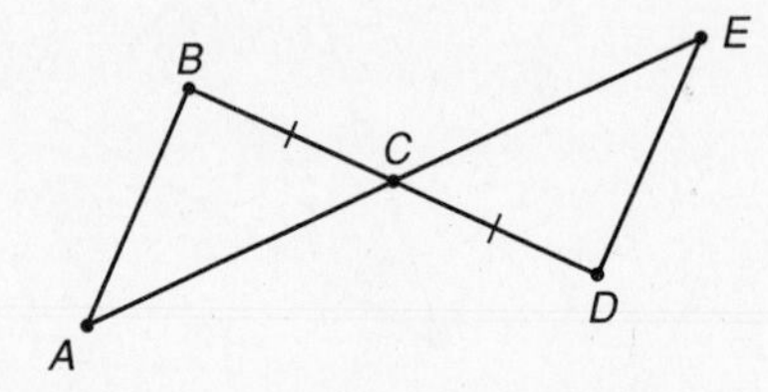

13. The measure of angle B is [?]. **13.** ______

a. 99°
b. 79°
c. 89°
d. 101°

130°
51°
B

Name ____________________

Date ____________________

14. *Given:* $\triangle ABC \cong \triangle DEF$ with $\overline{AB} \cong \overline{BC}$. Which statement of congruence is not provable? **14.** ______

a. $\triangle ABC \cong \triangle FED$ **b.** $\triangle DEF \cong \triangle CBA$
c. $\triangle ABC \cong \triangle FDE$ **d.** $\triangle ABC \cong \triangle CBA$

15. Which postulate or theorem can be used to determine the measure of $\overline{RT}$? **15.** ______

a. SAS Congruence Postulate
b. SSS Congruence Postulate
c. AAS Congruence Theorem
d. ASA Congruence Postulate

16. *Given:* $\overline{BD}$ bisects $\angle ABC$, **16.** ______

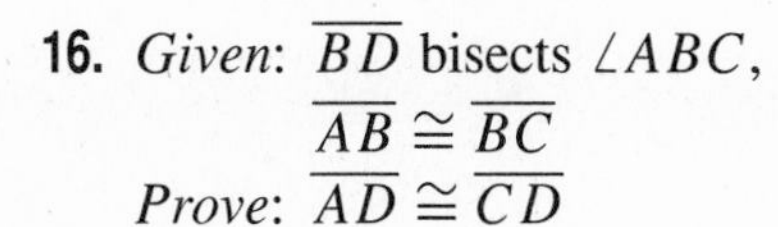

$\overline{AB} \cong \overline{BC}$

Prove: $\overline{AD} \cong \overline{CD}$

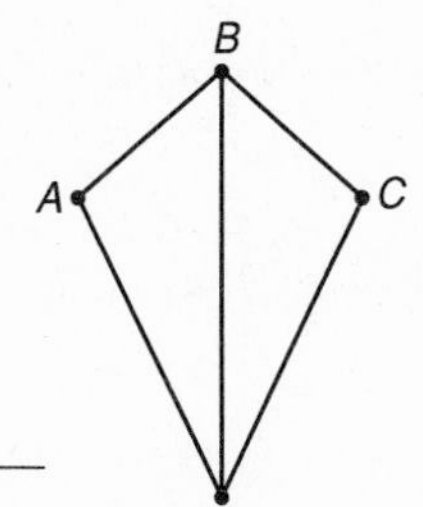

Proof:

Statements	*Reasons*
1. $\overline{AB} \cong \overline{BC}$	**1.** Given
2. $\overline{BD} \cong \overline{BD}$	**2.** Reflexive Prop. of $\cong$
3. $\overline{BD}$ bisects $\angle ABC$	**3.** Given
4. [?]	**4.** Def. of bisector
5. $\triangle ABD \cong \triangle CBD$	**5.** SAS Congruence Postulate
6. $\overline{AD} \cong \overline{CD}$	**6.** CPCTC

The missing step in the proof is [?].

a. $\angle ABC \cong \angle CBA$ **b.** $\angle ABD \cong \angle CBD$
c. $\angle BAD \cong \angle BCD$ **d.** $\angle BDA \cong \angle BDC$

17. *Given:* $\angle B \cong \angle E$ and $\angle C \cong \angle F$. What other piece of information is needed to show $\triangle ABC \cong \triangle DEF$ by ASA Congruence Postulate? **17.** ______

a. $\angle A \cong \angle D$ **b.** $\overline{BC} \cong \overline{EF}$
c. $\angle B = \angle F$ **d.** $\overline{EF} \cong \overline{FE}$

18. If a triangle has three lines of symmetry, then it is [?]. **18.** ______

a. scalene **b.** isosceles **c.** a right triangle **d.** equilateral

Chapter 4 Test **Form C**

(Page 1 of 3 pages)

Name ______________________

Date ______________________

1. If $\triangle RPQ \cong \triangle JKL$, then $\overline{LJ} \cong \boxed{?}$. **1.** ______________

2. Complete the statement using the words *always, sometimes,* or *never.* An isosceles triangle is $\boxed{?}$ an obtuse triangle. **2.** ______________

3. Do theorems require proofs? Explain. **3.**

4. Find the measure of $\angle A$. **4.** ______________

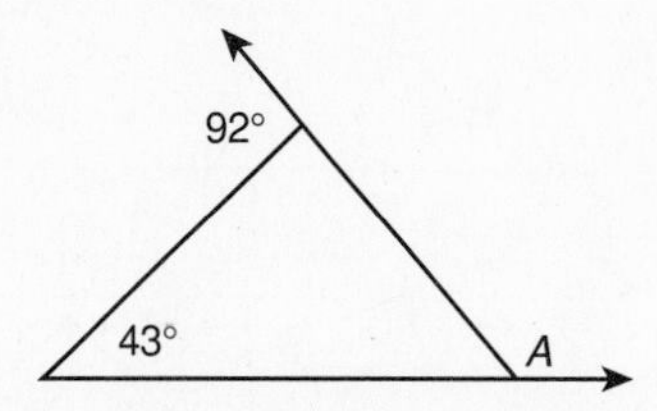

5. Solve for x, given that $\overline{AB} \cong \overline{BC}$. Is $\triangle ABC$ equilateral? **5.** ______________

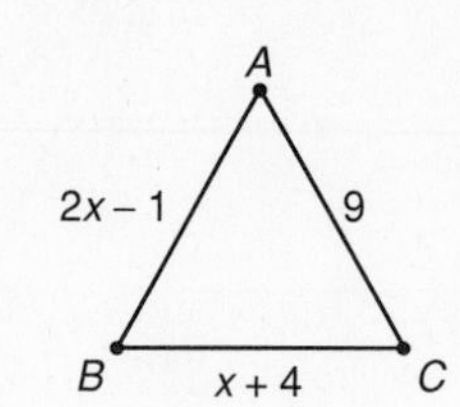

6. For the triangle at the right, $\triangle ABC \cong \triangle CAB$. What type of triangle must $\triangle ABC$ be? Explain. **6.**

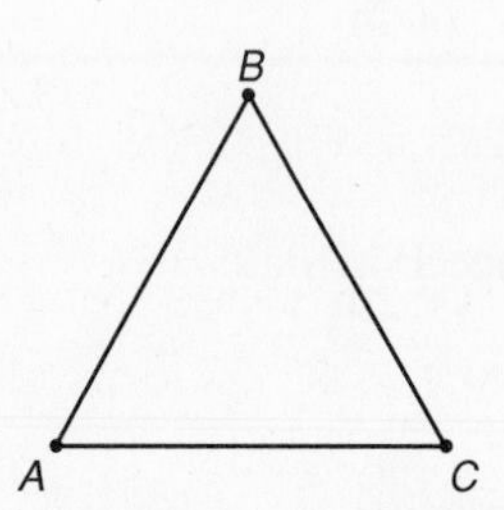

7. Consider a triangle, $\triangle DEF$. Must it be true that $\triangle DEF \cong \triangle FED$? Explain. **7.**

8. Construct and identify a triangle with angle measures of 45°, 45°, and 90°. **8.**

Name ______________________

Date ______________________

9. Find the measures of all three angles of the triangle.

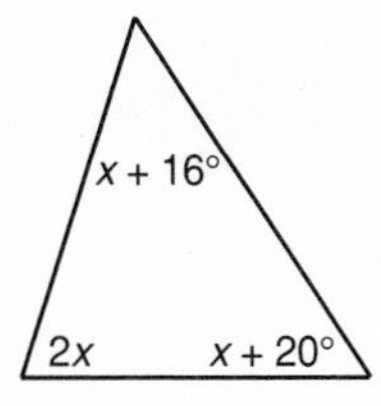

9. ______________________

10. A scalene triangle can have how many right angles? Explain your reasoning.

10. ______________________

11. State *two* postulates or theorems that can be used to conclude that $\triangle AOB \cong \triangle COD$.

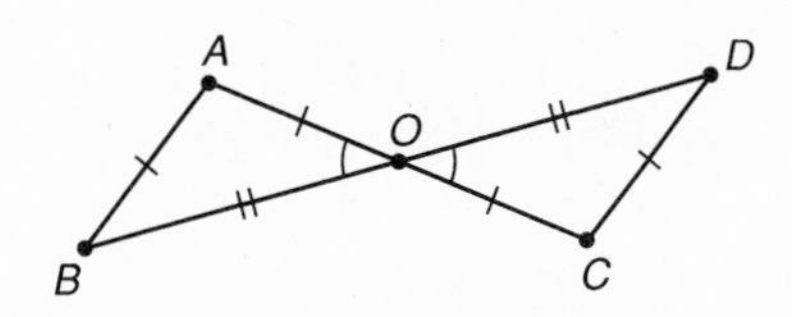

11. ______________________

12. Refer to the figure for Problem 11. Give a reason to justify the statement $\angle CDO \cong \angle COD$.

12. ______________________

13. Refer to the figure.

a. Is there enough information to know whether $\angle FEG$ is acute, obtuse, or a right angle? Explain.

b. If $m\angle F$ is less than 45°, what type of angle is $\angle FEG$? Explain.

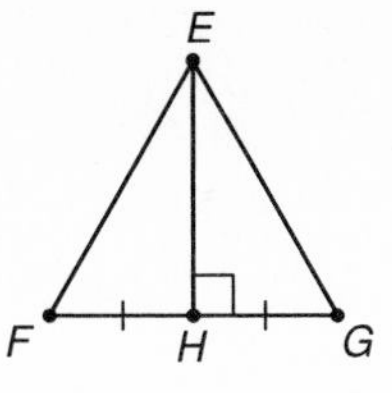

13. a. ______________________

b. ______________________

14. Find the measure of angle B.

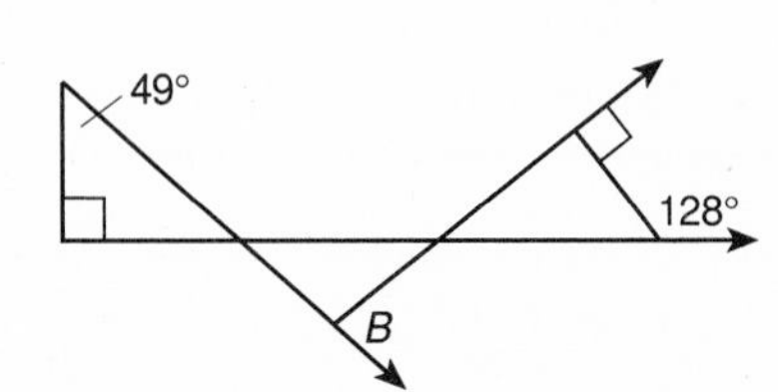

14. ______________________

Name ______________________

Date ______________________

15. $\triangle ABC \cong \triangle DEF$. Also $\overline{AB} \cong \overline{EF}$. What type of triangle is $\triangle ABC$? Explain.

15.

16. How would you determine the distance across the lake since it is not possible to measure it directly? Sketch and describe a method for determining the length of $\overline{AB}$.

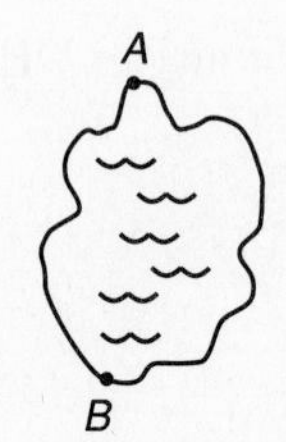

16.

17. *Given*: $\overline{AB} \cong \overline{DE}$

$\angle B \cong \angle E$

Prove: $\triangle ABC \cong \triangle DEC$

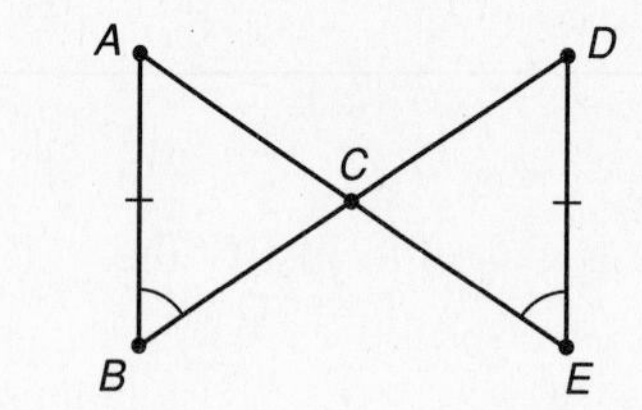

17.

18. *Given*: $\angle 1 \cong \angle 2$

$\overline{AB} \cong \overline{DE}$

$\overline{BC} \cong \overline{CD}$

Prove: $\triangle ABC \cong \triangle EDC$

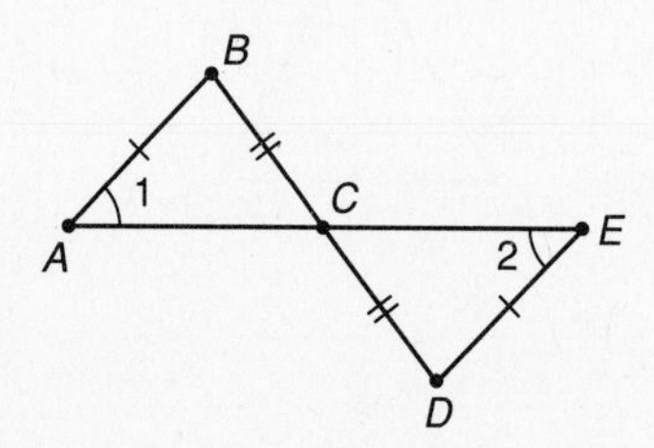

18.

19. Which triples (SSS, SAS, etc.) do not provide enough information to determine the exact size and shape of a triangle?

19. ______________________

5.2 Short Quiz

Name ______________________

Date ______________________

1. *Given:* $\overleftrightarrow{CD}$ is the perpendicular bisector of $\overline{AB}$.
   Complete: **a.** $\angle ACD \cong$ [?]
   b. $\overline{AC}$ [?] $\overline{BC}$

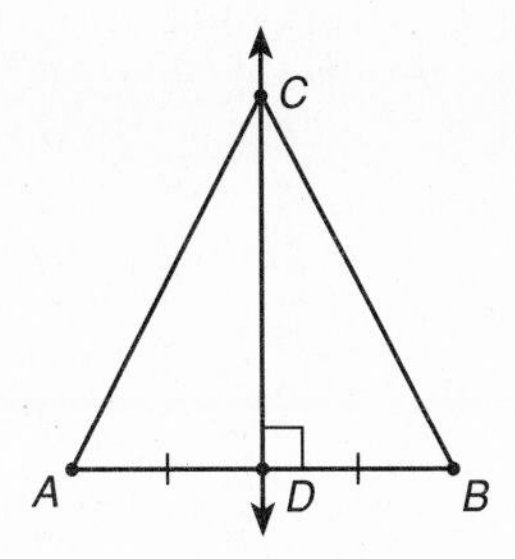

1. ______________________

2. *Given:* $\overrightarrow{OB}$ is the bisector of $\angle AOC$.
   $\overline{AB} \perp \overline{OA}$
   $\overline{CB} \perp \overline{OC}$
   Complete: **a.** $\angle ABO \cong$ [?]
   b. $\overline{BA}$ [?] $\overline{BC}$

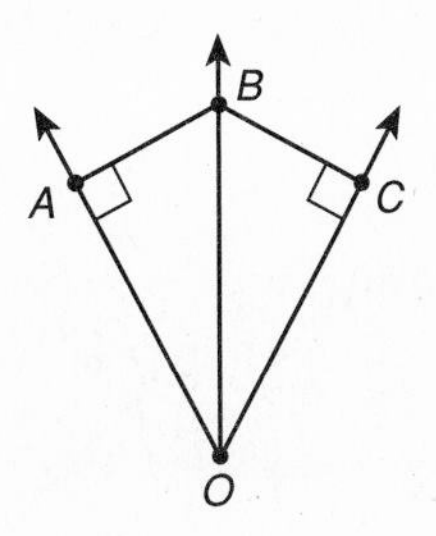

2. ______________________

3. A point is equidistant from points (0, 0), (10, 0), and (0, 10). Find its coordinates.

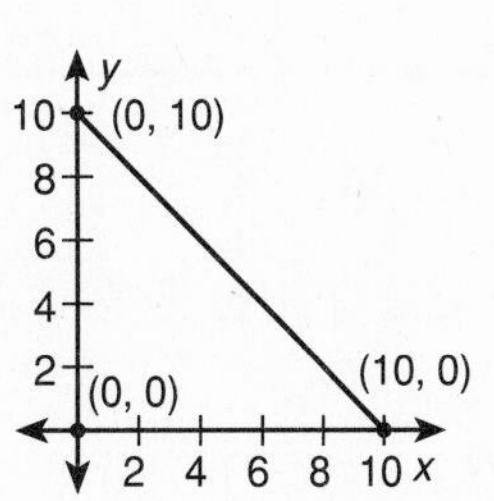

3. ______________________

4. *Given:* $\overleftrightarrow{AD}$ is the perpendicular bisector of $\overline{BC}$.
   $AD = 8$
   $\overline{BE} \cong \overline{EA}$
   $CE = 5$
   Find: **a.** Find AE.
   b. Find DE.

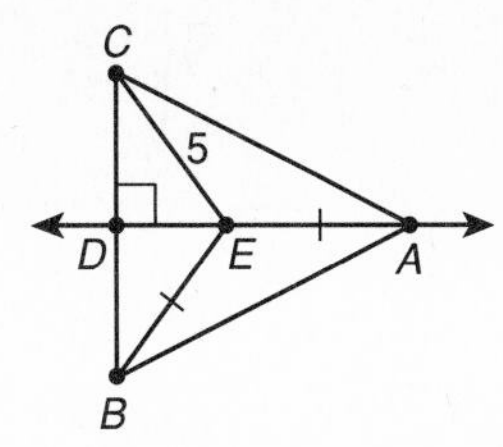

4. **a.** ______________________
   b. ______________________

5. What are the endpoints of a median?

5. ______________________

Mid-Chapter 5 Test Form A

(Use after Lesson 5.3)

Name ______________________

Date ______________________

1. The medians of a triangle are concurrent at a point called the [?].

1. ______________________

2. $\overrightarrow{OD}$ is the bisector of $\angle AOB$.
$\overline{DA} \perp \overline{OA}$; $\overline{DB} \perp \overline{OB}$
$DB =$ [?]

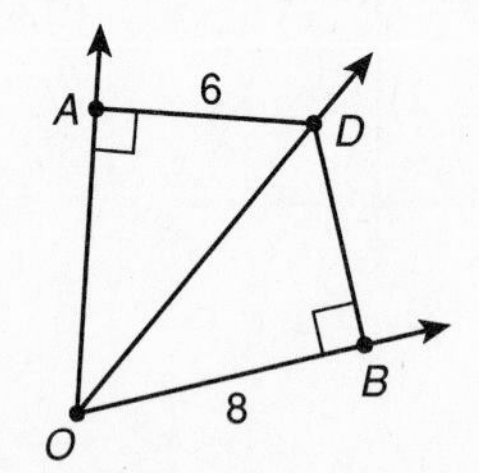

2. ______________________

3. Find the coordinates of the endpoints of the midsegments of $\triangle ABC$ as shown.

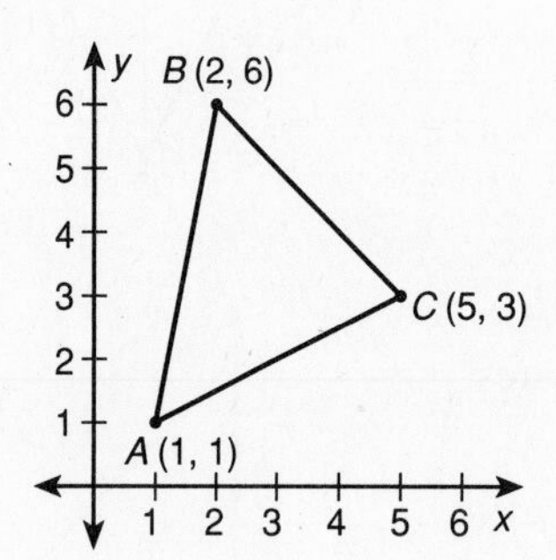

3. ______________________

4. The midpoints of the sides of a triangle are $R(0, -1)$, $S(3, 0)$, and $T(1, 2)$ as shown. Find the coordinates of the vertices of the triangle.

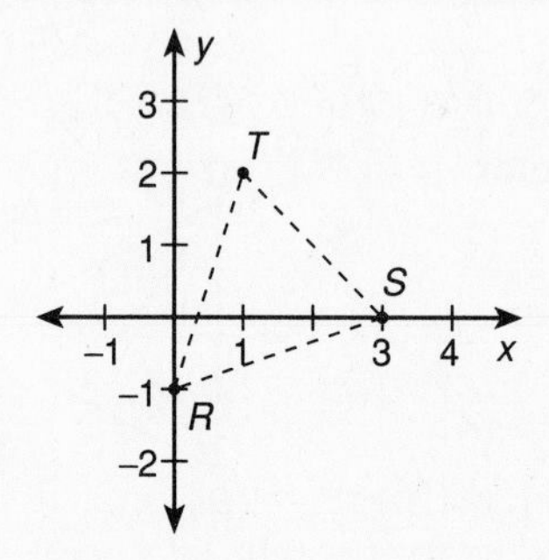

4. ______________________

5. Use a straightedge and compass to construct the incenter of the triangle shown. Then draw the inscribed circle.

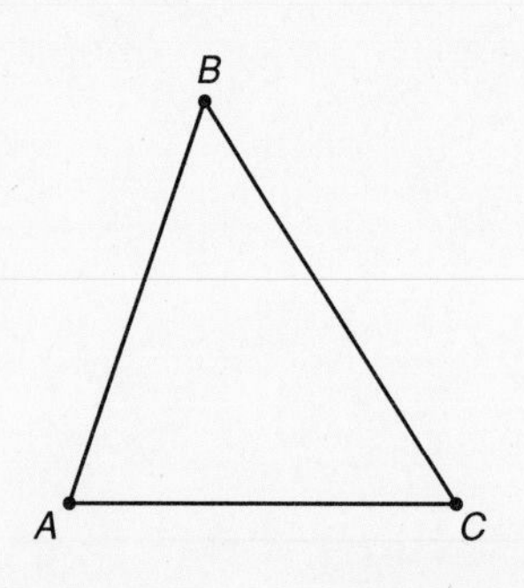

5.

Mid-Chapter 5 Test Form B

(Use after Lesson 5.3)

Name ______________________

Date ______________________

1. The angle bisectors of a triangle are concurrent at a point called the [?].

1. ______________________

2. $\overleftrightarrow{OA}$ is the $\perp$ bisector of $\overline{BC}$.
   $m\angle AOC = 25°$
   Find $m\angle OBC$.

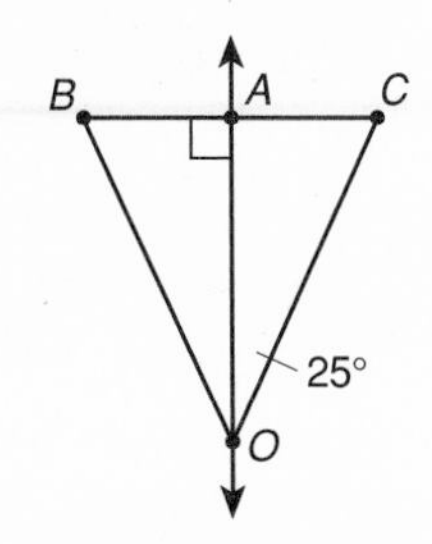

2. ______________________

3. **a.** What are the midsegments of a triangle?
   b. A triangle has how many midsegments?
   c. Name a special property of a midsegment.

3. a. ______________________
   b. ______________________
   c. ______________________

4. The midpoints of the sides of a triangle are $L(-1, 0)$, $M(2, -1)$, and $N(3, 2)$ as shown. Find the coordinate of the vertices of the triangle.

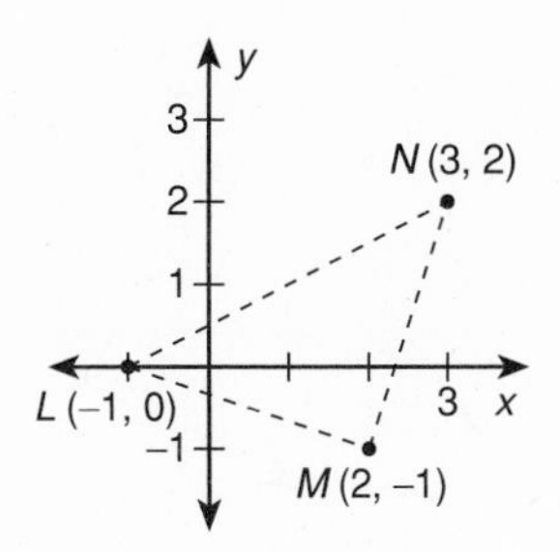

4. ______________________

5. Use a straightedge and compass to construct the circumcenter of the triangle shown and draw the circumscribed circle.

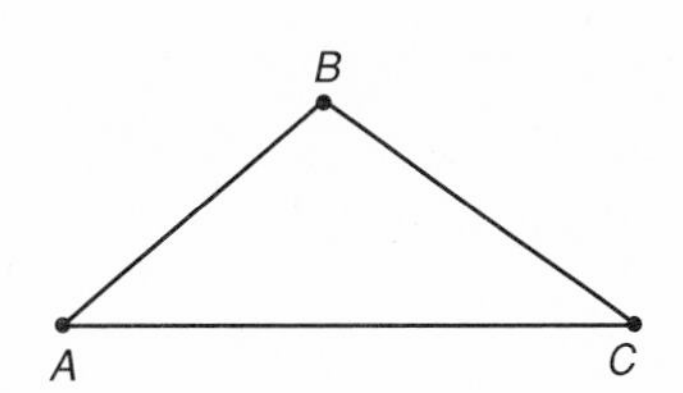

5.

5.4 Short Quiz

Name ______________________

Date ______________________

1. What are the endpoints of a midsegment? 1. ______________

2. For the triangle shown, if $EF = 7$, then $BD = \boxed{?}$. 2. ______________

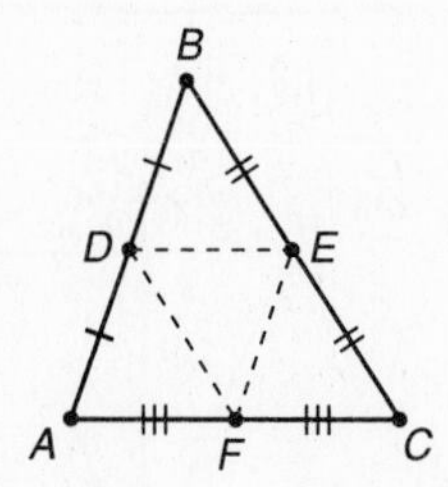

3. Refer to the triangle in Problem 2. If $AC = 4x + 6$ and $DE = 3x - 2$, then $AC = \boxed{?}$. 3. ______________

4. A triangle has side lengths of 5 and 8. What restrictions apply to the third side, x? 4. ______________

5. Name the shortest and longest sides of this triangle. 5. ______________

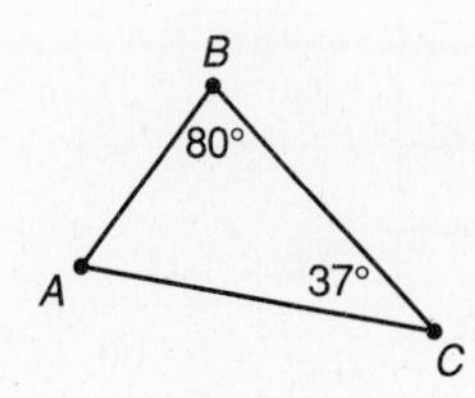

6. Name the smallest and largest angles of this triangle. 6. ______________

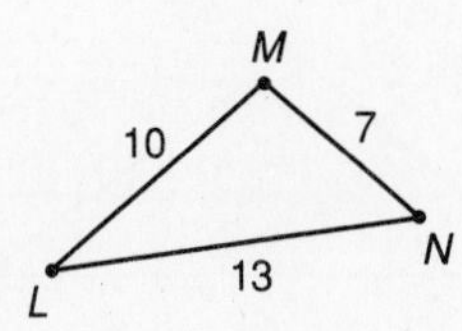

5.6 Short Quiz

Name ______________________

Date ______________________

In Problems 1–5, use the figure below.

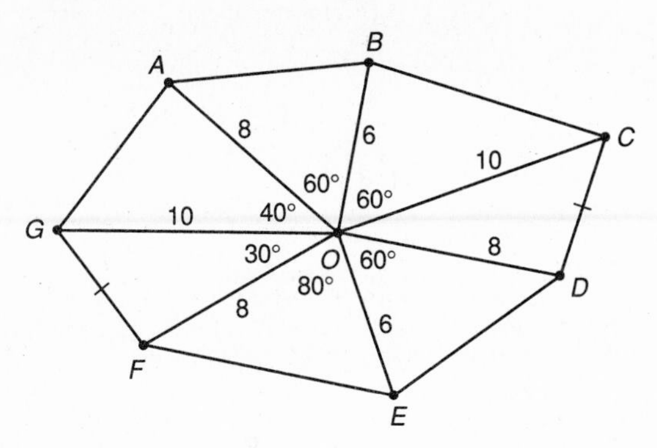

$\overline{CD} \cong \overline{GF}$ Complete each statement with <, >, or =.

1. AB [?] ED **1.** ______________

2. $m\angle GOF$ [?] $m\angle COD$ **2.** ______________

3. ED [?] BC **3.** ______________

4. AG [?] CD **4.** ______________

5. Find $m\angle COD$. **5.** ______________

Chapter 5 Test

Form A

(Page 1 of 3 pages)

Name ______________________

Date ______________________

1. The angle bisectors of a triangle are concurrent at a point called the [?].

1. ______________

2. The circumcenter of a triangle is equidistant from the three [?] of the triangle.

2. ______________

3. $\overleftrightarrow{NO}$ is the perpendicular bisector of $\overline{LM}$. If $OM = 4$ and $LN = 6$, then $LO =$ [?] and $MN =$ [?]. Explain your solutions.

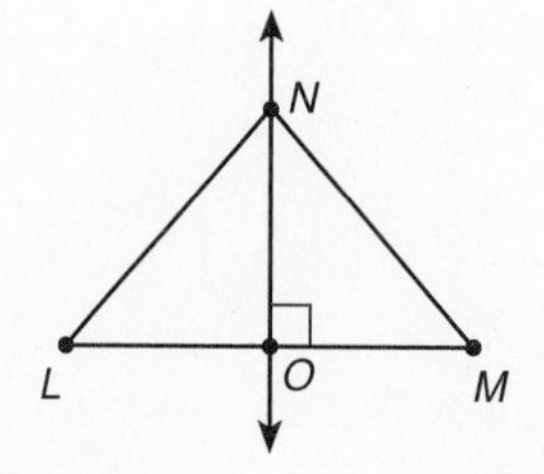

3.

4. Find the coordinates of the endpoints of the midsegments of $\triangle ABC$ as shown.

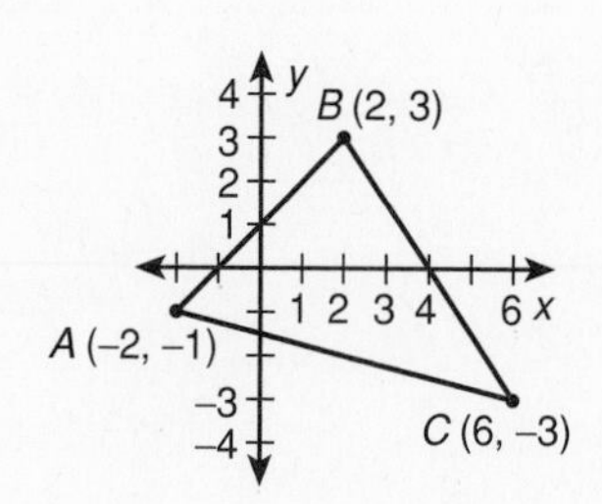

4. ______________

5. A median of a triangle connects what points of the triangle?

5. ______________

6. How many midsegments does a triangle have?

6. ______________

7. Use a straightedge and compass to construct the circumcenter of the triangle shown. Then draw the circumscribed circle.

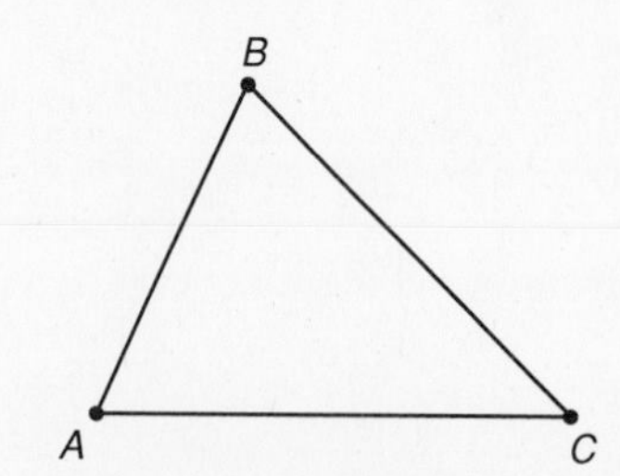

7.

8. A point is equidistant from points $A(2, 1)$, $B(2, -4)$, and $C(-3, 1)$. Find its coordinates. **8.** ______

9. Refer to the figure.
If $BC = 15$, then $LN = \boxed{?}$. **9.** ______

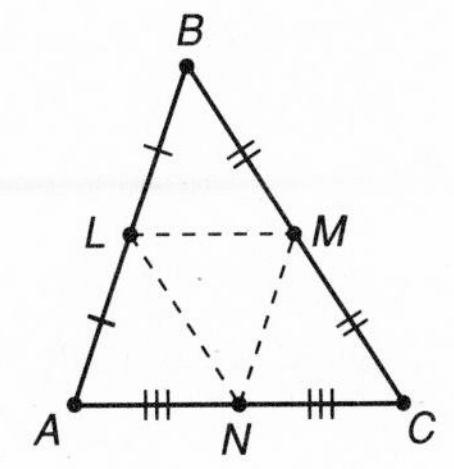

10. Refer to the triangle in Problem 9.
If $AB = 3x + 5$ and $NM = 2x + 1$, then $NM = \boxed{?}$. **10.** ______

11. Refer to the figure. Find $m\angle PQR$. **11.** ______

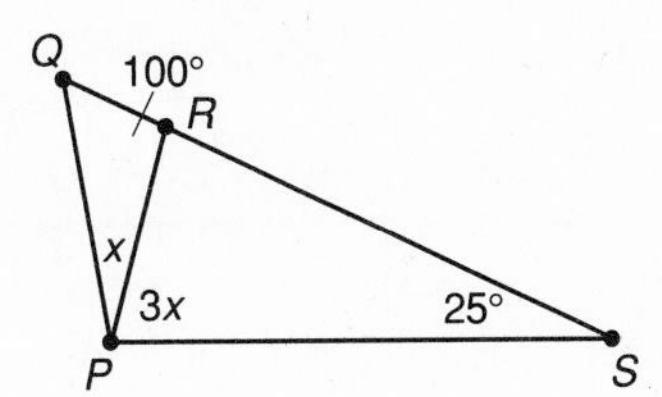

12. The medians of a triangle are concurrent. Their common point is called what? **12.** ______

In Problems 13 and 14, refer to the figure.

Given: $\overline{AF} \cong \overline{FC}$
$\angle ABE \cong \angle EBC$

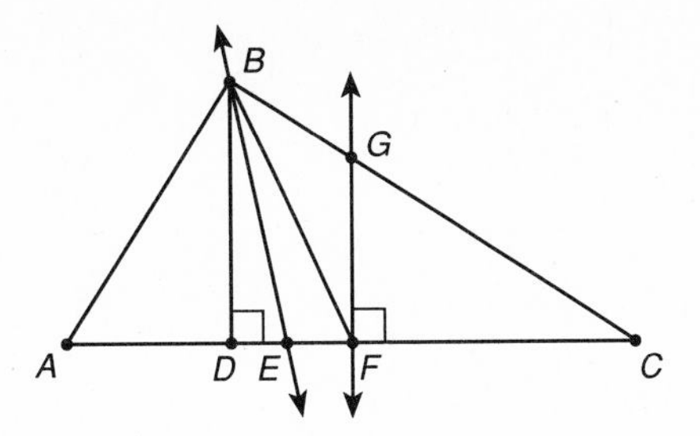

13. Which segment is a median of $\triangle ABC$? **13.** ______

14. Which line is a perpendicular bisector in $\triangle ABC$? **14.** ______

15. Refer to the figure. Which of the six angles in the two triangles is the largest?

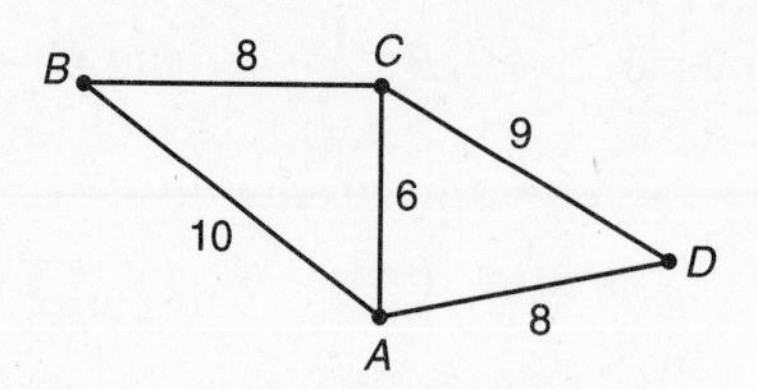

15. ______________

16. Two sides of a triangle are 8 and 11. What are the possible measures of the third side x?

16. ______________

17. Refer to the figure. Use the given information to complete the statement using <, >, or =.

Given: $\overline{AB} \cong \overline{AC}$,

$m\angle 1 < m\angle 2$

CD [?] DB

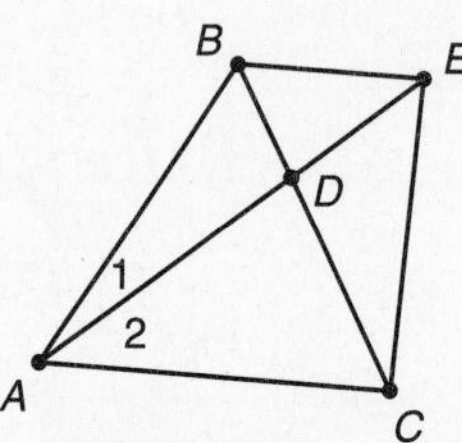

17. ______________

18. Using the Triangle Inequality Theorem, solve for all possible values of x.

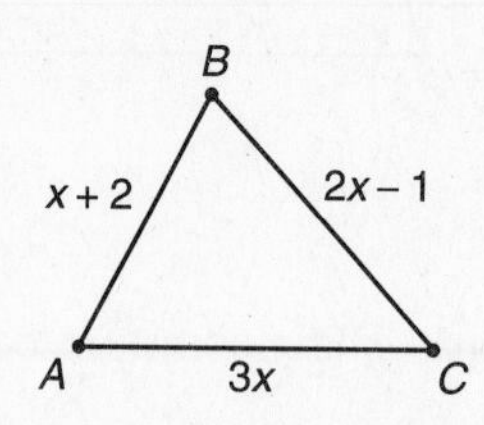

18. ______________

19. If $\overline{QR}$ is an altitude of $\triangle PQR$, what type of triangle is $\triangle PQR$?

19.

20. *Prove*: An angle bisector of an equilateral triangle is perpendicular to the opposite side.

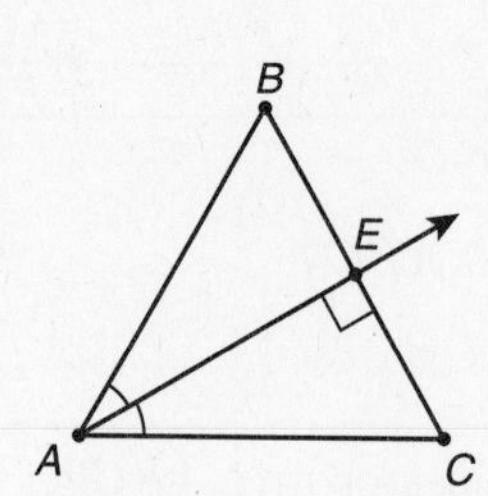

20.

Chapter 5 Test **Form B**

(Page 1 of 3 pages)

Name ______________________

Date ______________________

1. A point is equally distant from points (0, 0), (4, 0), and (4, 4). Its coordinates are $\boxed{?}$.

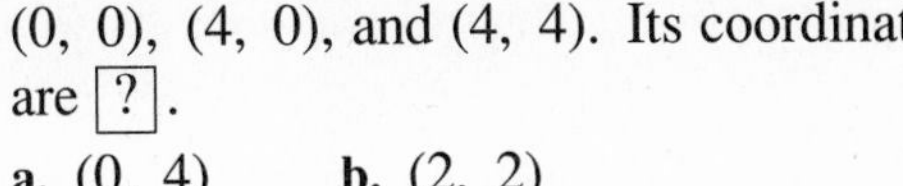

a. (0, 4) b. (2, 2)

c. (2, 1) d. (1, 3)

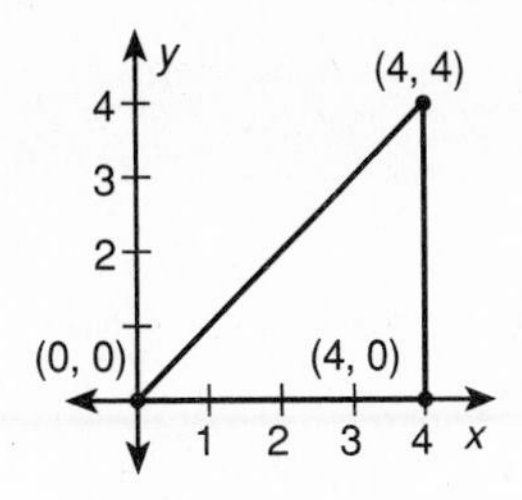

1. ________

2. $\overleftrightarrow{KF}$ is the perpendicular bisector of $\overline{GH}$. Then $\angle KGF \cong \boxed{?}$.

a. $\overline{KF}$ b. $\angle KFH$
c. $\angle KHF$ d. $\angle FKG$

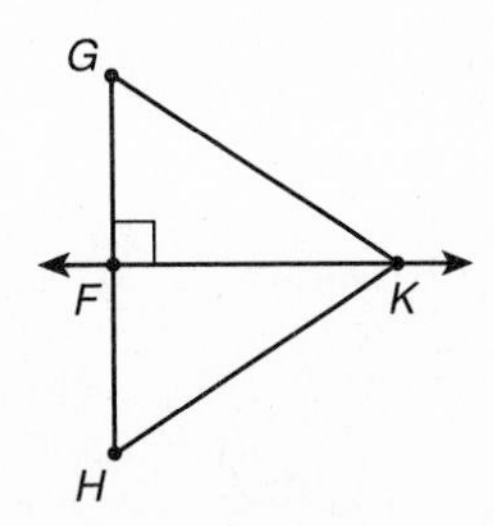

2. ________

3. $\overrightarrow{OE}$ bisects $\angle BOA$, $\overline{EA} \perp \overline{OA}$, and $\overline{EB} \perp \overline{OB}$. Which statement is *not* true?

a. $\angle AOE \cong \angle EAO$ b. $\overline{AE} \cong \overline{BE}$
c. $\overline{OA} \cong \overline{OB}$ d. $\angle AEO \cong \angle BEO$

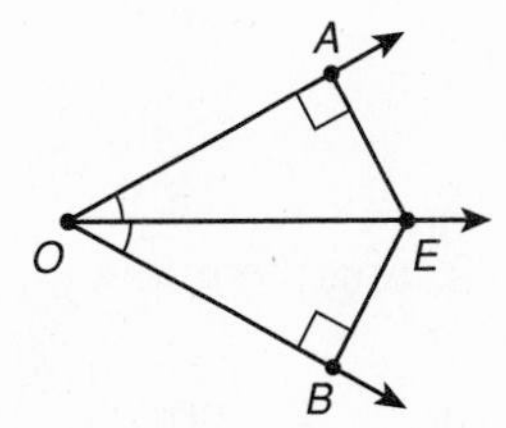

3. ________

4. Refer to the figure.
Given: $\overline{PQ} \perp \overline{QR}$.
Find $m\angle QSP$.

a. 50° b. 80°
c. 30° d. 100°

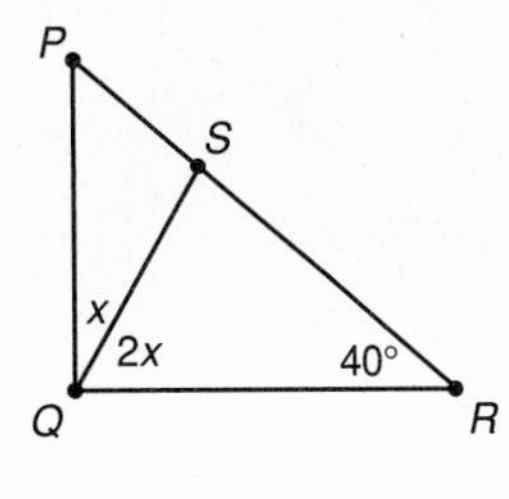

4. ________

In Problems 5 and 6, use the figure below.

Given: $\overline{AF} \cong \overline{FC}$, $\angle ABE \cong \angle EBC$

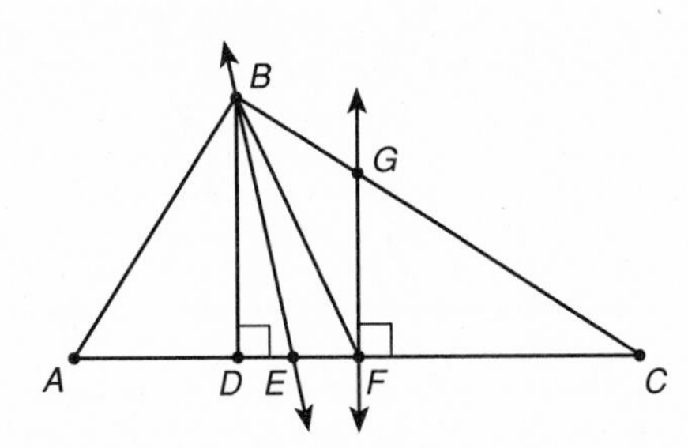

5. A median of $\triangle ABC$ is $\boxed{?}$.

a. $\overline{BD}$ b. $\overleftrightarrow{BE}$ c. $\overline{BF}$ d. $\overleftrightarrow{GF}$

5. ________

6. An altitude of $\triangle ABC$ is $\boxed{?}$.

a. $\overline{BD}$ b. $\overleftrightarrow{BE}$ c. $\overline{BF}$ d. $\overleftrightarrow{GF}$

6. ________

Name ______________________

7. The medians of a triangle are concurrent. Their common point is the [?].

a. incenter **b.** circumcenter **c.** orthocenter **d.** centroid

7. ________

8. The center of the inscribed circle of a triangle lies on all three of the circle's [?].

a. altitudes **b.** medians

c. angle bisectors **d.** perpendicular bisectors

8. ________

9. Which line is not a median of this triangle?

a. $x = 0$ **b.** $y = 0$

c. $y = \frac{1}{2}x + 2$ **d.** $y = -\frac{1}{2}x + 2$

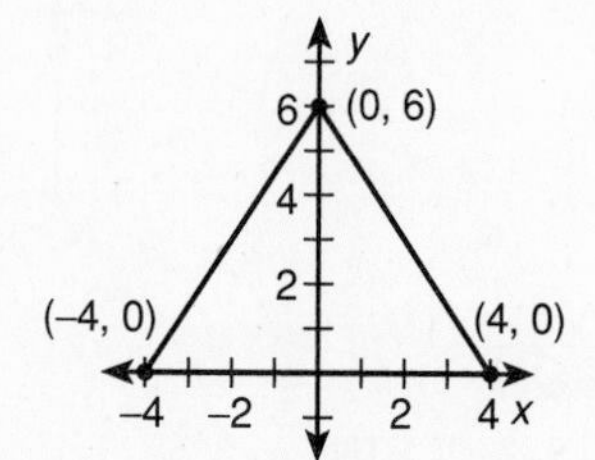

9. ________

10. In a triangle, a segment connecting the midpoints of two sides of the triangle is called a [?].

a. midsegment **b.** centroid **c.** vertrix **d.** shortcut

10. ________

11. For the triangle shown, $VS = 5$ and $VQ = 6$. Then $PQ =$ [?].

a. 11 **b.** 10

c. 12 **d.** 5

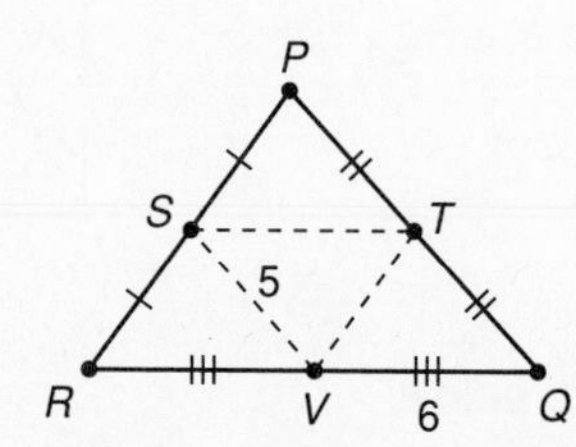

11. ________

12. Refer to the figure. The longest side is [?].

a. $\overline{NM}$ **b.** $\overline{LN}$

c. $\overline{ML}$ **d.** $\overline{MP}$

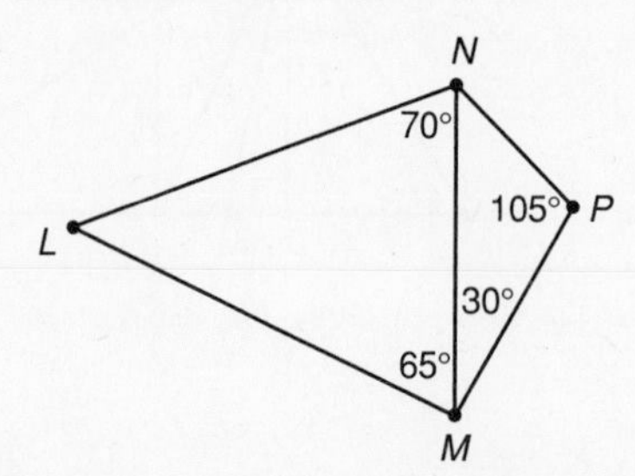

12. ________

Name ____________

Date ____________

13. Two sides of a triangle are 7 and 13. The third side is [?]. **13.** ______

a. > 20 **b.** < 20 and > 6

c. > 6 and < 13 **d.** < 6

14. Using the Triangle Inequality Theorem, solve for all possible values of x. **14.** ______

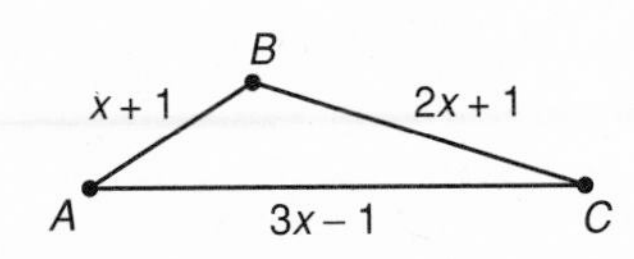

a. $x < \frac{1}{4}$ **b.** $x > \frac{1}{2}$

c. $x < \frac{1}{2}$ **d.** $x > \frac{1}{4}$

15. Refer to the figure. **15.** ______

Given: $AB = AD$,

$m\angle 1 > m\angle 2$

Then, [?].

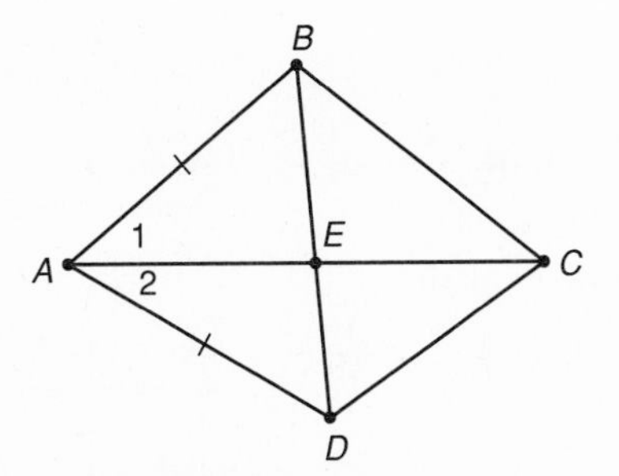

a. $BE > ED$ **b.** $BE = ED$

c. $BE < ED$ **d.** $AE = EC$

16. Refer to the figure. Choose the correct statement. **16.** ______

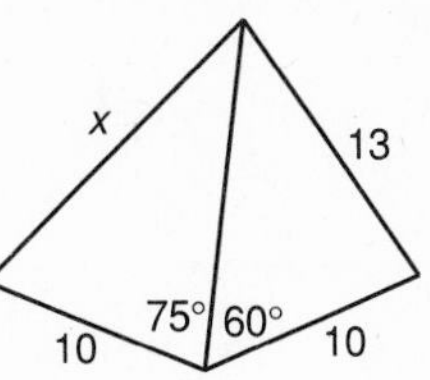

a. $10 < x < 13$ **b.** $x < 10$

c. $x = 13$ **d.** $x > 13$

17. $\overline{PR}$ is an altitude of $\triangle PQR$. Therefore, $\triangle PQR$ is [?]. **17.** ______

a. right **b.** isosceles **c.** acute **d.** equilateral

18. *Given:* $\overline{AB} \cong \overline{BC}$, **18.** ______

$AC > BC$

Then, [?].

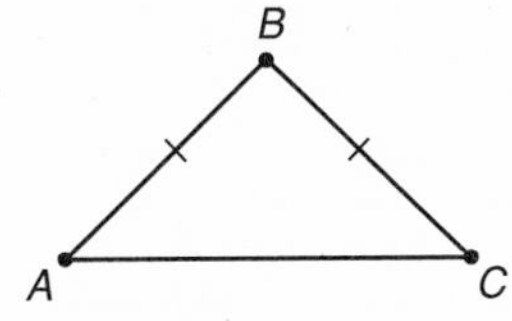

a. $m\angle A > m\angle B$ **b.** $AB > AC$

c. $m\angle B > m\angle C$ **d.** $AC = AB$

Chapter 5 Test

Form C

(Page 1 of 3 pages)

Name ______________________

Date ______________________

1. The perpendicular bisectors of a triangle all pass through what point?

1. ______________________

2. Complete the statement. The incenter of a triangle is equidistant from the three [?] of the triangle.

2. ______________________

3. $\overleftrightarrow{SU}$ is the bisector of $\angle RST$.
   $\overline{UR} \perp \overline{RS}$
   $\overline{UT} \perp \overline{ST}$
   Complete: $RS =$ [?], $UT =$ [?].

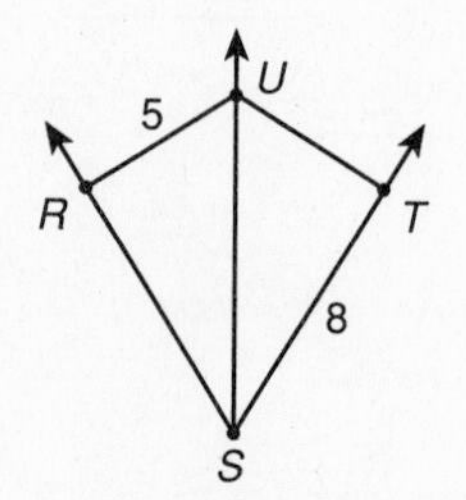

3. ______________________

4. $U(-2, -1)$, $V(2, 3)$, and $R(5, 0)$ are the midpoints of the sides of a triangle. Find the coordinates of the vertices of the triangle.

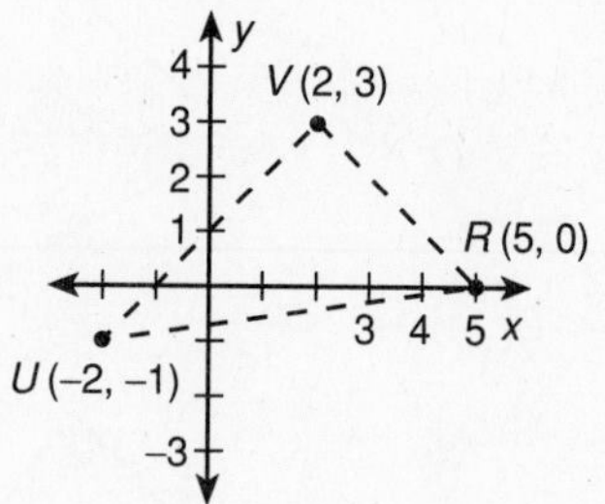

4. ______________________

5. What is a midsegment of a triangle?

5.

6. How many medians does a triangle have?

6. ______________________

7. Use a straightedge and compass to construct the incenter of the triangle shown and draw the inscribed circle.

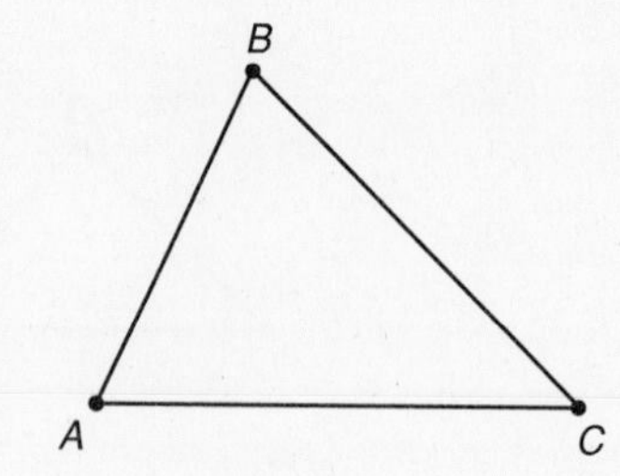

7. *Use figure at left.*

Name ____________________

Date ____________________

8. The vertices of a triangle are $A(4, -3)$, $B(-6, -3)$, and $C(-1, -8)$. Find its circumcenter.

8. 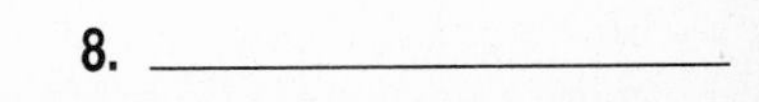

9. For the given triangle, state the relationships between $\overline{AB}$ and $\overline{DF}$.

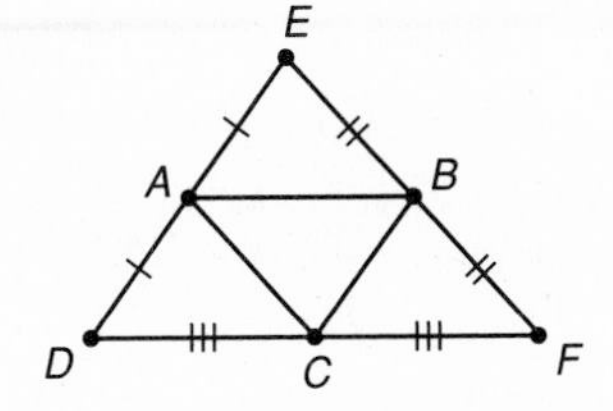

9. ____________________

10. Refer to the figure in Problem 9. If $EF = 5x + 6$ and $AC = 3x - 2$, then $BF = \boxed{?}$.

10. ____________________

11. Refer to the figure. Find $m\angle PRS$.

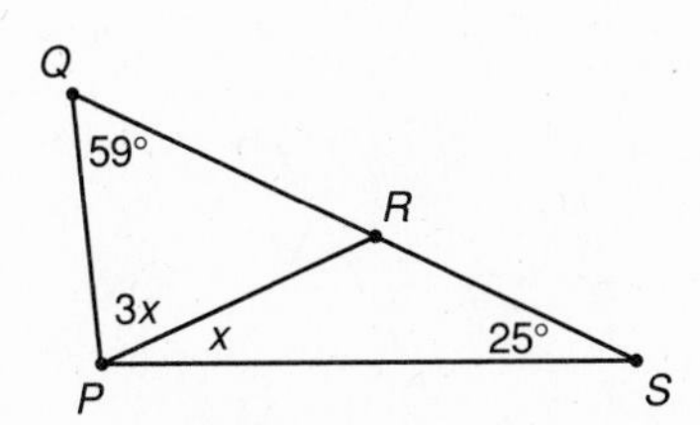

11. ____________________

12. The altitudes of a triangle are concurrent. What is the name of their common point?

12. ____________________

In Problems 13 and 14, use the sketch below.

Given: $\overline{AF} \cong \overline{FC}$
$\angle ABE \cong \angle EBC$

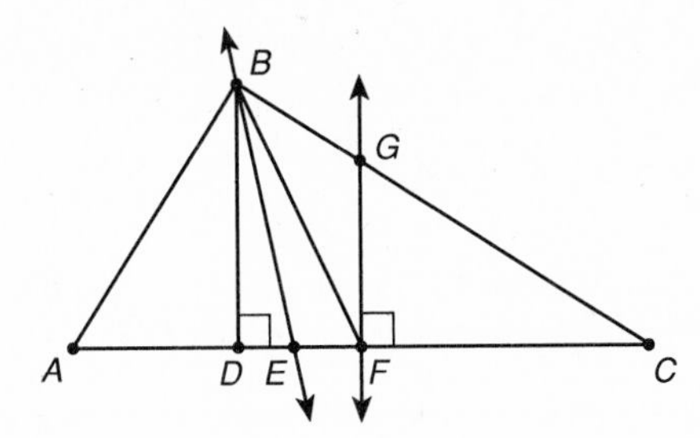

13. Which line is an angle bisector in $\triangle ABC$?

13. ____________________

14. Which segment is an altitude of $\triangle ABC$?

14. ____________________

Name ____________________

Date ____________________

15. Refer to the figure. Which of the six angles in the two triangles is the largest.

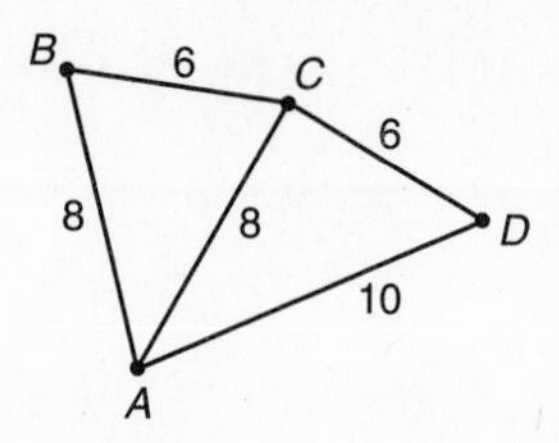

15. ____________________

16. Two sides of a triangle are 14 and 10. What are the possible measures of the third side, x?

16. ____________________

17. Refer to the figure.

Given: $\overline{AB} \cong \overline{AD}$, $BE > ED$

What is the relationship ($<$, $>$, or $=$) between $m\angle BAE$ and $m\angle DAE$?

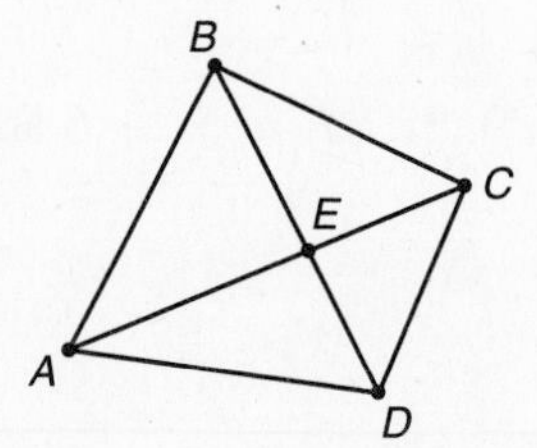

17. ____________________

18. Using the Triangle Inequality Theorem, solve for x (all possible values).

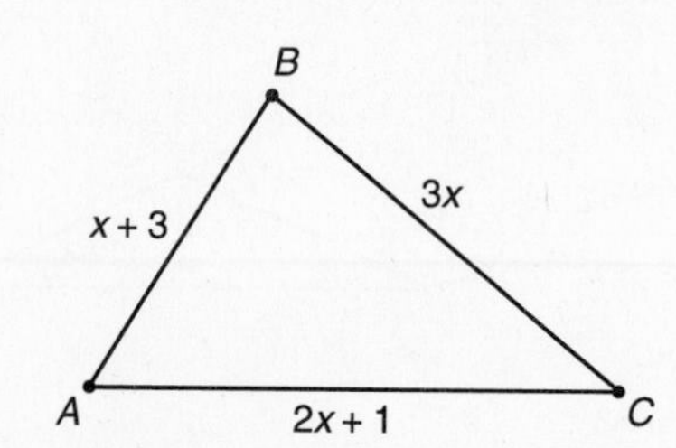

18. ____________________

19. If the incenter of a triangle is also its circumcenter, what type of triangle is it? Explain.

19. ____________________

20. *Prove*: An altitude from an angle of an equilateral triangle bisects that angle.

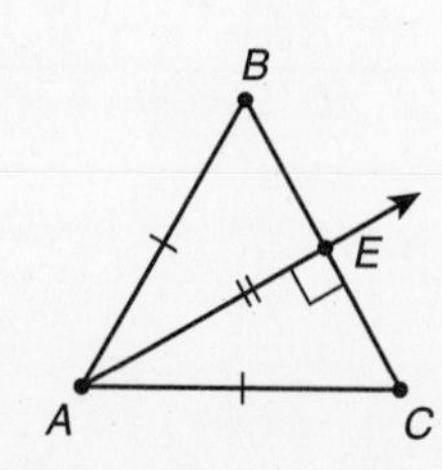

20. *Use space at left.*

6.2 Short Quiz

Name ________________

Date ________________

1. Sketch a convex polygon which has at least four sides. 1.

2. Sketch a polygon which is not convex. 2.

3. Define regular polygon. Sketch an example. 3.

4. How many diagonals does a convex quadrilateral have? 4. ________________

5. What is the measure of each interior angle in a regular pentagon? 5. ________________

6. What is the measure of each exterior angle in a regular hexagon? 6. ________________

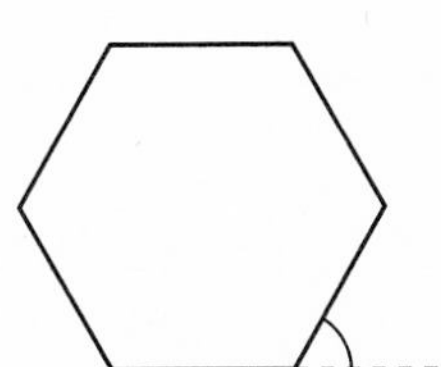

Mid-Chapter 6 Test Form A

(Use after Lesson 6.3)

Name ____________________

Date ____________________

1. Define convex. Then sketch an example of a convex polygon. 1.

2. Sketch, if possible, an equilateral hexagon that is not regular. 2.

3. How many diagonals does a convex pentagon have? 3. ____________

4. What is the sum of the measures of the interior angles of a hexagon? 4. ____________

5. Find the measure of $\angle A$. 5. ____________

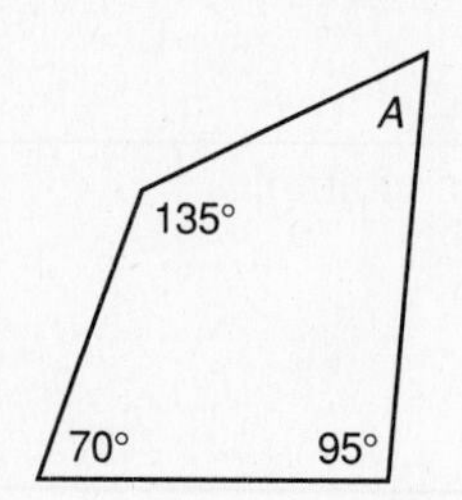

6. For the parallelogram shown, 6. ____________
   a. If $EC = 5$, find AC.
   b. If $m\angle ABC = 80°$, find $m\angle DCB$.

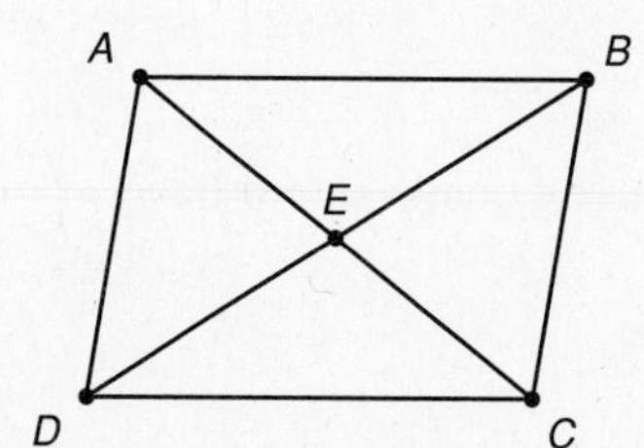

7. *Given:* $\overline{PQ} \parallel \overline{SR}$ 7.
   $\overline{PS} \parallel \overline{QR}$
   Prove Theorem 6.3: If a quadrilateral is a parallelogram, then its opposite sides are congruent. Justify each step.

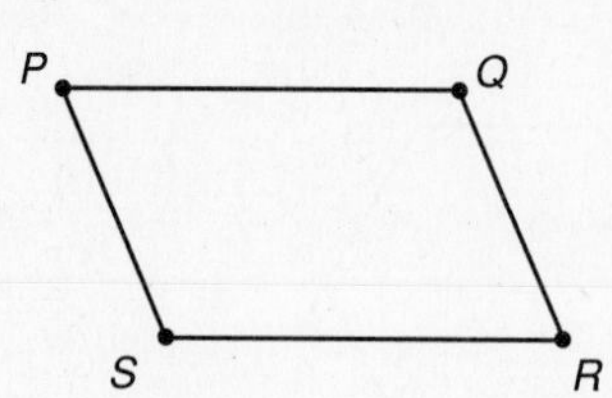

Mid-Chapter 6 Test Form B

(Use after Lesson 6.3)

Name ____________________

Date ____________________

1. Define what is meant by "not convex." Sketch an example of a polygon which is not convex.

1.

2. Sketch, if possible, an equiangular hexagon that is not regular.

2.

3. How many diagonals does a convex heptagon have?

3. ____________________

4. What is the sum of the measures of the interior angles of a pentagon?

4. ____________________

5. Find the measure of $\angle B$.

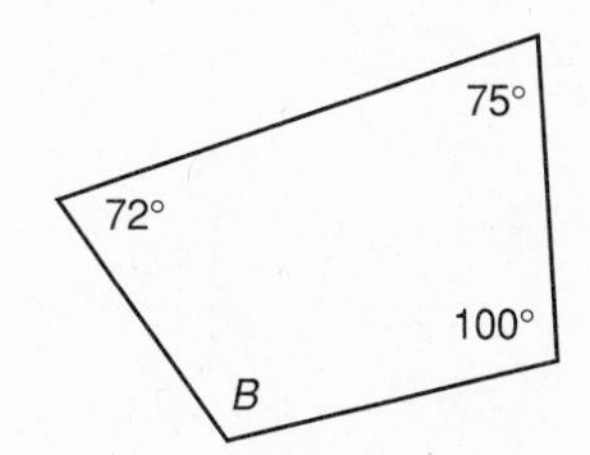

5. ____________________

6. For the parallelogram shown,
   a. If $SQ = 12$, find TQ.
   b. If $m\angle SRQ = 110°$, find $m\angle PSR$.

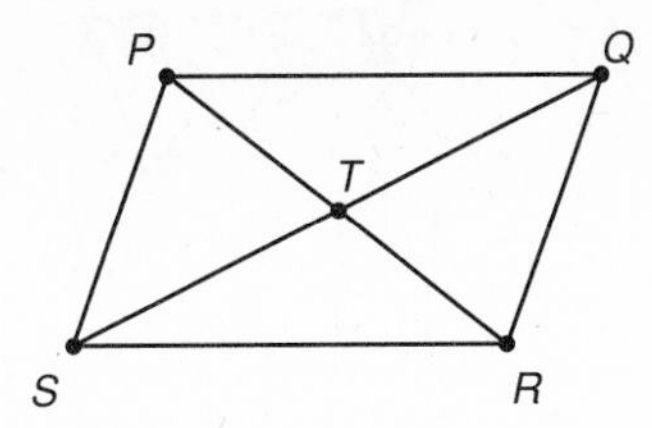

6. a. ____________________
   b. ____________________

7. *Given:* $ABCD$ is a parallelogram.
   Prove: $\angle BCA \cong \angle DAC$

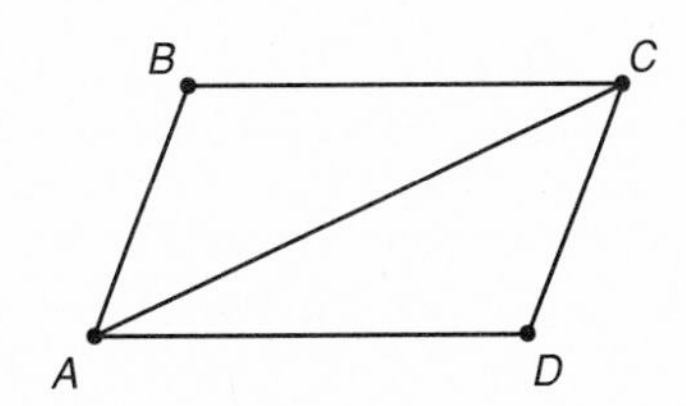

7.

6.4 Short Quiz

Name ____________________

Date ____________________

In Problems 1–4, use the figure below.

Given: $PQRS$ is a parallelogram.
$m\angle QPS = 60°$, $PQ = 15$, and $PS = 23$.

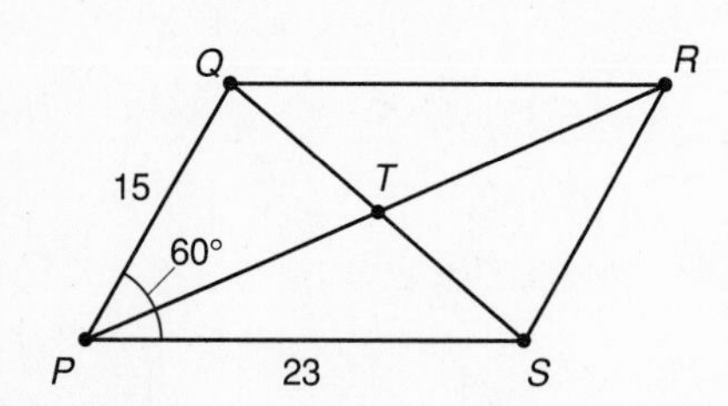

1. Find $m\angle PQR$. **1.** ____________________

2. Find the length of $\overline{RS}$. **2.** ____________________

3. Find $m\angle SRQ$. **3.** ____________________

4. Find the length of $\overline{QR}$. **4.** ____________________

5. Use the distance formula to determine whether $ABCD$ is a parallelogram. **5.** ____________________

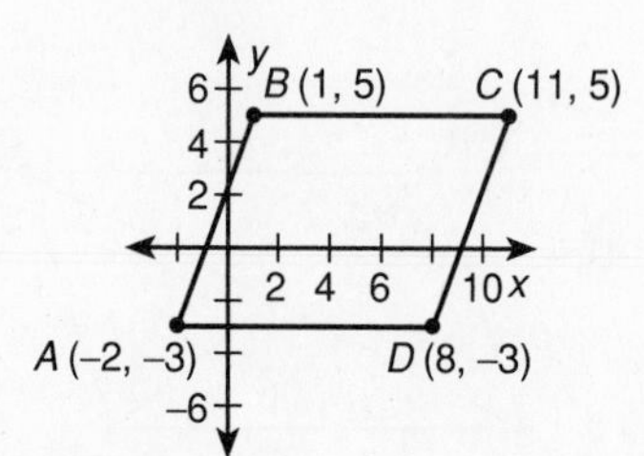

6. Sketch, if possible, a quadrilateral with two consecutive congruent sides that is not a parallelogram. **6.**

7. Sketch, if possible, a quadrilateral with two opposite congruent sides that is not a parallelogram. **7.**

6.6 Short Quiz

Name ______________________

Date ______________________

1. Use construction tools to construct a rhombus, one of whose angles has a measure of 60°. Show all construction lines.

1.

In Problems 2–4, refer by letter to the following properties.

A. Diagonals are perpendicular
B. Diagonals are congruent
C. Diagonals bisect opposite angles
D. All sides are congruent
E. All angles are congruent

2. Which of the properties does a nonspecial rhombus have?

2. ______________

3. Which of the properties does a nonsquare rectangle have?

3. ______________

4. Which of the properties does a square have?

4. ______________

5. **a.** Identify the type of quadrilateral shown.

   b. Find $m\angle B$.

   c. Find the measure of the midsegment.

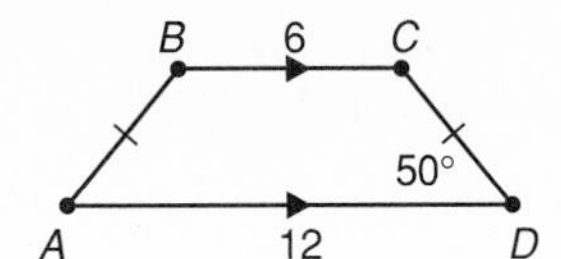

5. a. ______________

   b. ______________

   c. ______________

Chapter  Test

Form A

(Page 1 of 3 pages)

Name ______________________________

Date ______________________________

1. Sketch a five-sided polygon that is convex. **1.**

2. How many diagonals does a convex pentagon have? **2.** ____________________

3. What properties of a polygon make it regular? Sketch an example. **3.**

4. Explain why the figure shown does not satisfy the definition of a polygon.

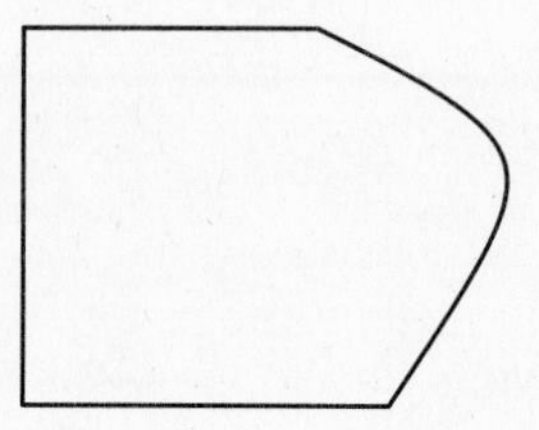

4.

5. What is the measure of each interior angle in a regular octagon? **5.** ____________________

6. A hexagon has six sides of various lengths. What is the sum of the measures of its interior angles? **6.** ____________________

7. What is the measure of each exterior angle in a regular pentagon? **7.** ____________________

Name ______________________
Date ______________________

In Problems 8–11, use the figure below.

Given: $FGHJ$ is a parallelogram.
$m\angle JHG = 68°$; $JH = 34$; and $GH = 19$.

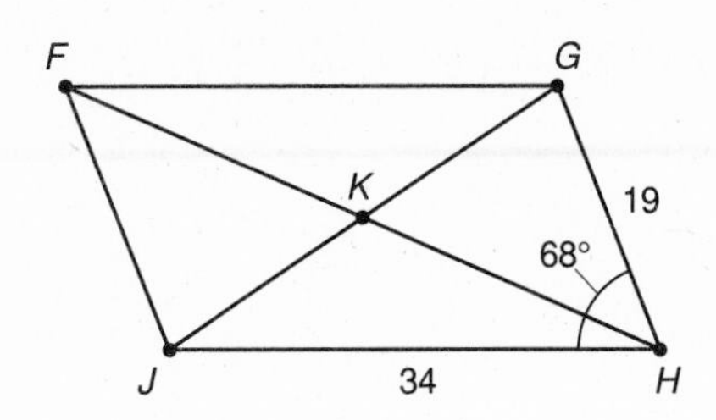

8. Find $m\angle FJH$. **8.** ______________

9. Find the length of $\overline{JF}$. **9.** ______________

10. Find $m\angle GFJ$. **10.** ______________

11. Find the length of $\overline{FG}$. **11.** ______________

12. Sketch, if possible, a quadrilateral with two consecutive supplementary angles that is not a parallelogram. **12.**

13. Sketch, if possible, a quadrilateral with three congruent sides but with no parallel sides. **13.**

14. Use the distance formula to determine whether $ABCD$ is a parallelogram. **14.** ______________

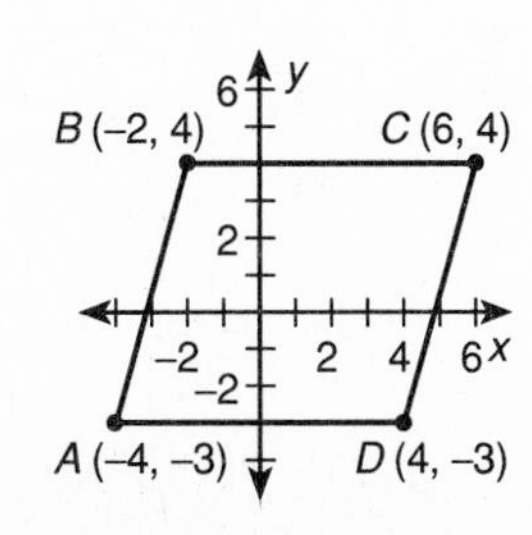

Name ______________________

Date ______________________

15. Consider the statement, "If a parallelogram is a square, then it is a rhombus."

a. Decide whether it is true or false.

b. Write the converse.

c. Decide whether the converse is true or false.

15.

16. If the diagonals of a parallelogram are equal then the parallelogram is also what type of figure?

16. ______________________

17. Draw a trapezoid whose parallel sides measure 1 inch and $1\frac{1}{2}$ inches. Find the length of the midsegment.

17.

18. In what type of trapezoid are the base angles congruent?

18. ______________________

19. Draw a kite in which one pair of congruent sides is twice the length of the other pair of congruent sides.

19.

20. What property must the diagonals of a quadrilateral have in order for it to be a kite?

20. ______________________

Chapter 6 Test

Form B

Name ______________________

Date ______________________

1. Which figure is *not* a polygon? **1.** ______

a. 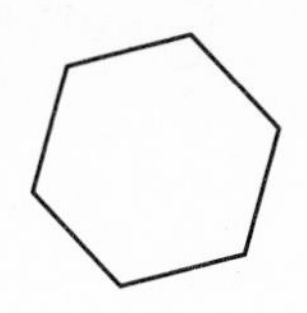 **b.** 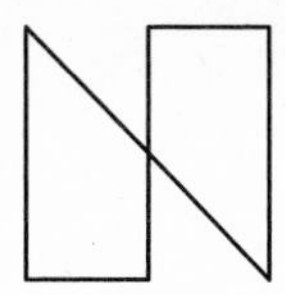 **c.** 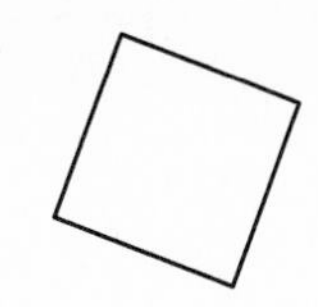 **d.** 

2. Which figure is *not* a convex polygon? **2.** ______

a. 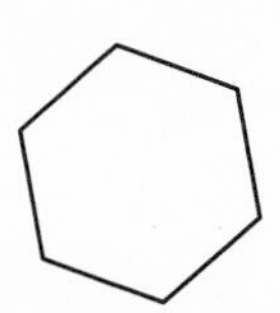 **b.** 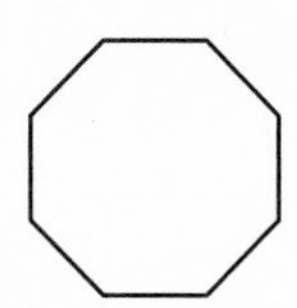 **c.** 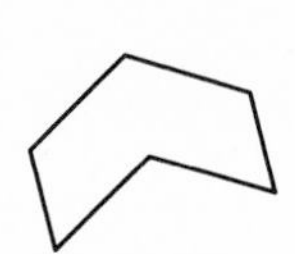 **d.** 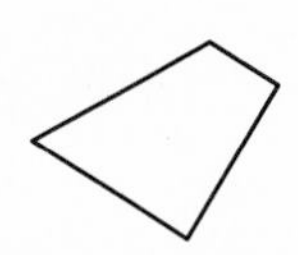

3. Which figure is *not* a regular polygon? **3.** ______

a.  **b.** 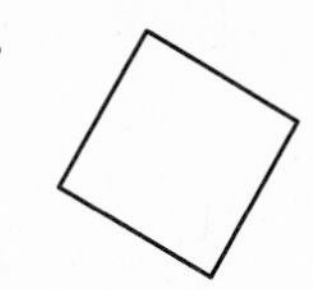 **c.**  **d.** 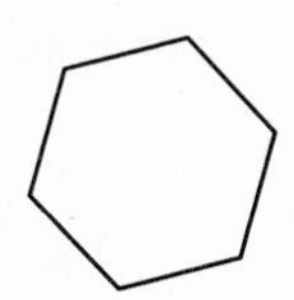

4. The figure is an example of a [?]. **4.** ______

a. hexagon **b.** heptagon
c. octagon **d.** nonagon

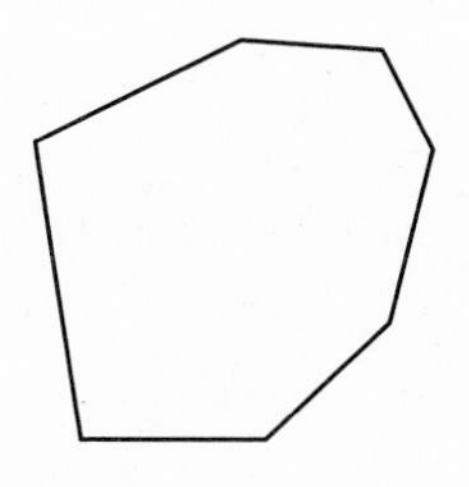

5. The sum of the measures of the interior angles of a convex quadrilateral is [?]. **5.** ______

a. 270° **b.** 180° **c.** 540° **d.** 360°

6. The measure of each interior angle of a regular hexagon is [?]. **6.** ______

a. 120° **b.** 60°
c. 30° **d.** 15°

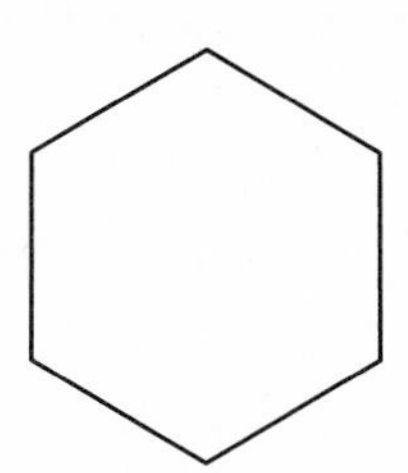

Name ______________________

Date ______________________

7. The measure of each exterior angle of a regular octagon is ?. **7.** ______

a. 135° b. 67.5°
c. 45° d. 22.5°

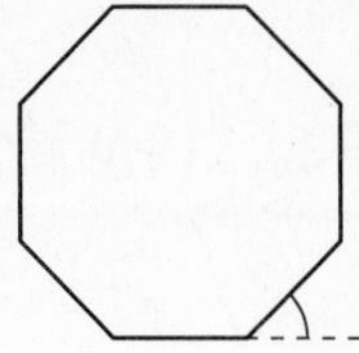

8. The diagonals of a nonspecial parallelogram ?. **8.** ______

a. are perpendicular b. are congruent
c. bisect each other d. are parallel

9. Consecutive angles in a nonspecial parallelogram are ?. **9.** ______

a. congruent angles b. supplementary angles
c. complementary angles d. vertical angles

10. For parallelogram $PQLM$, if $m\angle PML = 83°$, then $m\angle PQL =$?. **10.** ______

a. 83° b. 97°
c. $m\angle PQM$ d. $m\angle QLM$

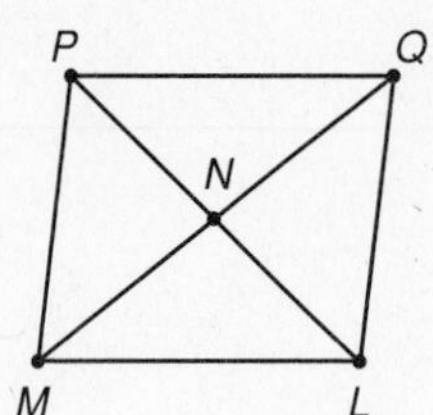

11. If all four sides of a quadrilateral are congruent, then the quadrilateral is always ?. **11.** ______

a. a trapezoid b. a nonsquare rectangle
c. a rhombus d. a kite

12. (2, 3) and (3, 1) are opposite vertices in a parallelogram. If (0, 0) is the third vertex, then the fourth vertex is ?. **12.** ______

a. $(-1, 2)$ b. $\left(\frac{5}{2}, 2\right)$ c. $(1, -1)$ d. $(5, 4)$

13. Choose the statement that is *not always* true. For any parallelogram ?. **13.** ______

a. opposite sides are congruent b. opposite angles are congruent
c. the diagonals are perpendicular d. the diagonals bisect each other

Name ______________________

Date ______________________

14. Choose the statement that is *not always* true.
For a rhombus [?].
a. the diagonals are perpendicular
b. each diagonal bisects a pair of opposite angles
c. the diagonals are congruent
d. all four sides are congruent

14. ______

15. The coordinates of quadrilateral $PQRS$ are $P(-3, 0)$, $Q(0, 4)$, $R(4, 1)$, and $S(1, -3)$. Which *best* describes the quadrilateral?
a. a rectangle **b.** a rhombus **c.** a parallelogram **d.** a square

15. 

16. Choose the statement which is *not* always true.
For an isosceles trapezoid [?].
a. the diagonals are congruent
b. the diagonals are perpendicular
c. the legs are congruent
d. the base angles are congruent

16. ______

17. For the trapezoid shown, the measure of the midsegment is [?].
a. 25 **b.** 30
c. 58 **d.** 29

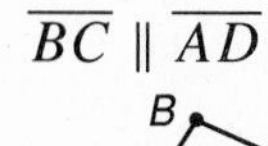

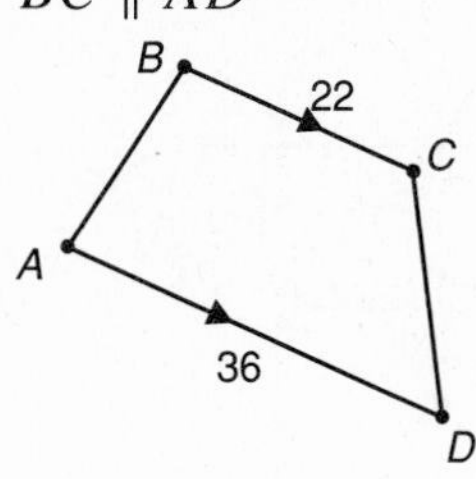

17. ______

18. Choose the figure which satisfies the definition of a kite.

a. 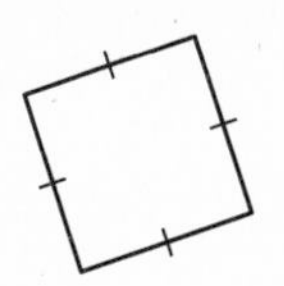

b. 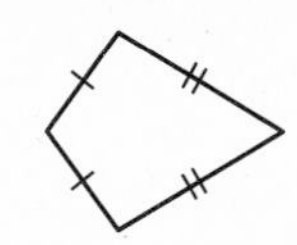

c. 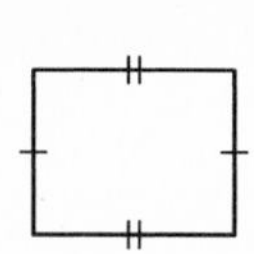

d. 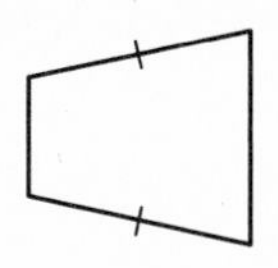

18. 

19. Which type of quadrilateral has no parallel sides?
a. trapezoid **b.** rhombus **c.** kite **d.** rectangle

19. ______

Chapter 6 Test

Form C

(Page 1 of 3 pages)

Name ____________________

Date ____________________

1. Sketch a polygon which is *not* convex and has five sides. 1.

2. Sketch a convex hexagon. How many diagonals does it have? 2. ____________

3. What properties of a polygon make it regular? Sketch an example. 3.

4. Explain why the figure shown does not satisfy the definition of a polygon. 4.

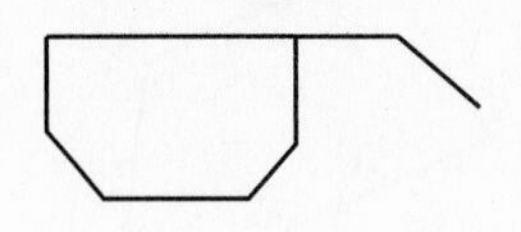

5. A regular pentagon has five congruent interior angles. What is the measure of each angle? 5. ____________

6. An octagon has eight sides of varying lengths. What is the sum of the measures of its interior angles? 6. ____________

7. Determine the number of sides of a regular polygon if each interior angle measure is 135°. 7. ____________

Name ______________________

Date ______________________

In Problems 8–11, refer to the figure below.

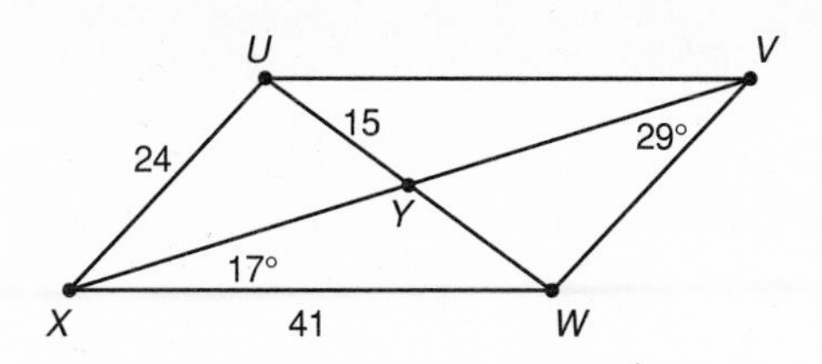

Given: $UVWX$ is a parallelogram.
$m\angle WXV = 17°$; $m\angle WVX = 29°$; $XW = 41$; $UX = 24$; $UY = 15$

8. Find $m\angle WVU$. **8.** ______________

9. Find the length of $\overline{WV}$. **9.** ______________

10. Find $m\angle XUV$. **10.** ______________

11. Find the length of $\overline{UW}$. **11.** ______________

12. Sketch, if possible, a quadrilateral with two consecutive angles congruent which is not a parallelogram. **12.**

13. Sketch, if possible, a quadrilateral with two opposite angles congruent which is not a parallelogram. **13.**

14. Use the distance formula to determine whether $ABCD$ is a parallelogram. **14.** ______________

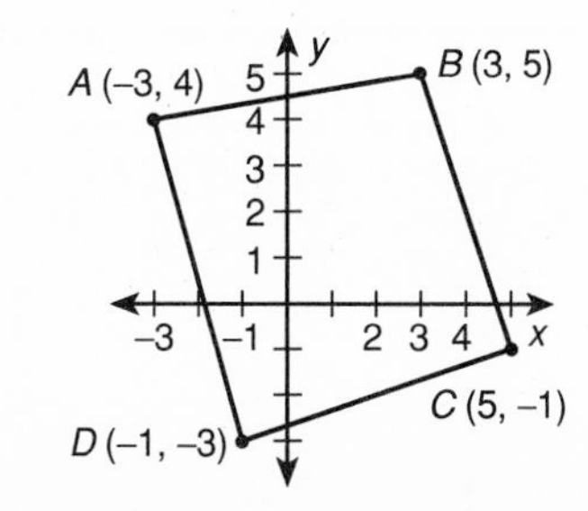

Name ______________________

Date ______________________

15. Draw a Venn diagram showing the relationship between squares, rectangles, rhombuses, parallelograms, and quadrilaterals.

15.

16. If the diagonals of a parallelogram are perpendicular, then the parallelogram is also what type of figure?

16. ______________

17. Sketch a trapezoid whose parallel sides measure $\frac{7}{8}$ inch and $1\frac{1}{8}$ inches. Find the length of the midsegment.

17. ______________

18. Sketch, if possible, an isosceles trapezoid with exactly two lines of symmetry.

18.

19. Sketch, if possible, a kite with exactly one right angle.

19.

20. Explain why a kite can have no sides parallel.

20. ______________

Cumulative Test

Chapters 1–6

(Page 1 of 8 pages)

Name ______________________________

Date ______________________________

1. Sketch, if possible, a quadrilateral that has exactly one line of symmetry.

1.

2. Identify any symmetry that the figure appears to have.

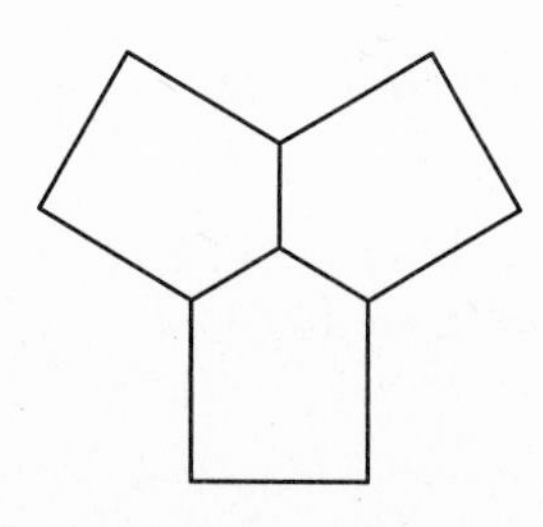

2. ______________________

3. Draw two straight lines that divide the shaded figure into four congruent figures.

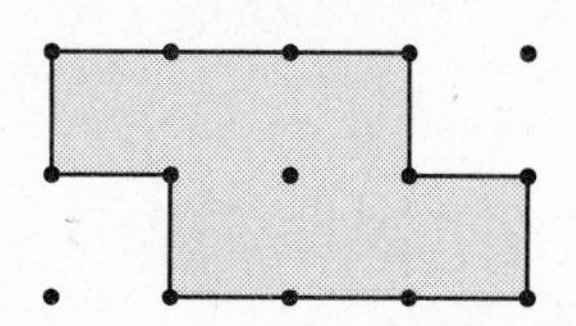

3. *Use figure at left.*

4. What is the relationship between a photographic transparency and its properly projected image on a screen?

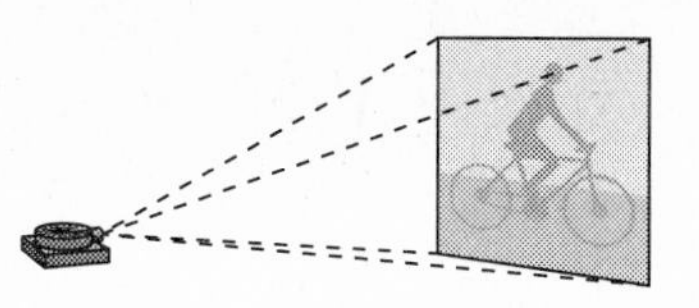

4. ______________________

5. Consider the points: $A(-7, 4)$ and $B(-3, -2)$. If C is the midpoint of $\overline{AB}$, find the coordinates of C.

5. ______________________

6. Find the slope of the line passing through the points $(-5, 6)$ and $(2, 7)$.

6. ______________________

Name ____________________

Date ____________________

7. Use construction tools to construct the perpendicular bisector of the segment $\overline{AB}$. Show all construction lines.

7. *Use figure at left.*

8. State the relationship between the lines $y = \frac{3}{4}x + 1$ and $y = -\frac{4}{3}x - 1$. Explain.

8. ____________________

9. What does the symbol $\overrightarrow{PN}$ represent? Make a sketch of $\overrightarrow{PN}$.

9. ____________________

10. $\overleftrightarrow{AB}$ intersects $\overline{ST}$ at point Q (distinct from S and T). Describe the position of Q with respect to S and T.

10. ____________________

11. C is between A and B. D is between C and B. B is between C and E. $AE = 20$, $CB = 8$, and $AC = CD = DB$. Find BE.

11. ____________________

Name ______________________

Date ______________________

12. B is in the interior of $\angle AOC$. C is in the interior of $\angle BOD$. D is in the interior of $\angle COE$. $m\angle AOE = 154°$, $m\angle COE = 62°$ and $m\angle AOB = m\angle COD = m\angle DOE$. Find $m\angle AOD$.

12. ______________________

13. Find the length of $\overline{QT}$ if $\overleftrightarrow{PR}$ bisects $\overline{QT}$ at N and $NT = 37$.

13. ______________________

14. Consider the statement: "If a quadrilateral is a square, then it is a rhombus."

a. What is the conclusion?
b. Is the statement true? Explain.
c. What is the converse statement?
d. Is the converse true? Explain.

14.

15. Write the contrapositive for the statement, "If I vote for his bill, then he will vote for mine."

15.

16. *Given:* $\angle A$ and $\angle B$ are a linear pair and $m\angle B = 111°$. Find $m\angle A$.

16. ______________________

17. Define supplementary angles.

17.

18. If ℓ_1 and ℓ_2 are parallel lines, they must be coplanar. If ℓ_3 is also parallel to ℓ_1, explain whether ℓ_3 must also lie in that same plane and if it is parallel to ℓ_2.

18.

19. What are two lines that do not lie in the same plane called? **19.** ________

20. Solve the system and check your answer algebraically. **20.** ________

$$\begin{cases} 2x + 3y = 7 \\ x + 2y = 6 \end{cases}$$

21. Write the equation of a line that is parallel to $y = -\frac{2}{3}x + 2$ and passes through point (3, 4). **21.** ________

22. Sketch an example of alternate exterior angles. **22.**

23. Use the figure and the given information to determine which lines are parallel.
Given: $m\angle 6 = m\angle 2$ **23.** ________

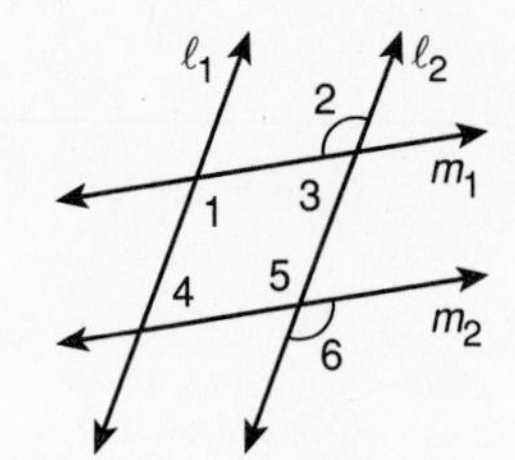

24. Use construction tools to construct a line through P perpendicular to m. Show all construction lines. **24.** *Use figure at left.*

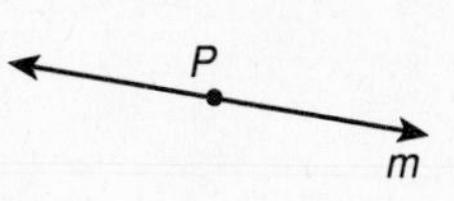

In Problems 25–27, use the vectors $\vec{u} = \langle 2, -7 \rangle$ and $\vec{v} = \langle 1, 3 \rangle$.

25. Find the length of $\vec{u}$. **25.** ________

26. Find $\vec{u} + \vec{v}$. **26.** ________

27. Find $\vec{u} \cdot \vec{v}$. **27.** ________

28. An airplane is flying northeast at a speed of $400\sqrt{2}$ (about 566) miles per hour. Its velocity is represented by $\vec{v} = \langle 400, 400 \rangle$. The airplane encounters a wind of 90 miles per hour blowing from the west, which is represented by $\vec{u} = \langle 90, 0 \rangle$. What is the new speed of the airplane?

28. ______________

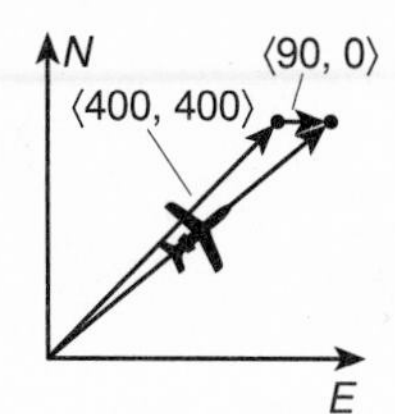

29. If $\triangle DEF \cong \triangle PQR$ and $\triangle PQR \cong \triangle UVW$, is $\triangle DEF \cong \triangle UVW$? Explain your reasoning.

29. ______________

30. Find the measure of the exterior angle $\angle BCD$.

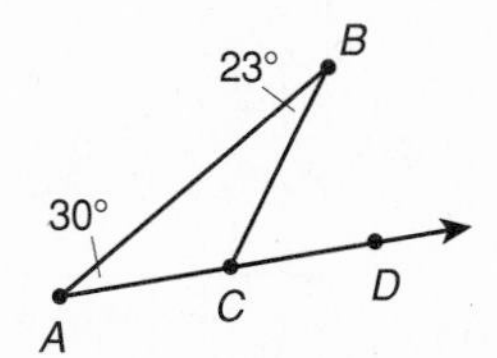

30. ______________

31. Find the measures of angles A, B, and C.

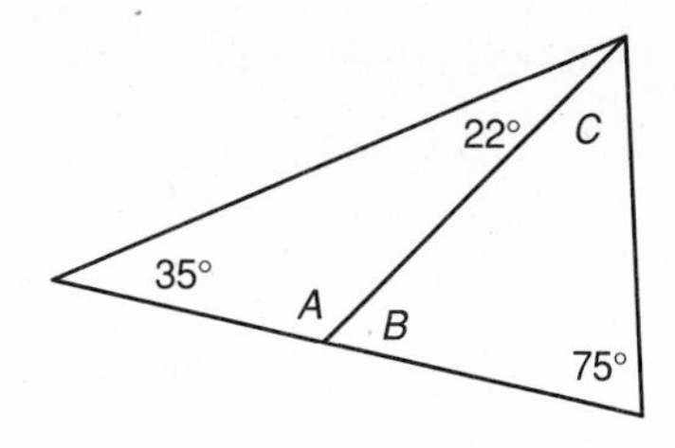

31. ______________

32. How many acute angles can an isosceles triangle have? Explain.

32.

33. If $\triangle PQR \cong \triangle UVW$, does it follow that $\triangle RQP \cong \triangle WVU$? Explain.

33.

Name ______________________

Date ______________________

34. Find the measure of $\overline{AB}$. State the argument you used to do this.

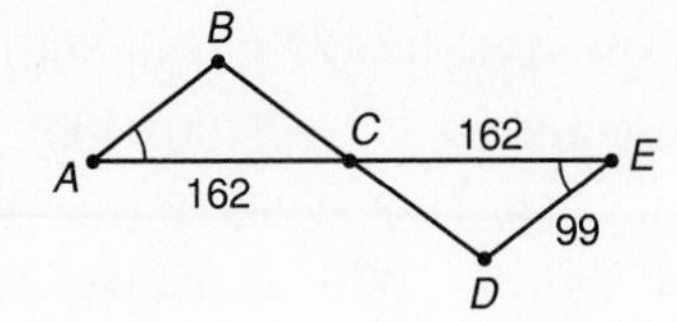

34.

35. Write a proof. Justify each statement.
Given: $\overline{AB} \cong \overline{CD}$, $\overline{AB} \parallel \overline{CD}$
Prove: $\triangle ABD \cong \triangle CDB$

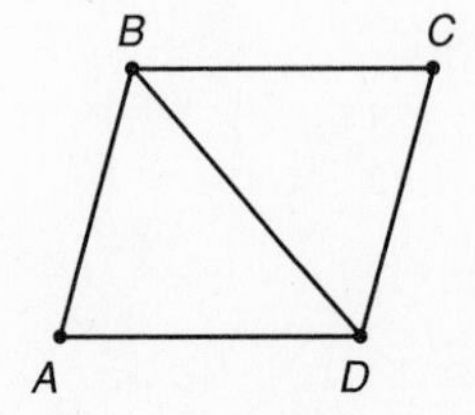

35.

36. A point is equidistant from points (0, 0), (8, 0), and (0, 6). Find its coordinates.

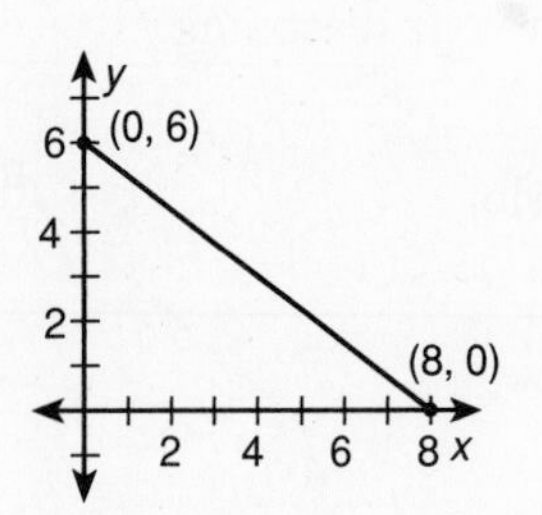

36. ______________________

37. Sketch the three medians of an obtuse isosceles triangle. Label the point of concurrency X. What is the name for point X?

37.

38. Define what a midsegment of a triangle is. Draw an example to illustrate your definition.

38.

39. Use construction tools to construct the circumcenter of the triangle shown. Then draw the circumscribed circle.

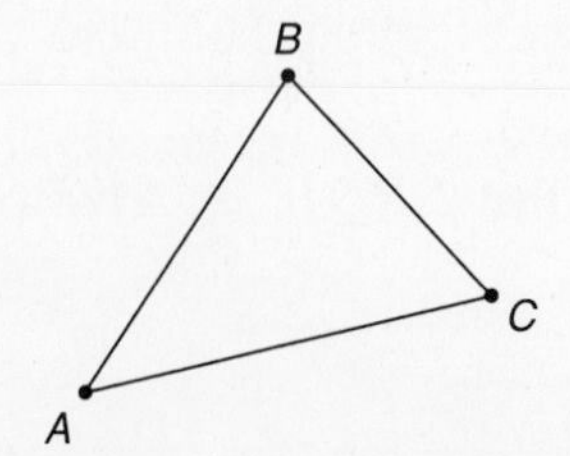

39.

Name ______________________

Date ______________________

40. For the triangle shown, if $QR = 8.5$, what is the measure of $\overline{LM}$?

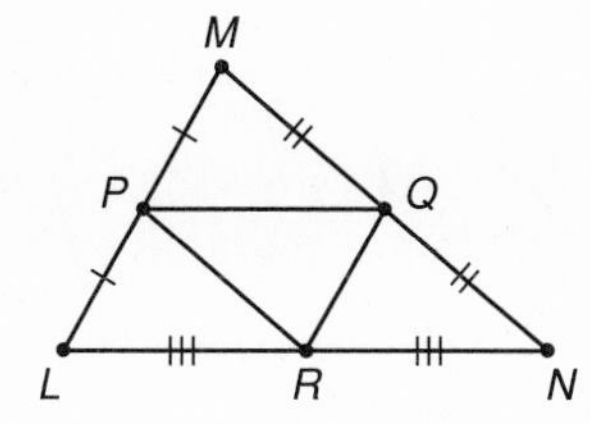

40. ______________________

41. For the figure shown, what is $m\angle GDE$?

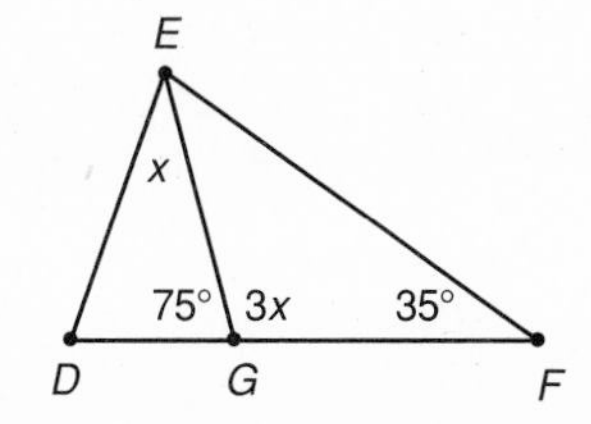

41. ______________________

42. *Given:* $\overline{AF} \cong \overline{FC}$, $\angle ABE \cong \angle EBC$

a. Give the name of each special segment in $\triangle ABC$.

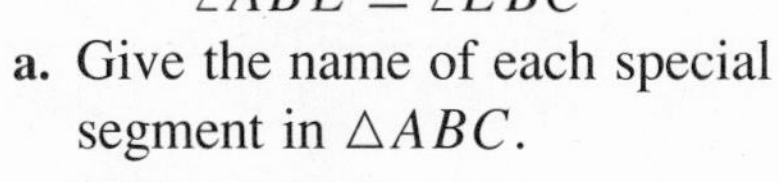

$\overline{BD}$ $\quad$ $\overline{BE}$

$\overline{BF}$ $\quad$ $\overline{GF}$

b. On which line segment does the center of the inscribed circle of the triangle lie?

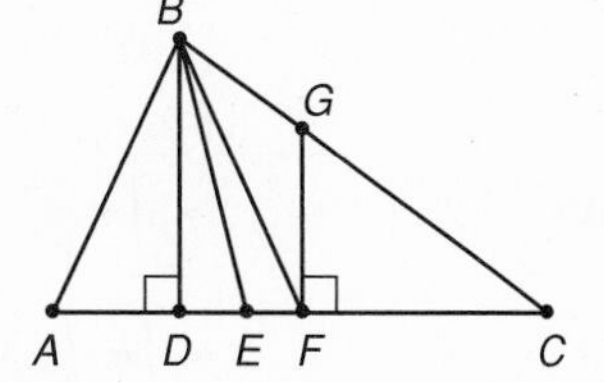

42.

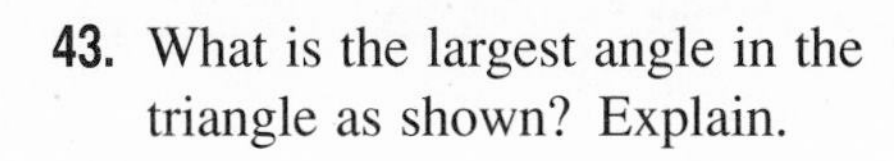

43. What is the largest angle in the triangle as shown? Explain.

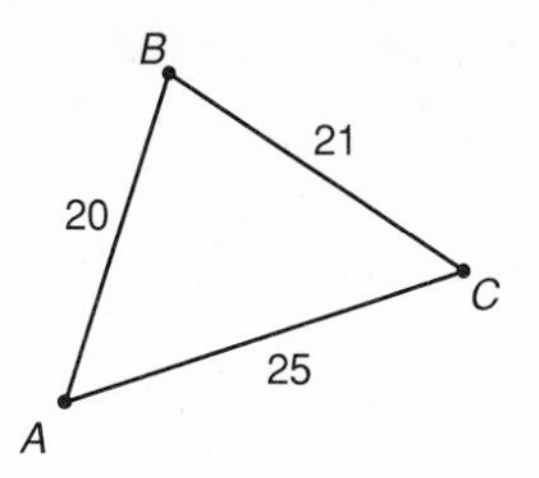

43. ______________________

44. If $\overline{PT}$ is an altitude of $\triangle PST$, what type of triangle is $\triangle PST$?

44. ______________________

Name ______________________

Date ______________________

45. *Given:* $\overline{BC} \cong \overline{DA}$
$\overline{AB} \cong \overline{CD}$
Prove: $ABCD$ is a parallelogram.

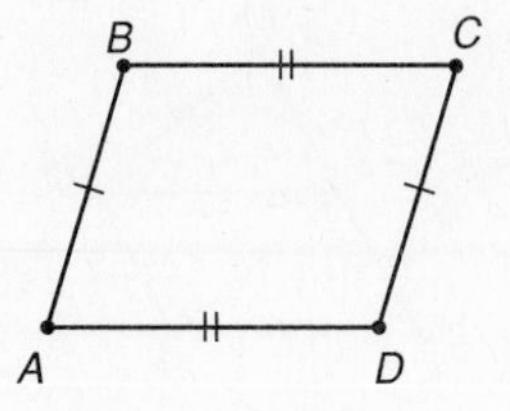

45. ______________________

46. Two sides of a triangle are 6 and 14. What are the possible measures of the third side x?

46. ______________________

47. How many diagonals does a heptagon have?

47. ______________________

48. What is the sum of the measures of the interior angles of a heptagon?

48. ______________________

49. Determine whether or not $ABCD$ is a parallelogram. Explain.

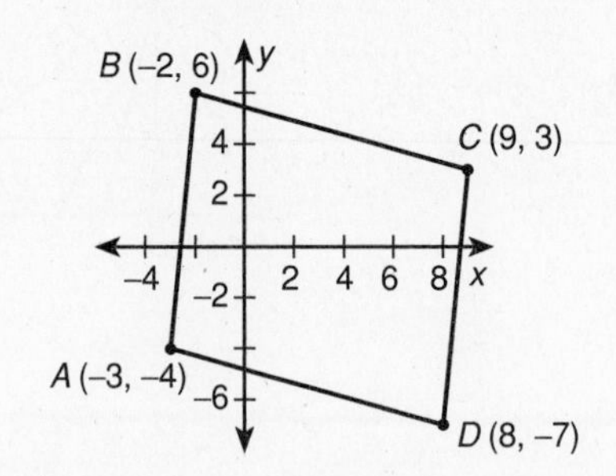

49. ______________________

50. If the diagonals of a rhombus are congruent then it is also what type of quadrilateral?

50. ______________________

51. For the trapezoid shown:

a. Find the length of the midsegment.

b. What is required for $ABCD$ to be an isosceles trapezoid?

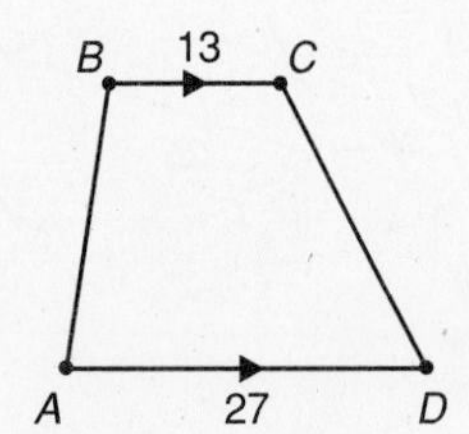

51. ______________________

7.2 Short Quiz

Name ______________________

Date ______________________

1. Identify each transformation shown.

a. 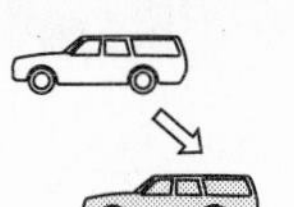 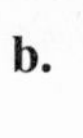

b. 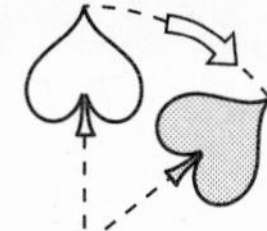

c. 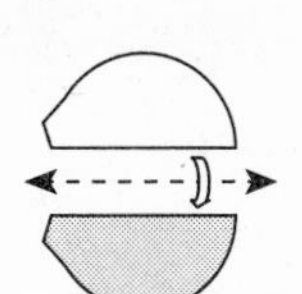

d. 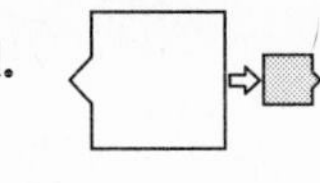

1. a. ______________________
b. ______________________
c. ______________________
d. ______________________

2. In each case, describe those attributes that remain the same and those that change.

a. 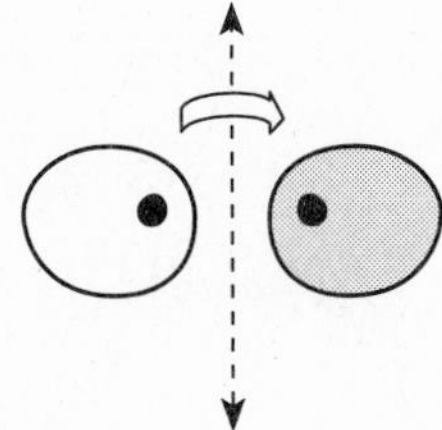

b. 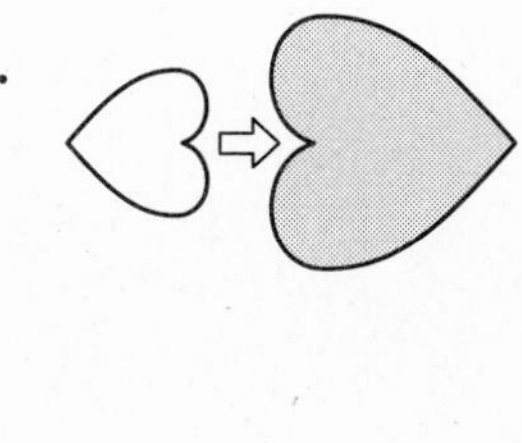

2.

3. Explain why the transformation is not an isometry.

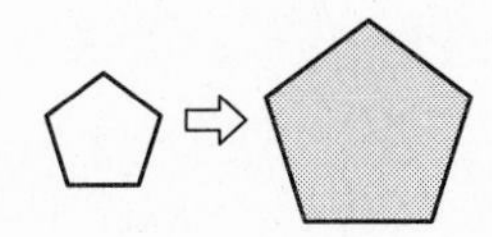

3.

4. How many lines of symmetry does a regular octagon have? Sketch them on the figure at right.

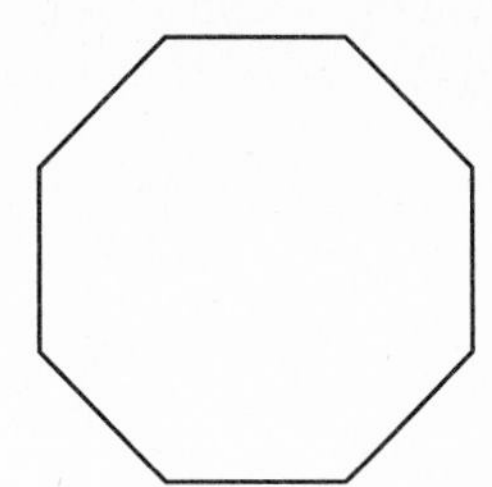

4. ______________________
Use figure at left.

5. Draw the reflection of the figure in line ℓ.

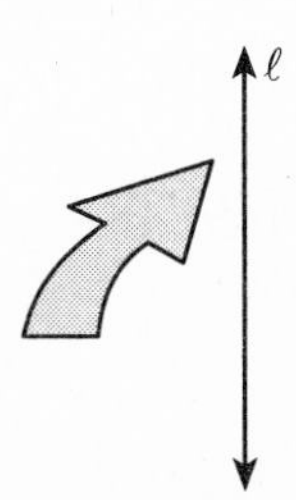

5. *Use figure at left.*

Mid-Chapter 7 Test Form A

(Use after Lesson 7.3)

Name ______________________

Date ______________________

1. Explain what is meant by a *rigid* transformation. **1.**

2. Which of the following are isometries? **2.** ____________

a. 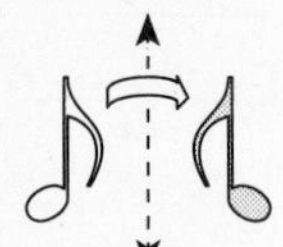 **b.** **c.** 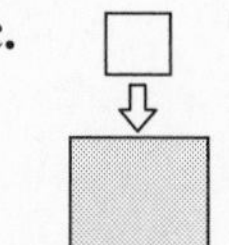 **d.** 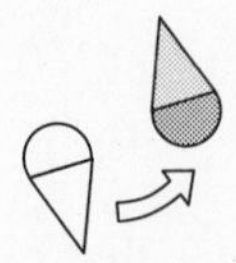

3. How many lines of symmetry does the figure have? Sketch them on the figure at right. **3.** ____________ *Use figure at left.*

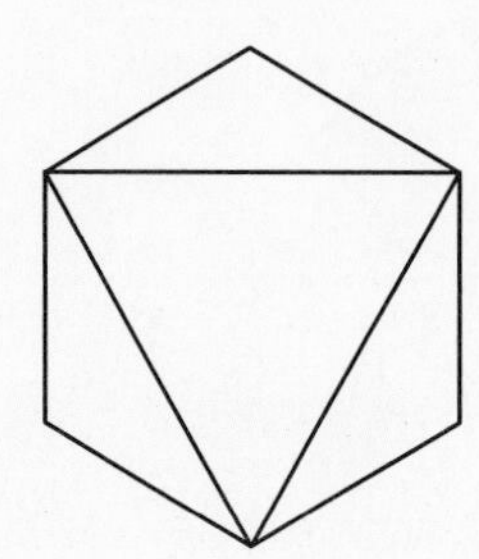

4. Draw the reflection of $\triangle ABC$ in line ℓ. **4.** *Use figure at left.*

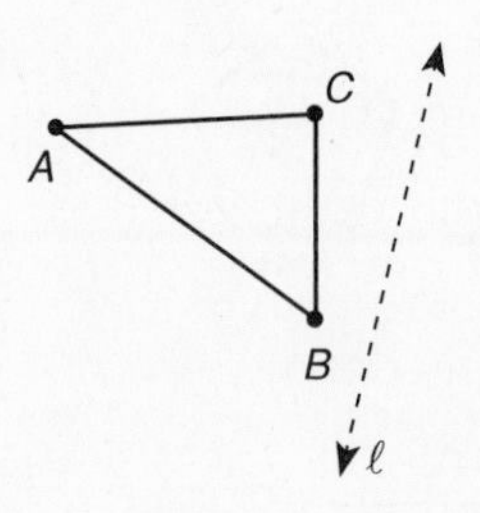

5. $\triangle ABC$ is rotated 90° clockwise about point (0, 0). Find A', B', and C', the coordinates of the image of its vertices $A(-1, -1)$, $B(-4, 1)$, and $C(0, 5)$. **5.** ____________

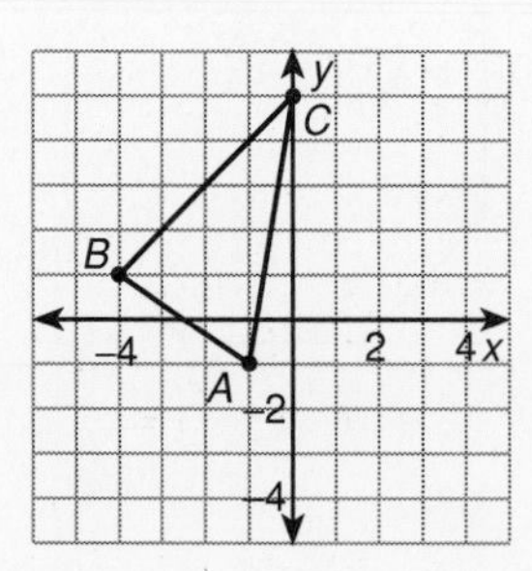

6. Lines ℓ and m intersect at point O. Consider a reflection of $\triangle RST$ in line ℓ, followed by a reflection in line m. If the angle between ℓ and m is 45°, what is the angle of rotation about O? **6.** ____________

Mid-Chapter 7 Test **Form B** (Use after Lesson 7.3)

Name ______________________

Date ______________________

1. Explain what is meant by a transformation that is an *isometry*.

1.

2. Which of the following transformations are rigid?

a. 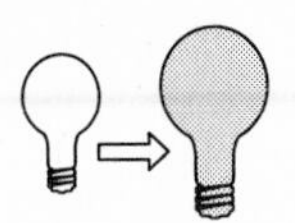

b.

c. 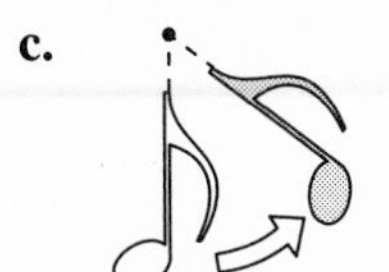

d. 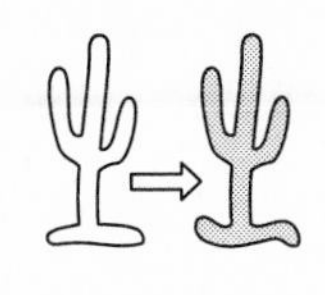

2. ______________________

3. How many lines of symmetry does the figure have? Sketch them on the figure at the right.

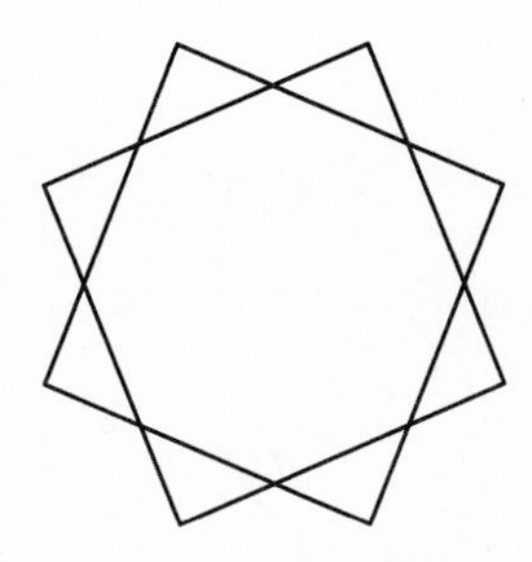

3. ______________________ *Use figure at left.*

4. Draw the reflection of $\triangle ABC$ in line ℓ.

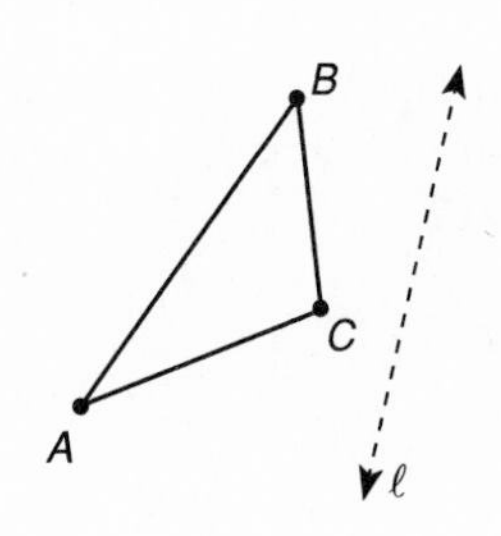

4. *Use figure at left.*

5. $\triangle ABC$ is rotated 90° clockwise about point (0, 0). Find A', B', and C', the coordinates of the image of its vertices $A(-1, -2)$, $B(-4, 2)$, and $C(2, 3)$.

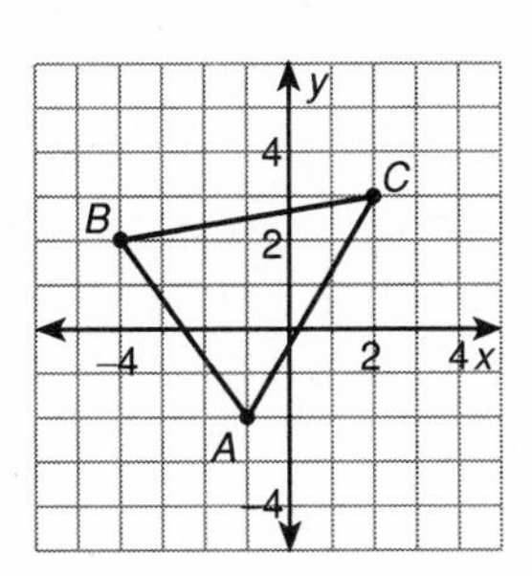

5. ______________________

6. Lines ℓ and m intersect at point O. Consider a reflection of $\triangle PQR$ in line ℓ, followed by a reflection in line m. If the angle between ℓ and m is 30°, what is the angle of rotation about O?

6. ______________________

7.4 Short Quiz

Name ______________________

Date ______________________

1. Name three capital letters in the alphabet that have a rotational symmetry of 180°. **1.** ______________

2. Lines ℓ and m intersect at point O. Consider a reflection of $\triangle PQR$ in line ℓ, followed by a reflection in line m. If the angle between ℓ and m is 48°, what is the angle of rotation about O? **2.** ______________

3. Describe the symmetry of the figure. **3.**

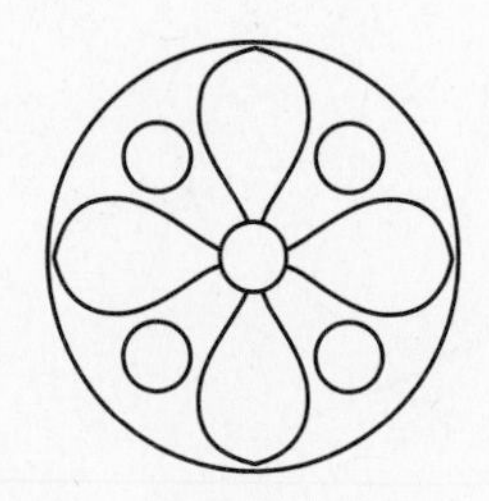

4. $\triangle ABC$ is translated by the vector $\overrightarrow{u} = \langle 5, -2 \rangle$. Find the coordinates of its image A', B' and C'. **4.** ______________

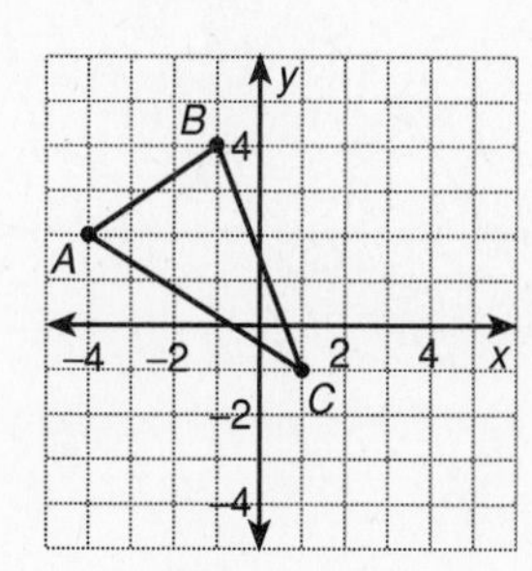

5. ℓ is parallel to m and $\overline{PQ}$ is reflected in ℓ. The result, $\overline{P'Q'}$, is then reflected in m resulting in $\overline{P''Q''}$. The distance between ℓ and m is 3.5 units. Find the distance of QQ''. **5.** ______________

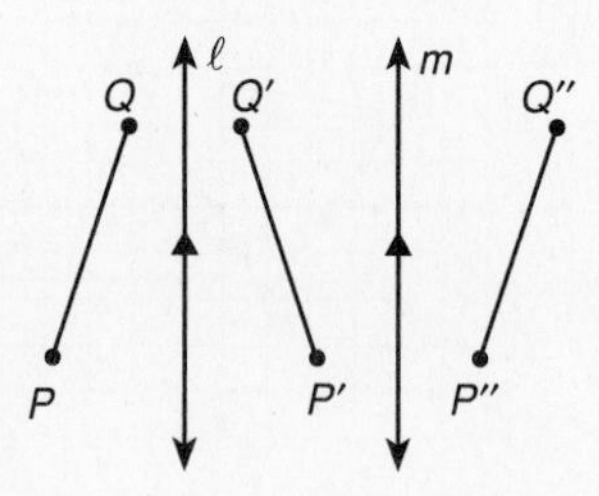

7.6 Short Quiz

Name ____________________

Date ____________________

1. What is a glide reflection?

1.

2. $\triangle PQR$ is translated by the vector $\overrightarrow{v} = \langle 5, 4 \rangle$ and then reflected in the y-axis. Find the resulting coordinates of P', Q' and R'.

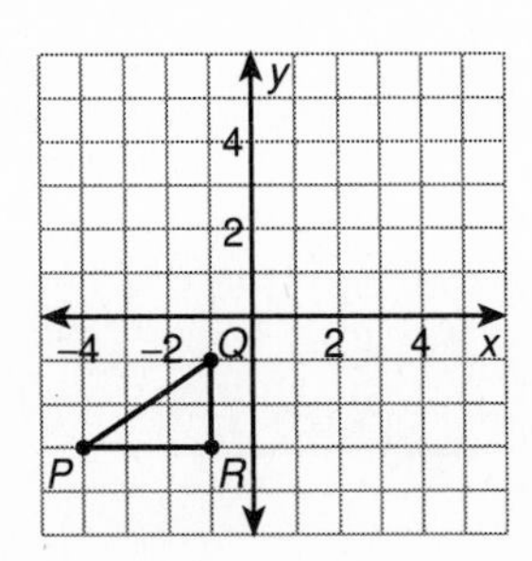

2. ____________________

3. In Problem 2, does the order in which the translation and reflection are carried out matter?

3. ____________________

4. Point $(-3, 2)$ is translated by the vector $\langle 0, -5 \rangle$ and then reflected in the line $x = 0$. Find the coordinates of its image.

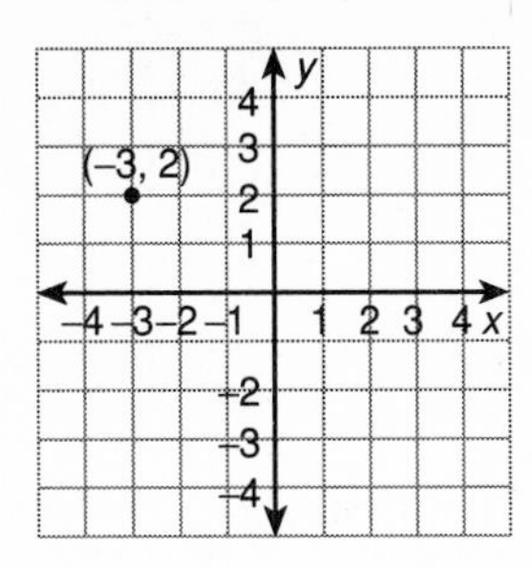

4. ____________________

5. Classify the frieze pattern.

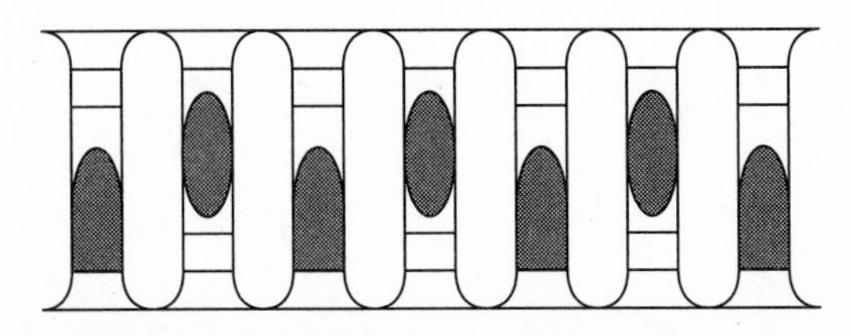

5. ____________________

Chapter 7 Test

Form A (Page 1 of 3 pages)

Name ______________________

Date ______________________

1. Explain what is meant by a *rigid* transformation. 1.

In Problems 2 and 3, name the transformation. (Preimages are unshaded; images are shaded.)

2. 2. ______________________

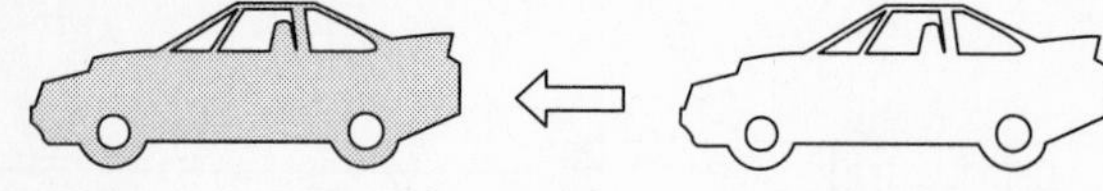

3. 3. ______________________

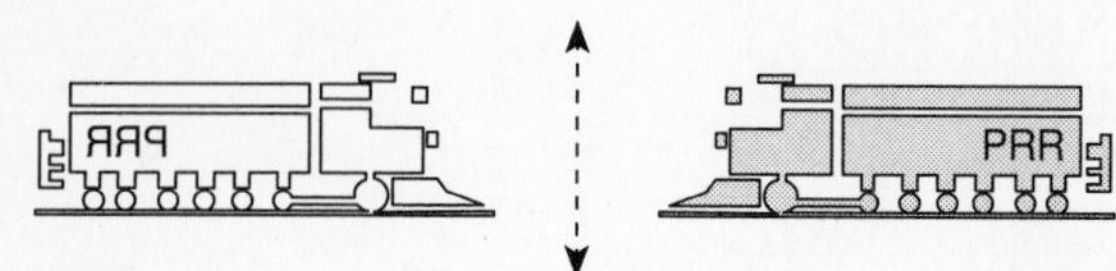

4. Is the transformation an isometry? Explain. 4.

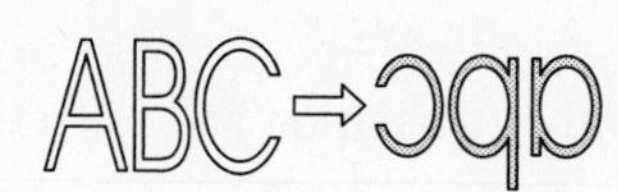

5. Segment AB is translated by the motion rule $(x, y) \rightarrow (x + 4, y - 3)$. Find the coordinates of the endpoints of the image $\overline{A'B'}$. 5. ______________________

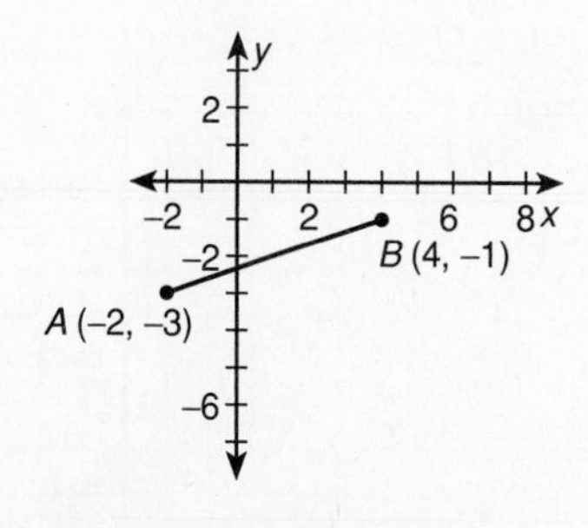

6. The points in a coordinate plane are reflected in the y-axis. In general, every point (x, y) is mapped onto what point? 6. ______________________

7. Draw the reflection of $\overline{PQ}$ in the line ℓ. 7. *Use figure at left.*

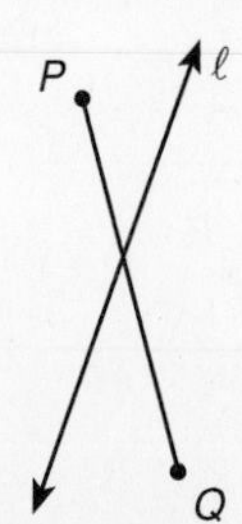

Name ______________________

Date ______________________

8. How many lines of symmetry does this equiangular hexagon have? Sketch them on the figure at the right.

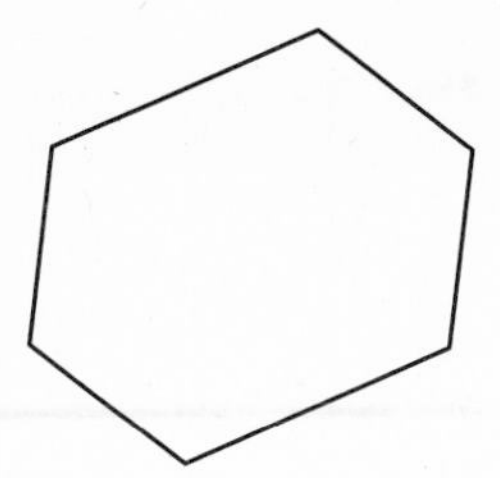

8. ______________________ *Use figure at left.*

9. Draw, if possible, a quadrilateral with exactly one line of symmetry.

9.

10. $\triangle PQR$ is rotated clockwise $90°$ about the origin. Find the coordinates of the vertices of the image $\triangle P'Q'R'$.

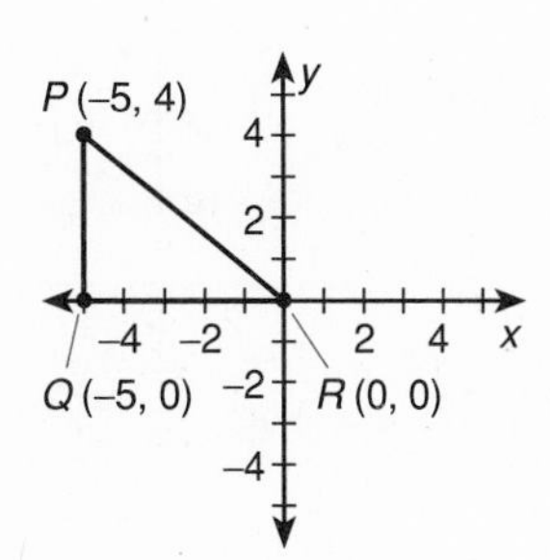

10. ______________________

11. Lines ℓ and m intersect at point O forming an $80°$ angle. Point P is reflected in ℓ, followed by a reflection in m. Describe the location of the image P''.

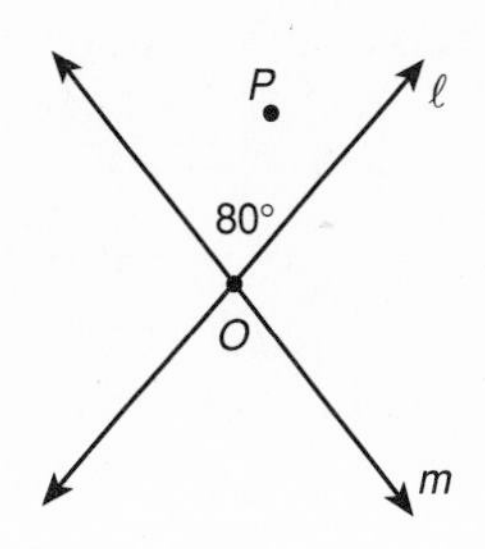

11. ______________________

12. Does the clock face shown have any rotational symmetry? If so, list any angles of rotation, $180°$ or less, that can map it onto itself.

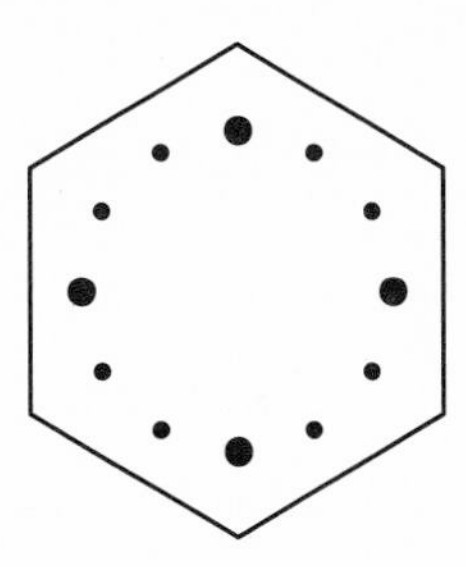

12. ______________________

13. You are given that $\ell \parallel m$ and the distance between the lines is d. Point P is reflected in ℓ onto image P' which is then reflected in m onto image P''. Then $\overline{PP''}$ is perpendicular to ℓ. Find PP''.

13. ______________________

Name ______________________

Date ______________________

14. Find the translation vector, $\overrightarrow{v}$, that maps $\triangle ABC$ onto $\triangle A'B'C'$.

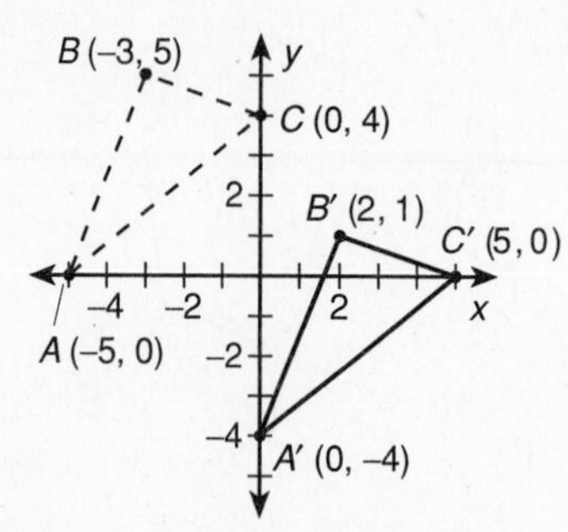

14. ______________________

15. The translation vector is $\overrightarrow{u} = \langle -7, 4 \rangle$. If the image of A is $A'(6, -4)$, find the coordinates of point A.

15. ______________________

16. The point $A(2, 5)$ is translated by the vector $\overrightarrow{v} = \langle -5, -3 \rangle$ and then reflected in the y-axis. Find the coordinates of its image, A'.

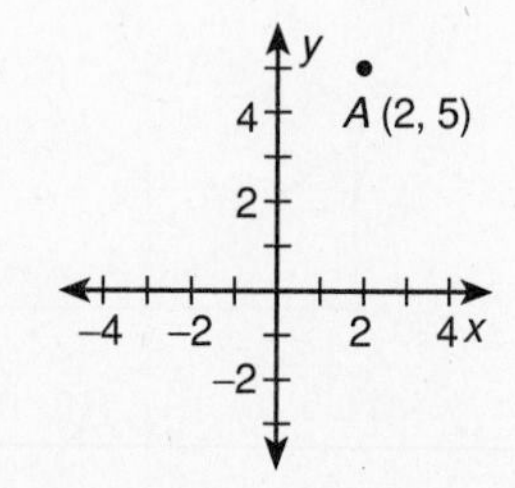

16. ______________________

17. Explain whether or not, in a composition of two or more isometries, the order of applying them matters.

17.

18. Classify the frieze pattern.

18. ______________________

Chapter 7 Test

Form B

(Page 1 of 3 pages)

Name ______________________

Date ______________________

1. A *rigid transformation* always maps a figure onto [?].

a. itself **b.** a similar figure
c. a congruent figure **d.** its mirror image

1. ______

2. An *isometry* is a transformation which does not usually preserve [?].

a. position **b.** betweeness **c.** length **d.** angle measure

2. ______

3. Refer to this stenciled border pattern.

3. ______

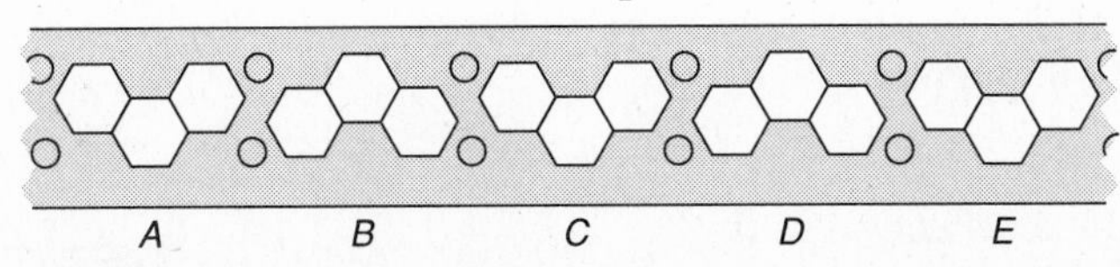

Which statement is *not* true?

a. Figures C and E are translations of figure A.
b. Figure D is a translation of figure B.
c. Figure C is a reflection of figure D.
d. Figure E is a rotation of figure B.

4. The motion rule for this transformation of $\triangle ABC$ onto $\triangle A'B'C'$ is [?].

a. $(x, y) \rightarrow (x - 6, y - 5)$
b. $(x, y) \rightarrow (x - 5, y - 6)$
c. $(x, y) \rightarrow (x + 6, y + 5)$
d. $(x, y) \rightarrow (x + 5, y + 6)$

4. ______

5. A reflection is always [?].

a. a translation **b.** a rotation **c.** reflexive **d.** an isometry

5. ______

6. The hexagon shown is equiangular. How many lines of symmetry does it have?

a. 2 **b.** 1 **c.** 3 **d.** 6

6. ______

Name ____________________

Date ____________________

In Problems 7 and 8, use the figure below.

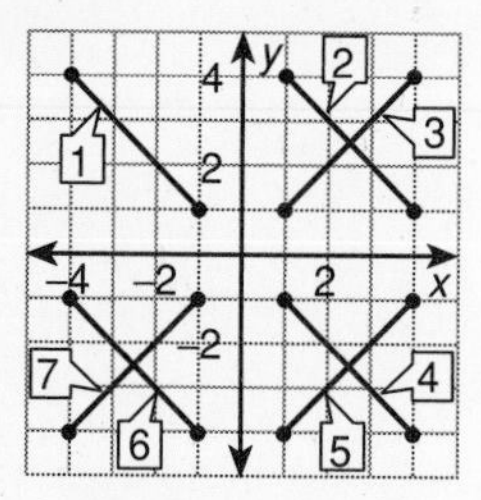

7. If segment 1 is reflected in the line $y = x$, its image is [?]. 7. ______
   a. segment 2 **b.** segment 4 **c.** segment 5 **d.** segment 7

8. Segment 1 is reflected in the x-axis, followed by a reflection in the y-axis, followed by another reflection on the x-axis. Its final image is [?]. 8. ______
   a. segment 1 **b.** segment 2 **c.** segment 3 **d.** segment 4

In Problems 9 and 10, lines ℓ and m intersect at point O. The acute or right angle between ℓ and m has a measure of $x°$. A figure is reflected in ℓ followed by a reflection in m.

9. The result is [?]. 9. ______
   a. a translation **b.** a reflection
   c. a rotation **d.** no overall change

10. The overall effect is [?]. 10. ______
    a. a rotation about O of $x°$
    b. a reflection in the bisector of the angle between ℓ and m
    c. no change
    d. a rotation about O of $2x°$

11. The degree of rotation that maps parallelogram $PQRS$ onto parallelogram $P'Q'R'S'$ is [?]. 11. ______
    a. 90° **b.** 180°
    c. 270° **d.** 360°

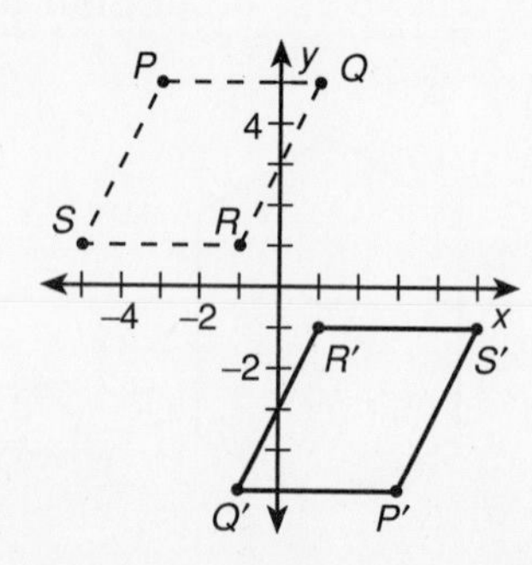

Name ______________________

Date ______________________

12. The point $A(-7, 3)$ is translated onto A' by the vector $\overrightarrow{u} = \langle 5, -4 \rangle$. The coordinates of A' are [?].

a. $(-2, -1)$ **b.** $(5, -4)$ **c.** $(-12, 7)$ **d.** $(12, -7)$

12. ______

13. Which of the following is *not* true?

a. A parallelogram has rotational symmetry and may have line symmetry.

b. A rectangle has rotational symmetry and always has line symmetry.

c. A triangle has rotational symmetry and always has line symmetry.

d. A regular hexagon has rotational symmetry and always has line symmetry.

13. ______

14. One of the following represents a glide reflection. Which one?

a. $A \Rightarrow D$ **b.** $A \Rightarrow B$

c. $A \Rightarrow C$ **d.** $A \Rightarrow E$

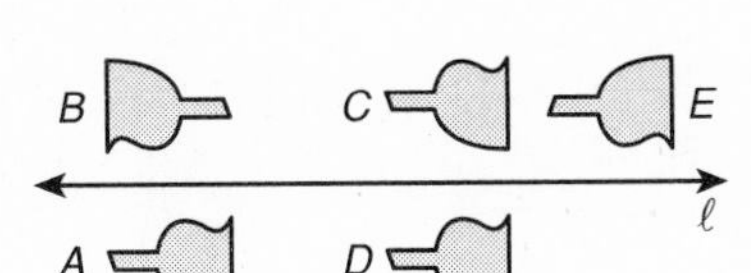

14. ______

15. The composition of two (or more) isometries is always [?].

a. a reflection **b.** a translation **c.** a rotation **d.** an isometry

15. ______

16. Point $A(-4, 2)$ is translated by the vector $\overrightarrow{v} = \langle 7, -4 \rangle$ and then reflected in the line $y = x$ onto A'. The coordinates of its image A' are [?].

a. $(-2, 3)$ **b.** $(3, -2)$

c. $(-3, -2)$ **d.** $(3, 2)$

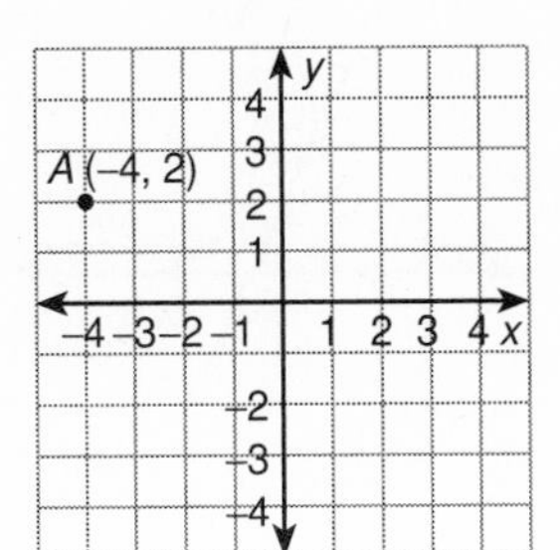

16. ______

17. Point $A(-5, -3)$ is translated by the vector $\overrightarrow{v} = \langle 8, 0 \rangle$ and then reflected in the line $y = 0$. The coordinates of its image, A', are [?].

a. $(3, -3)$ **b.** $(3, 3)$ **c.** $(13, 3)$ **d.** $(-5, 5)$

17. ______

18. Classify the frieze pattern.

a. TG **b.** TV **c.** TR **d.** THG

18. ______

Chapter 7 Test

Form C

(Page 1 of 3 pages)

Name ______________________

Date ______________________

1. When is a transformation in the plane an *isometry*? **1.**

In 2 and 3, name the transformation.

2. **2.** ______________

Preimage Image

3. **3.** ______________

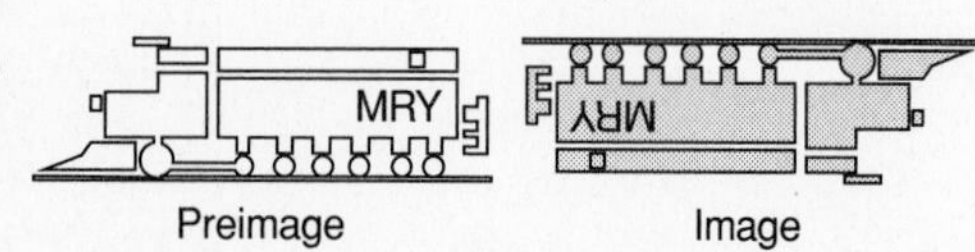

Preimage Image

4. Explain whether or not the transformation is an isometry. **4.**

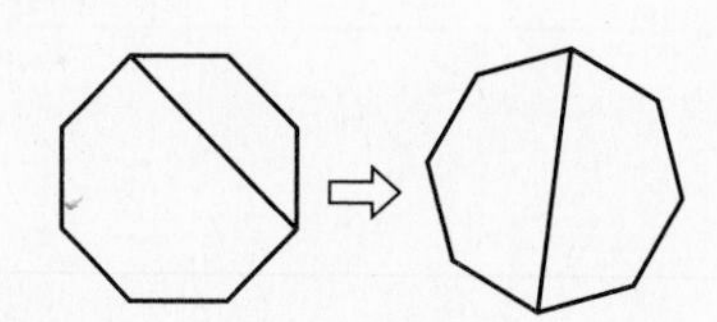

5. Segment AB is translated by the motion rule $(x, y) \rightarrow (x - 4, y + 5)$. Find the coordinates of the endpoints of the image $\overline{A'B'}$. **5.** ______________

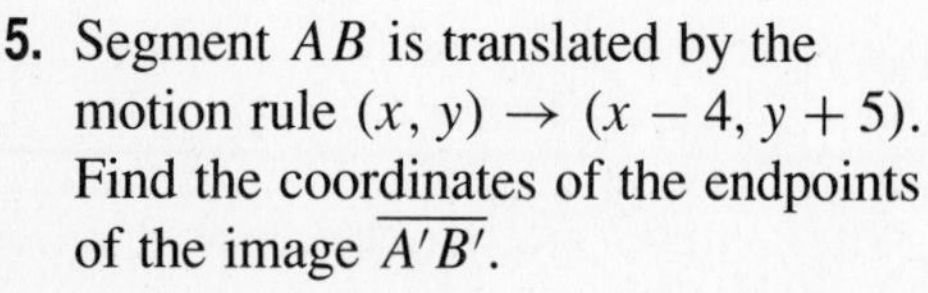

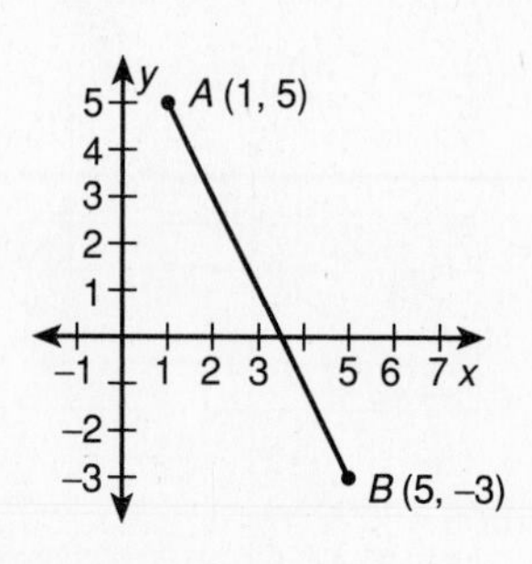

6. The points in a coordinate plane are reflected in the line $y = x$. In general, every point (x, y) is mapped onto what point? **6.** ______________

7. If point P is not on line ℓ and P is reflected in ℓ to the point P', what is the relationship between line ℓ and $\overline{PP'}$? Make a sketch to support your answer. **7.** ______________

Name ______________________

Date ______________________

8. How many lines of symmetry does a regular hexagon have? Sketch the symmetry lines.

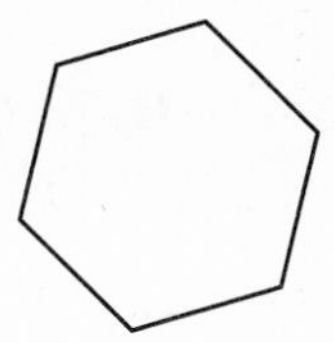

8. ______________ *Use figure at left.*

9. A water line, ℓ, passes near two houses, as shown. Show, by construction, where to place a tap-in so the least amount of pipe is used to supply water to both houses. Explain your solution.

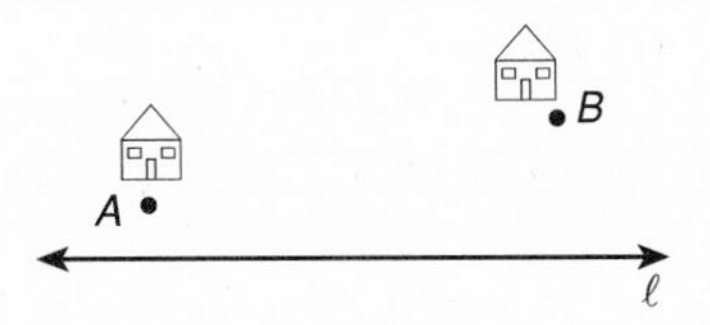

9. *Use figure at left.*

10. $\overline{PQ}$ is rotated counterclockwise 90° about the origin. Find the coordinates of the endpoints of the image $\overline{P'Q'}$.

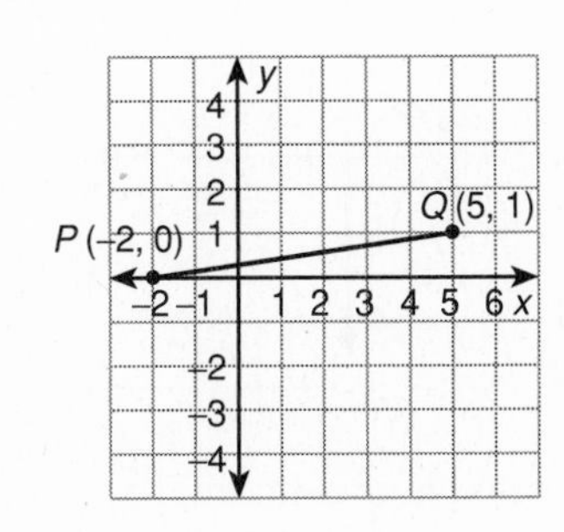

10. ______________

11. Draw lines ℓ and m so that their intersection is point O and any reflection in ℓ followed by a reflection in m results in a rotation about point O of 136°. What is the acute angle between ℓ and m?

11. ______________ *Use back of page for constructions.*

12. Does the clock face shown have any rotational symmetry? If so, list any angles of rotation, 180° or less, that can map it onto itself.

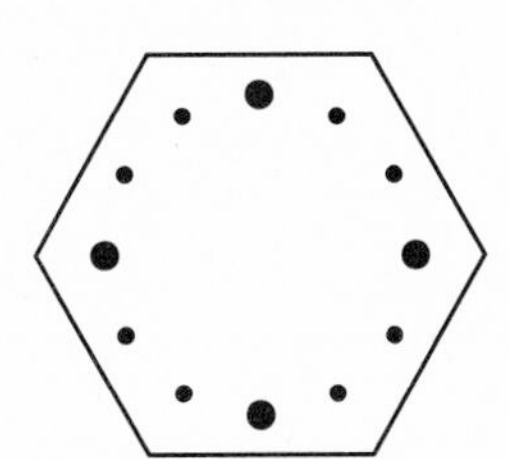

12. ______________

13. *Given:* $\ell \parallel m$. Point P is reflected in ℓ onto image P' which is then reflected in m onto image P''. Then PP'' is perpendicular to ℓ. If $PP'' = 16$, what is the distance between the lines ℓ and m? Make a sketch to support your solution.

13. ______________ *Use back of page for constructions.*

Chapter 7 Test

Form C

(Page 3 of 3 pages)

Name ______________________

Date ______________________

14. Find the vector, $\overrightarrow{v}$, which will translate $\triangle ABC$ onto $\triangle A'B'C'$.

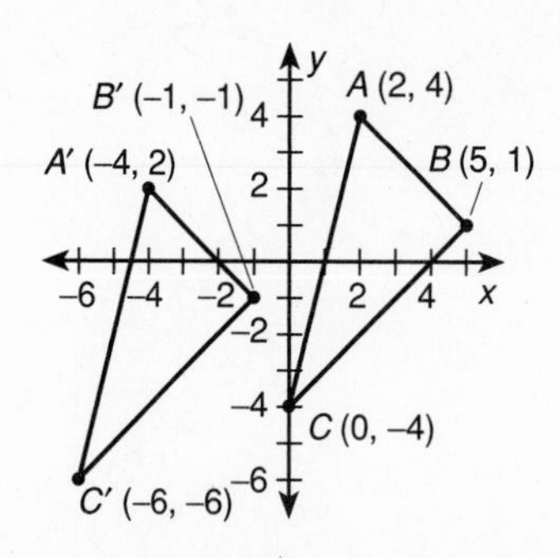

14. ______________

15. The translation vector is $\overrightarrow{u} = \langle 7, -3 \rangle$. The image of point A is $A'(5, -7)$. Find the coordinates of A.

15. ______________

16. Consider the glide reflection composed of the translation by vector $\overrightarrow{v} = \langle 5, 0 \rangle$, followed by a reflection in the x-axis. Find the coordinates of the endpoints (A' and B') of the image.

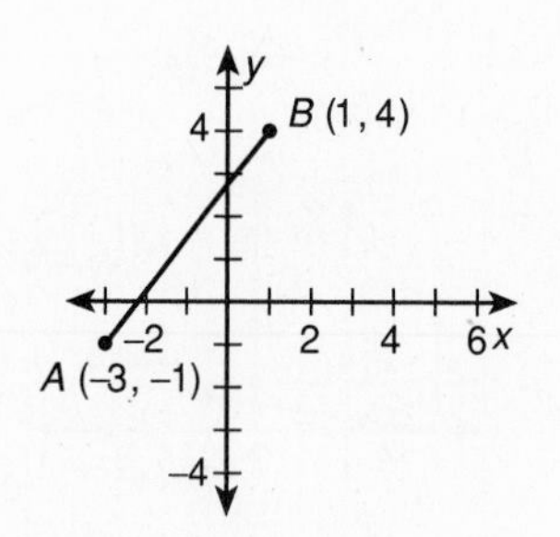

16. ______________

17. Prove that in any rigid transformation a triangle and its image is congruent.

Given: $\triangle ABC \Rightarrow \triangle A'B'C'$

The transformation is an isometry.

17.

18. Explain whether or not the composition of two or more isometries is an isometry.

18.

19. Classify the frieze pattern.

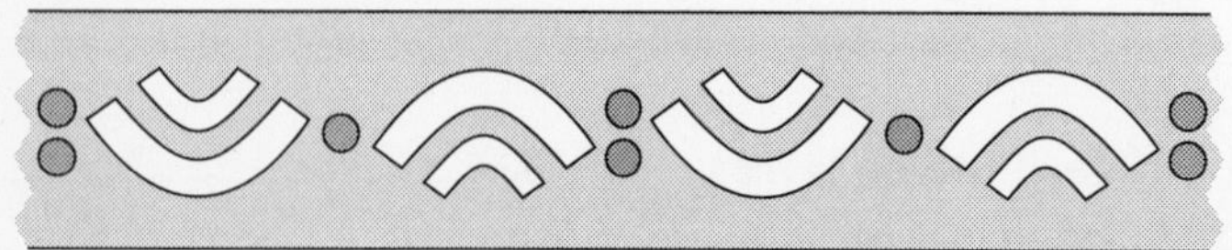

19. ______________

8.2 Short Quiz

Name ______________________

Date ______________________

1. Simplify the ratio.

$$\frac{112 \text{ people}}{88 \text{ people}}$$

1. ______________

2. Solve the proportion.

$$\frac{8}{d+2} = \frac{3}{d}$$

2. ______________

3. Rewrite the fraction so that the numerator and denominator have the same units. Then simplify.

$$\frac{3 \text{ ft}}{48 \text{ in.}}$$

3. ______________

4. As of August 28, 1993, the Pittsburgh Pirates baseball team had reported 60 wins in 129 games played. Write the ratio of wins to losses.

4. ______________

5. Given that $\frac{a}{b} = \frac{c}{d}$, decide whether it is true or not that $\frac{a+b}{a} = \frac{c+d}{c}$. Explain your reasoning.

5.

6. A wheelchair ramp has a slope of $\frac{1}{12}$. If its rise is $3\frac{1}{2}$ feet, what is its run?

6. ______________

7. The ratio of Gear A to Gear B is equal to the ratio of Gear B to Gear C. Gear B has 18 teeth and Gear C has 36 teeth. Find the number of teeth for Gear A.

7. ______________

Mid-Chapter 8 Test Form A

(Use after Lesson 8.3)

Name ______________________

Date ______________________

1. Rewrite the fraction so that the numerator and denominator have the same units. Then simplify.

$$\frac{3 \text{ hr}}{90 \text{ min}}$$

1. ______________

2. Consider this true proportion $\frac{x}{y} = \frac{w}{z}$. If the product of the means increases, what must happen to the product of the extremes? Explain your reasoning.

2.

3. Solve the proportion.

$$\frac{5}{b+2} = \frac{2}{b}$$

3. ______________

4. The points (2, 1), (3, −2), and (5, y) are collinear. Find y by solving the proportion

$$\frac{-2-1}{3-2} = \frac{y+2}{5-3}.$$

Explain why this proportion yields the correct value of y.

4.

5. Find the geometric mean of 6 and 15.

5. ______________

6. Assume the exchange rate of Canadian dollars to American dollars is 1 to 0.78. If a camera costs $55.00 in Canadian dollars, then what is its price in American dollars?

6. ______________

7. Determine whether or not a 4 in. × 5 in. photograph and an 8 in. ×10 in. photograph represent similar rectangles. Explain.

7.

8. If possible, sketch two isosceles triangles that have a common angle measure but are not similar.

8.

Mid-Chapter 8 Test Form B

(Use after Lesson 8.3)

Name ____________________

Date ____________________

1. Rewrite the fraction so that the numerator and denominator have the same units. Then simplify.
   $$\frac{2 \text{ pounds}}{40 \text{ ounces}}$$

1. ____________________

2. Consider this true proportion $\frac{x}{y} = \frac{w}{z}$. If the product of the extremes increases, what must happen to the product of the means? Explain your reasoning.

2.

3. Solve the proportion.
   $$\frac{7}{b+2} = \frac{3}{b}$$

3. ____________________

4. The points (6, 3), (9, y), and (−3, −9) are collinear. Find y by solving the proportion
   $$\frac{y-3}{9-6} = \frac{-9-3}{-3-6}.$$
   Explain why this proportion yields the correct value of y.

4.

5. Find the geometric mean of 4 and 20.

5. ____________________

6. Assume the exchange rate of Canadian dollars to American dollars is 1 to 0.78. If a VCR costs $198.50 in Canadian dollars, then what is its price in American dollars?

6. ____________________

7. Determine whether or not a 16 in. × 20 in. photograph and an 11 in. × 14 in. photograph represent similar rectangles. Explain.

7.

8. If possible, sketch two rhombuses that have the same side lengths but are not similar.

8.

8.4 Short Quiz

Name ____________________

Date ____________________

1. Given that $\triangle ABC \sim \triangle A'B'C'$. Find x.

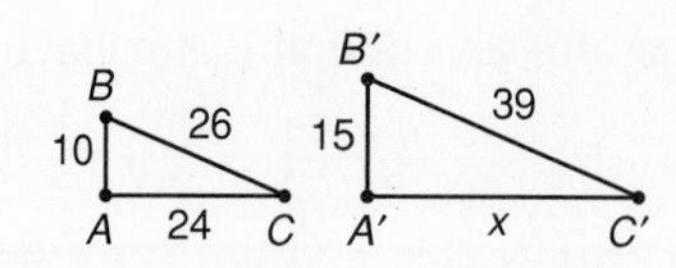

1. ____________________

2. When are two polygons similar?

2.

3. Are any of the following rectangles similar? If so, which ones?
   a. 7 ft by 13 ft **b.** 14 in. by 39 in.
   c. 1 ft by 3 ft **d.** $2\frac{1}{3}$ ft by $6\frac{1}{2}$ ft

3. ____________________

4. Decide whether or not an 8 in. × 10 in. photograph is similar to an 11 in. × 14 in. photograph. Explain your reasoning.

4.

5. If pentagon A is similar to pentagon B and pentagon B is similar to pentagon C, then what conclusions can be drawn. What property does this illustrate?

5.

6. Calculate the slope of the line. Does it matter which points are used? Why?

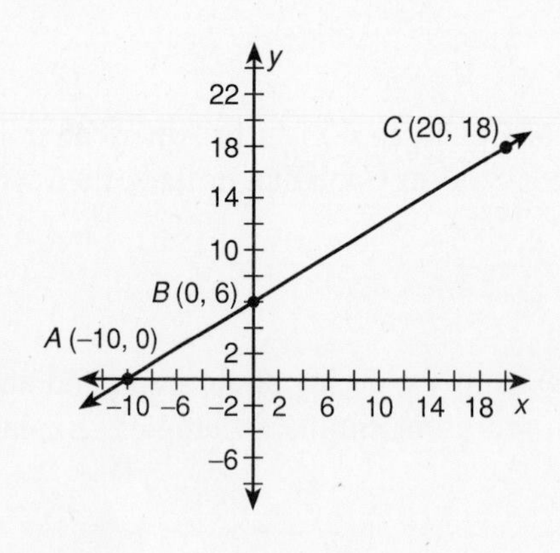

6.

7. Are all equilateral triangles similar? Explain.

7.

8.6 Short Quiz

Name ______________________

Date ______________________

1. $\triangle ABC$ has side lengths of 50, 90, and 130. $\triangle DEF$ has side lengths of 60, 108 and 156. Determine whether or not the two triangles are similar. Explain your reasoning.

1.

2. Are any of the following triangles similar? If so, which ones?

a. 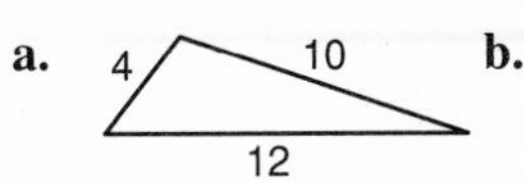

b. 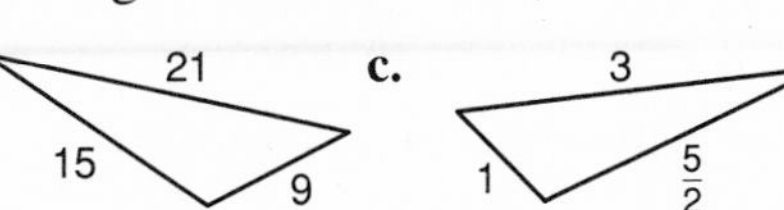

c.

2. ______________________

3. *Given:* $\triangle LMN \sim \triangle PQR$.
Find LM and LN.

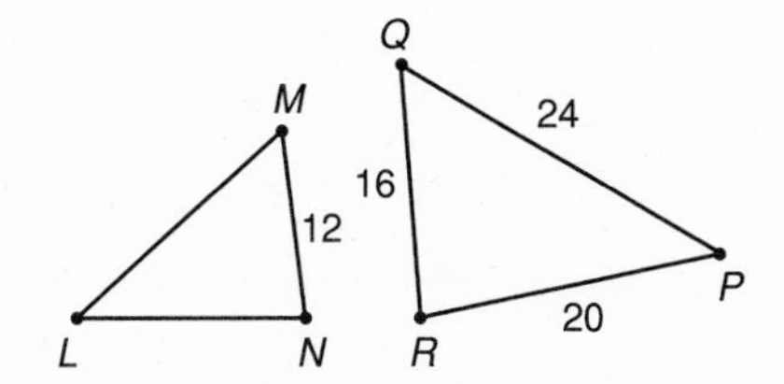

3. ______________________

In Problems 4 and 5, state the postulate or theorem that can be used to prove that the two triangles are similar.

4.

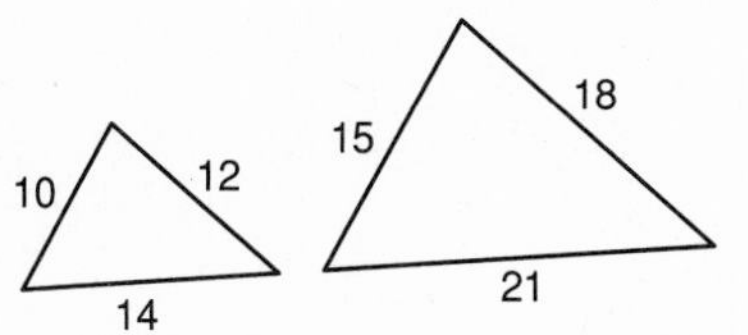

4. ______________________

5.

5. ______________________

6. Find VS, given that $\overline{RT} \parallel \overline{UV}$, $RU = 16$, $TV = 12$, and $US = 18$.

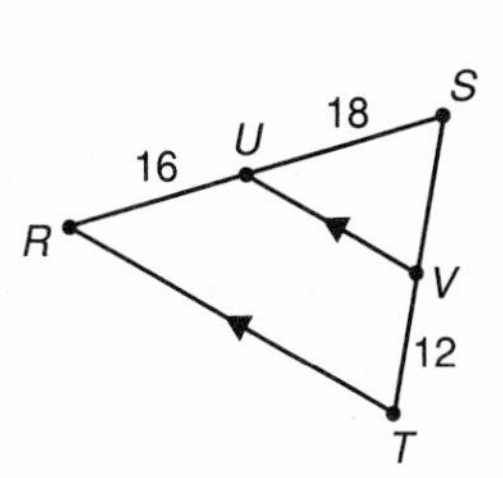

6. ______________________

7. Given that $\triangle PBQ \sim \triangle ABC$, what conclusion can you make about $\angle P$ and $\angle A$?

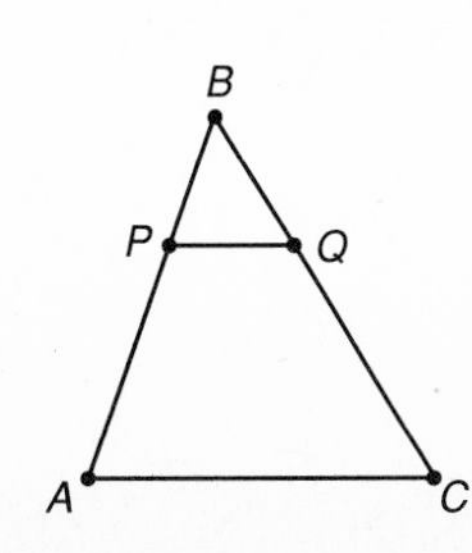

7. ______________________

Chapter 8 Test

Form A

(Page 1 of 3 pages)

Name ____________________

Date ____________________

1. Rewrite the fraction so that the numerator and denominator have the same units. Then simplify. 1. ____________

   $$\frac{7 \text{ min}}{400 \text{ sec}}$$

2. Consider this true proportion $\frac{R}{W} = \frac{S}{T}$. If the product of the means increases, what must happen to the product of the extremes? Explain your reasoning. 2.

3. Solve the proportion. 3. ____________

   $$\frac{3}{2x} = \frac{7}{5}$$

4. A student takes a Geometry test worth 200 points. How many correct answers did she have if her percent of correct answers on the test was 79%? 4. ____________

5. Find the geometric mean of 6 and 24. 5. ____________

6. Given that $\frac{a}{b} = \frac{c}{d}$, decide whether it is true or not that $\frac{a}{c} = \frac{b}{d}$. Explain your reasoning. 6.

7. The official height-to-width ratio of the United States flag is 1 : 1.9. If a United States flag is 9.5 feet wide, how high should it be? 7. ____________

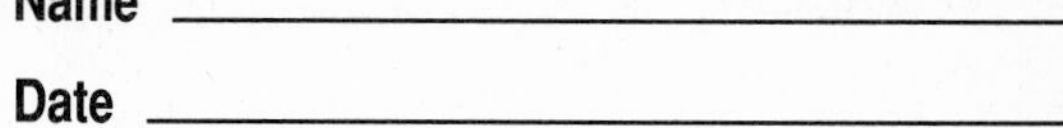

Name ______________________

Date ______________________

8. If hexagon A is similar to hexagon B and hexagon B is similar to hexagon C, then what conclusion can be drawn? What property does this illustrate?

8.

9. A wheelchair ramp has a slope of $\frac{1}{10}$. If its rise is $5\frac{1}{2}$ feet, what is its run?

9. ______________________

10. Calculate the slope of the line. Does it matter which points are used? Why?

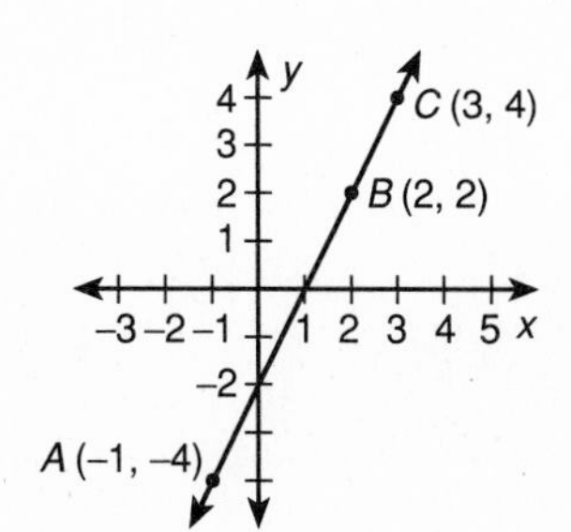

10.

11. If possible, draw two isosceles triangles in which all the equal-length sides in both triangles are congruent, but the two triangles themselves are not similar.

11.

12. Are all regular hexagons similar? Explain.

12.

13. Which triangle is not similar to any of the others?

13. ______________________

a.

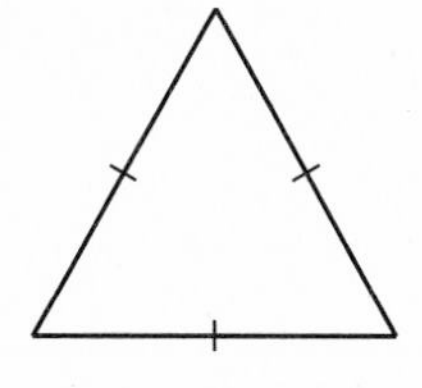

b.

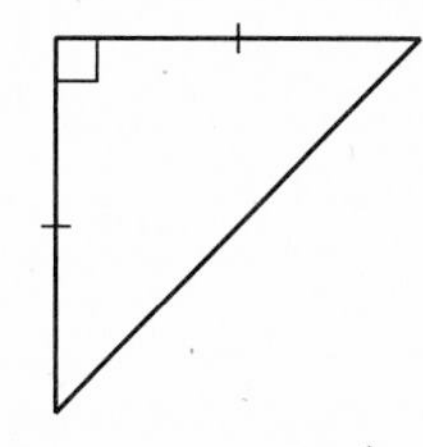

d.

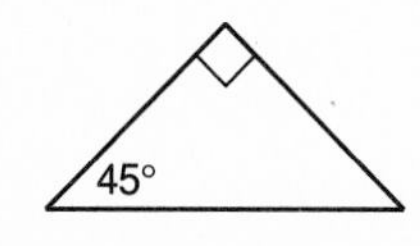

d.

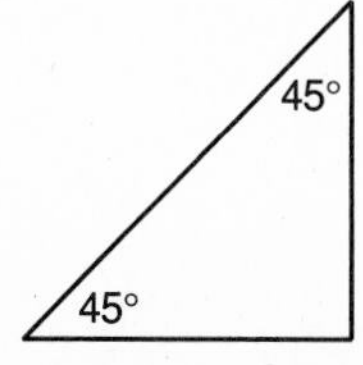

Name ____________________

Date ____________________

14. What value of x will make the two triangles similar?

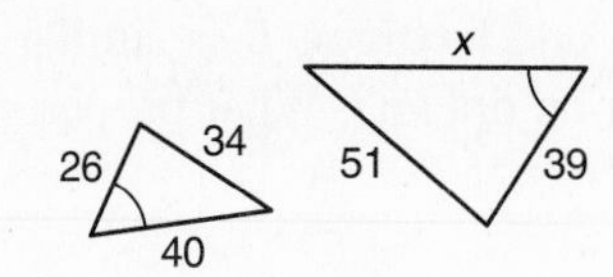

14. ____________

In Problems 15 and 16, state the postulate or theorem that can be used to prove that the two triangles are similar.

15.

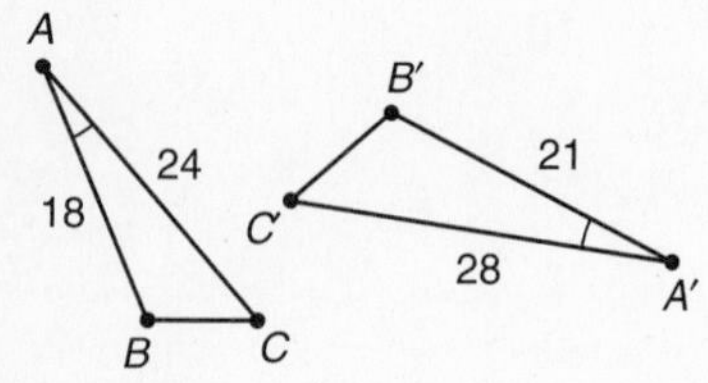

15. ____________

16.

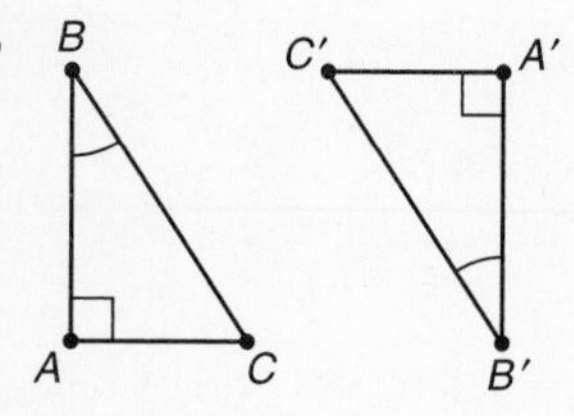

16. ____________

17. Given that $\triangle PQR \sim \triangle PST$, explain why $\overline{QR} \parallel \overline{ST}$.

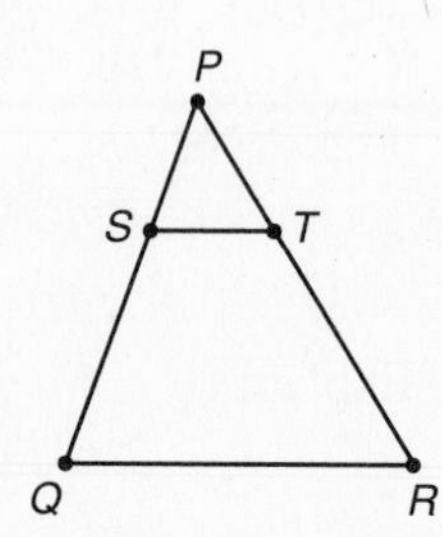

17.

18. Find the value of x to one decimal place.

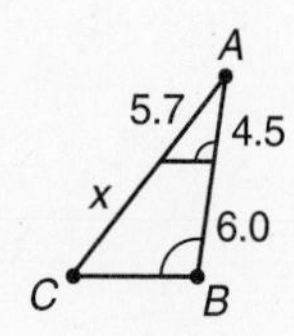

18. ____________

19. What is the scale factor of this dilation?

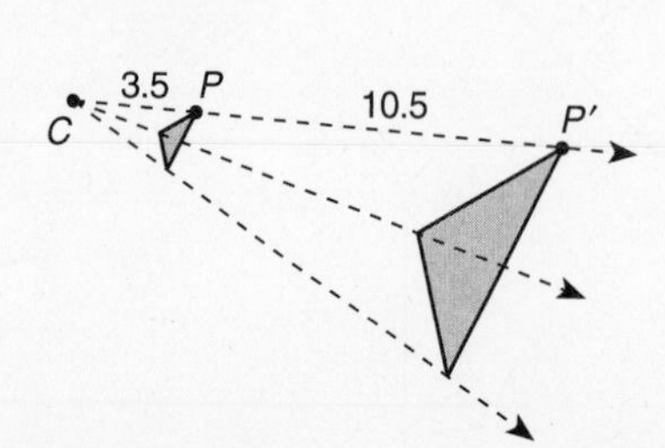

19. ____________

Name ______________________

Date ______________________

1. According to a recent survey, 20 out of every 25 students do not walk to school. Which of the following represents the ratio of walkers to total students? 1. ______

   a. $\frac{4}{5}$ b. $\frac{1}{4}$ c. $\frac{1}{5}$ d. 5

2. Which of the following is a proportion? 2. ______

   a. $XY = AB$ b. $\frac{5 \text{ in.}}{3 \text{ in.}} = \frac{15 \text{ lb}}{9 \text{ lb}}$ c. $\frac{\text{grams}}{\text{cubic cm}}$ d. A is to B

3. Consider this true proportion $\frac{x}{y} = \frac{u}{v}$. If the product of the means increases, then the product of the extremes [?]. 3. ______

   a. decreases b. increases c. remains the same

4. If $\frac{3}{x-4} = \frac{7}{x}$, then [?]. 4. ______

   a. $x = 3$ b. $x = 4$ c. $x = 7$ d. $x = \frac{7}{3}$

5. If $\frac{P}{Q} = \frac{R}{S}$, which of the following is *not* true? 5. ______

   a. $\frac{Q}{P} = \frac{S}{R}$ b. $PR = SQ$ c. $PS = RQ$ d. $\frac{R}{S} = \frac{P}{Q}$

6. Assume the exchange rate of Canadian dollars to American dollars is 1 to 0.77. If a stove costs $529.50 in Canada dollars, then what would its price be in American dollars? 6. ______

   a. $407.72 b. $687.66 c. $452.50 d. $506.50

7. If $\frac{a}{b} = \frac{c}{d}$, then [?]. 7. ______

   a. $\frac{a+b}{b} = \frac{c+b}{d}$ b. $ac = bd$ c. $\frac{a+b}{b} = \frac{c+d}{d}$ d. $\frac{a}{b} = \frac{a+c}{b+d}$

Name ______________________

Date ______________________

8. The geometric mean of 5 and 15 is [?]. **8.** ______

a. $5\sqrt{3}$ **b.** 10 **c.** 75 **d.** $\frac{15}{5} = 3$

9. If two polygons are *similar*, then their corresponding angles must be [?]. **9.** ______

a. supplementary **b.** linear pairs **c.** complementary **d.** congruent

10. If two polygons are *similar*, then their corresponding sides must be [?]. **10.** ______

a. parallel **b.** proportional **c.** similar **d.** congruent

11. One standard photograph size is a 4 in. × 5 in. rectangle. Which of these other standard rectangular sizes is similar to it? **11.** ______

a. $2\frac{1}{2}$ in.×$3\frac{1}{2}$ in. **b.** 5 in.×7 in. **c.** 8 in.×10 in. **d.** 11 in.×14 in.

12. One way to show that two triangles are similar is to show that [?]. **12.** ______

a. an angle of one is congruent to an angle of the other
b. a side of one is congruent to a side of the other
c. two sides of one are proportional to two sides of the other
d. two angles of one are congruent to two angles of the other

13. Which triangle is *not* similar to any of the others? **13.** ______

a.

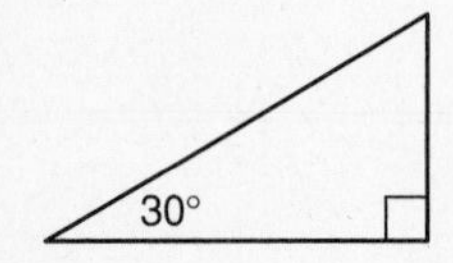

b.

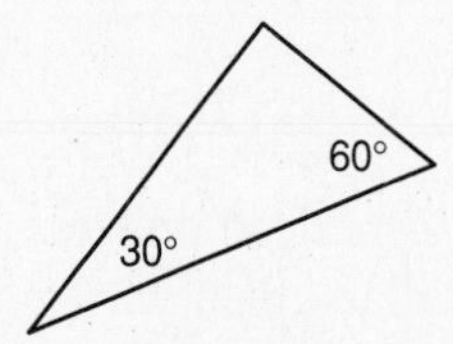

c.

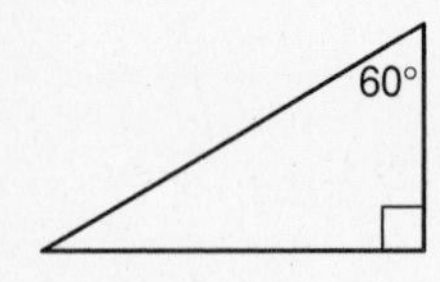

d.

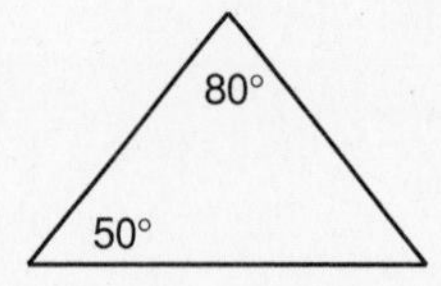

14. If $\triangle ABC \sim \triangle DEF$ and $\triangle DEF \sim \triangle GHI$, then [?]. **14.** ______

a. $\triangle ABC \sim \triangle GHI$ **b.** $\triangle ABC \cong \triangle GHI$
c. $AB = GH$ **d.** $\angle BCA \cong \angle GHI$

Chapter 8 Test

Form B (Page 3 of 3 pages)

Name ______________________

Date ______________________

15. If the corresponding sides of two triangles are proportional, then [?].

a. the triangles are congruent **b.** the triangles are right triangles
c. corresponding side lengths are equal **d.** the triangles are similar

15. ______

16. Shown at the right is an illustration of the [?].

a. AA Similarity Postulate
b. SAS Similarity Theorem
c. SSS Similarity Theorem
d. SAS Congruence Theorem

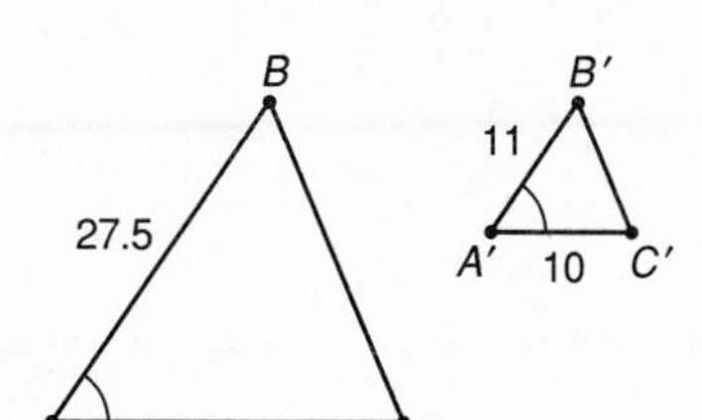

16. ______

17. If $\triangle ABC \sim \triangle PBQ$, then which of the following proportions is *not* true?

a. $\dfrac{PB}{AB} = \dfrac{PQ}{AC}$ **b.** $\dfrac{AC}{PQ} = \dfrac{CB}{QB}$

c. $\dfrac{AP}{PB} = \dfrac{AC}{PQ}$ **d.** $\dfrac{AP}{PB} = \dfrac{CQ}{QB}$

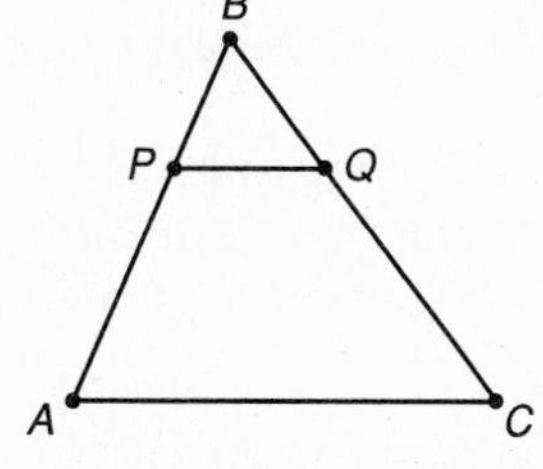

17. ______

18. The postulate or theorem that can be used to prove that the two triangles are similar is [?].

a. SSS Similarity Theorem
b. SAS Similarity Theorem
c. AA Similarity Postulate
d. ASA Congruence Theorem

18. ______

19. Find the value of x to one decmical place.

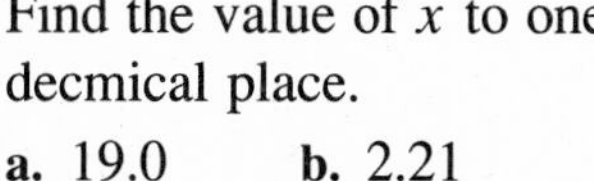

a. 19.0 **b.** 2.21
c. 0.53 **d.** 22.5

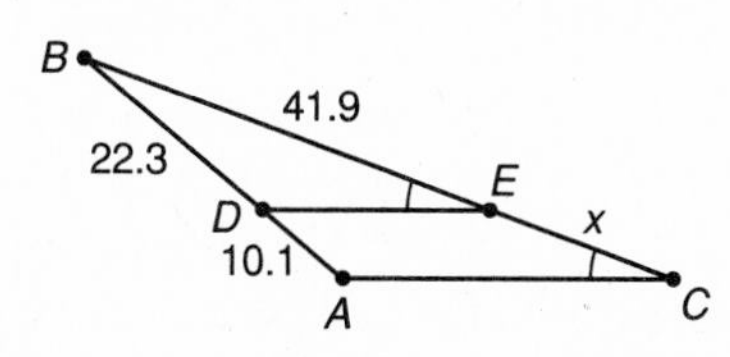

19. ______

20. The scale factor for the dilation shown is [?].

a. $\frac{13}{15}$ **b.** $\frac{28}{15}$
c. $\frac{13}{28}$ **d.** $\frac{15}{13}$

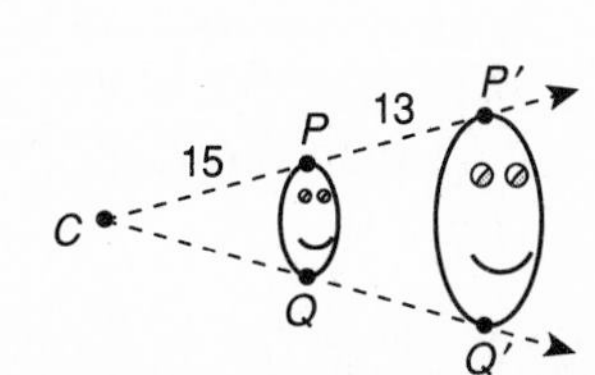

20. ______

Chapter 8 Test

Form C

(Page 1 of 3 pages)

Name ____________________

Date ____________________

1. Rewrite the fraction so that the numerator and denominator have the same units. Then simplify.

$$\frac{3 \text{ yards}}{48 \text{ inches}}$$

1. ____________________

2. The ratios of the side lengths of triangle ABC are 7:9:12 ($AB{:}AC{:}BC$). Solve for x.

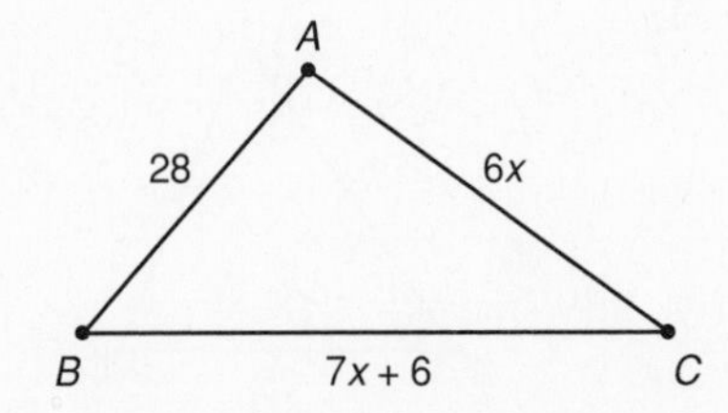

2. ____________________

3. Solve the proportion $\dfrac{5}{x-1} = \dfrac{7}{x}$.

3. ____________________

4. The batting average is the ratio of the number of hits to the number of at bats. A baseball player with 512 at bats needs how many hits to have a batting average of at least .275?

4. ____________________

5. Is it possible for the geometric mean and the arithmetic mean of two numbers, a and b, to be the same? Explain.

5.

6. If $\dfrac{a}{b} = \dfrac{c}{d}$, show that $\dfrac{a-b}{b} = \dfrac{c-d}{d}$.

6.

7. Calculate the slope of the line. Does it matter which points are used? Why or why not.

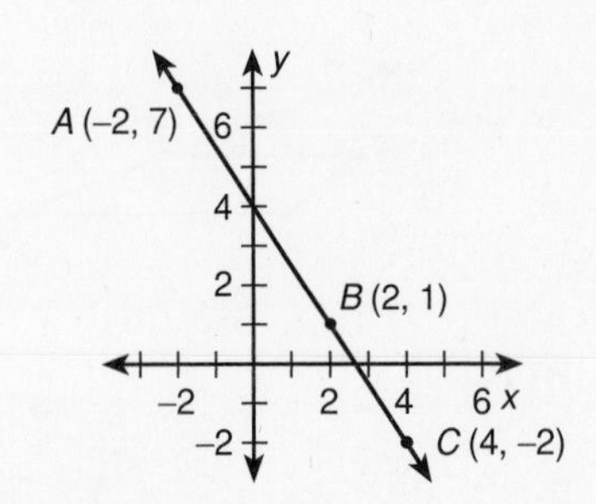

7.

Name ______________________

Date ______________________

8. *Given*: $\triangle ABC \sim \triangle A'B'C'$
$\triangle A'B'C' \sim \triangle A''B''C''$
What is the relationship between $\triangle ABC$ and $\triangle A''B''C''$? Find the values of x and y to the nearest tenth.

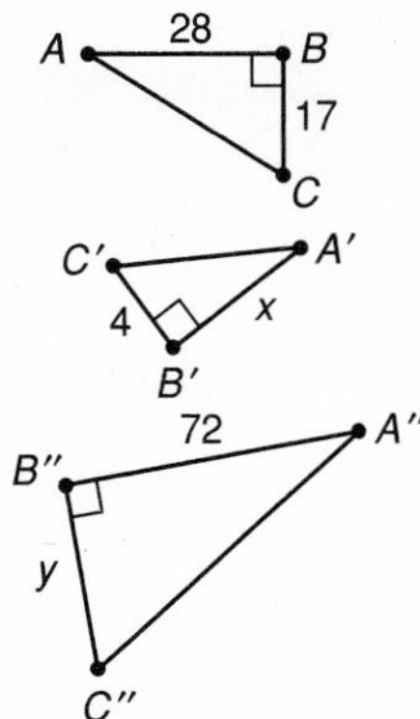

8.

9. Are the values determined for x and y in Problem 8 unique? Explain.

9.

10. Standard sizes of photo enlargements are not usually similar. Assume that all sizes were similar to the 5 in. × 7 in. size, where 5 in. is the width. What would be the corresponding length of an 8-in. wide enlargement? (Note: 8 in. × 10 in. is the standard offering.)

10. ______________________

11. If possible, draw two parallelograms in which corresponding sides are congruent but the figures are not similar.

11.

12. Are all isosceles trapezoids similar? Explain.

12. ______________________

13. Which triangle is not similar to any of the others?

13. ______________________

a.

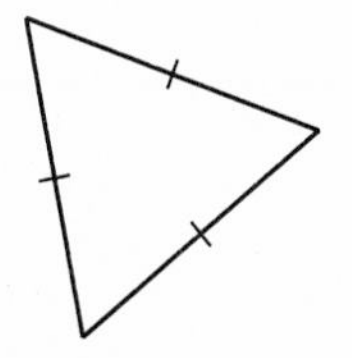

b.

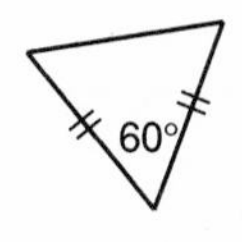

c.

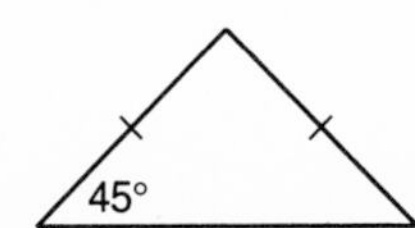

d.

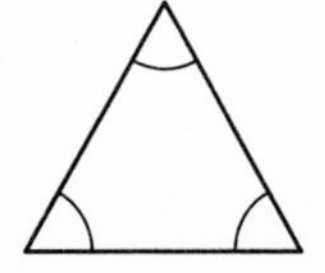

Name ______________________

Date ______________________

14. Find x so that the two triangles are similar.

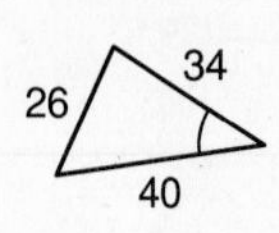

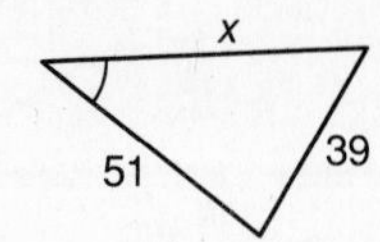

14. ______________

15. *Given*: $\overline{BF} \parallel \overline{AE}$, $\overline{AC} \parallel \overline{EF}$
Prove: $\triangle ACE \sim \triangle FED$

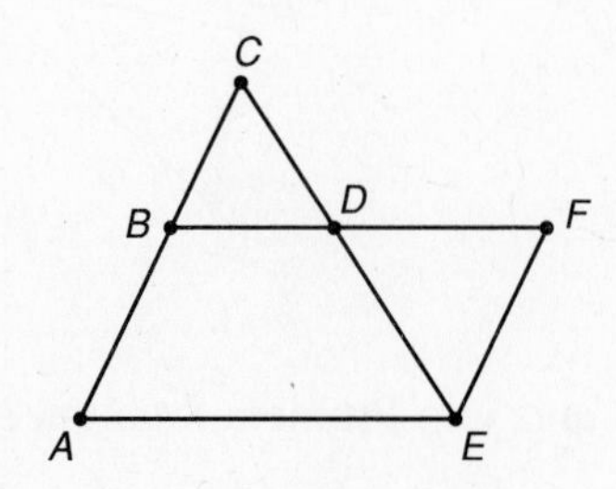

15.

16. Given that $\dfrac{RU}{UT} = \dfrac{RV}{VS}$, what is the relationship between $\overline{UV}$ and $\overline{TS}$?

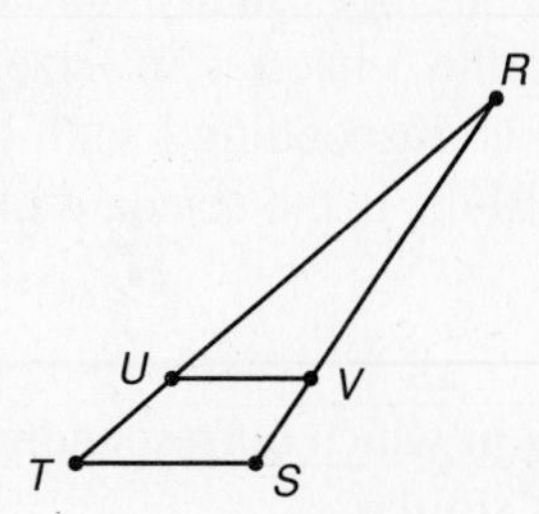

16. ______________

17. Find the value of x to one decimal place.

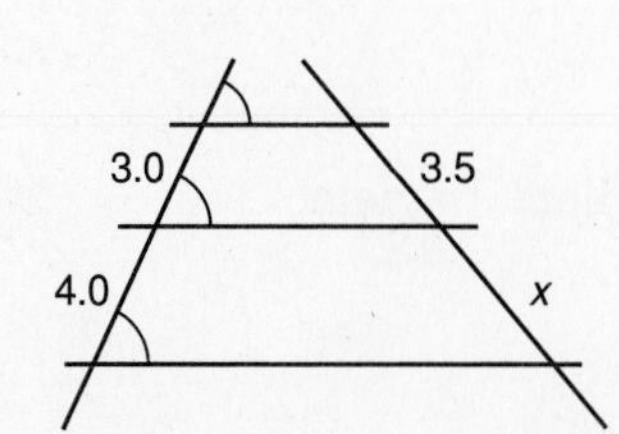

17. ______________

18. Find the scale factor to two decimal places for the dilation shown.

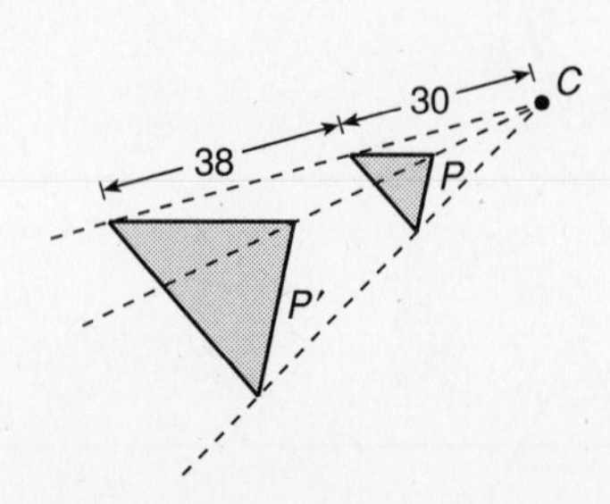

18. ______________

9.2 Short Quiz

Name ______________________

Date ______________________

In Problems 1–3, refer to the figure.

1. Name two triangles that are similar to $\triangle RST$.

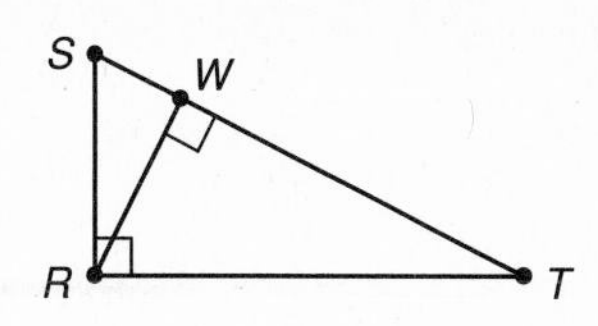

1. ______________________

2. Complete the proportion.

$$\frac{RT}{RS} = \frac{\boxed{?}}{WS}$$

2. ______________________

3. Which segment length is the geometric mean of SW and WT?

3. ______________________

4. Find the geometric mean of 5 and 80.

4. ______________________

5. Find the length of the altitude, x.

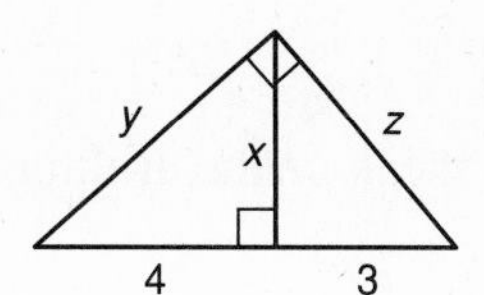

5. ______________________

6. Find the values of x and y.

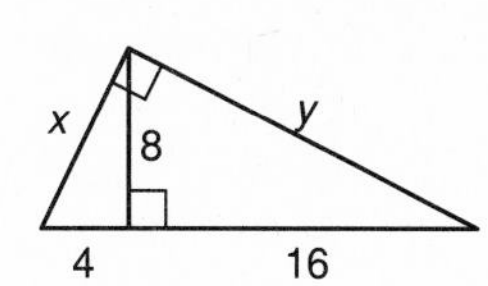

6. ______________________

Mid-Chapter  Test Form A

(Use after Lesson 9.3)

Name ______________________

Date ______________________

1. Find BD.

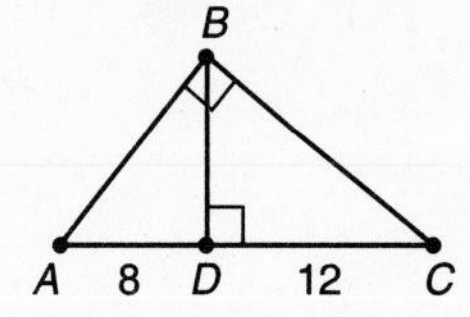

1. ______________

2. Find the value of PQ.

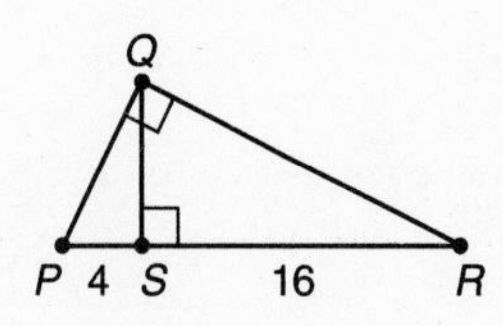

2. ______________

3. Find the value of x.

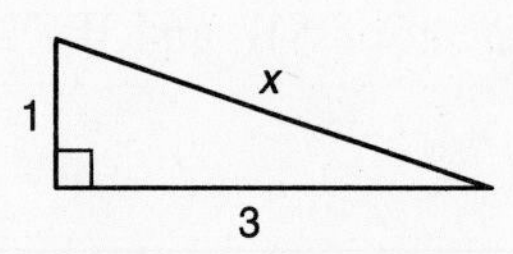

3. ______________

4. Choose the sets that are Pythagorean triples.

a. 12, 16, 20 **b.** 6, 7, 9 **c.** 10, 24, 26 **d.** 5, 12, 15

4. ______________

5. Which is greater, the geometric mean or the arithmetic mean of 4 and 40? Explain.

5.

6. The sides of a triangle are 6, 7 and 9. Is the triangle right, acute, or obtuse? Explain.

6.

7. Find the length of the diagonal of a square whose side lengths are 8.

7. ______________

8. The diagonal of a rectangle measures 10. If the rectangle is twice as long as it is wide, then find its dimensions.

8. ______________

Mid-Chapter 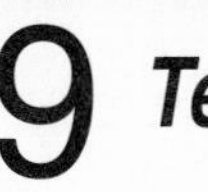 Test Form B

(Use after Lesson 9.3)

Name ______________________

Date ______________________

1. Find AB.

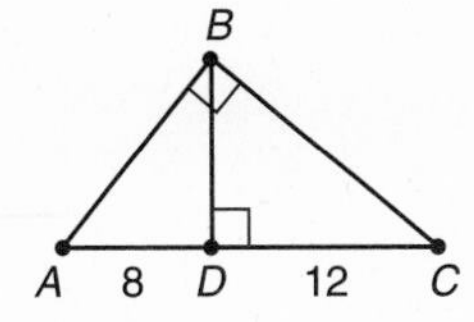

1. ______________

2. Find the value of QR.

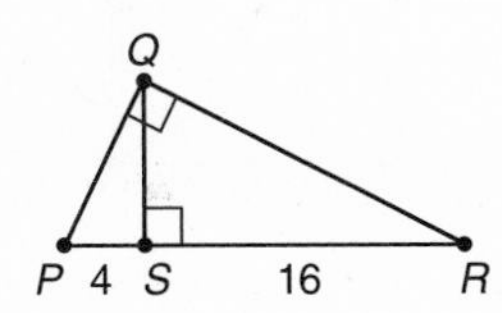

2. ______________

3. Find the value of x.

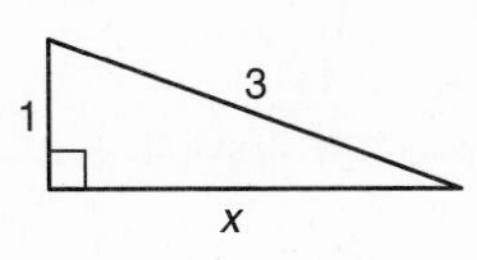

3. ______________

4. Choose the sets that are Pythagorean triples.

a. 4, 8, 10 **b.** 15, 20, 25 **c.** 3, 11, 12 **d.** 5, 12, 13

4. ______________

5. Which is greater, the geometric mean or the arithmetic mean of 6 and 54? Explain.

5.

6. The sides of a triangle are 7, 8, and 11. Is the triangle right, acute, or obtuse? Explain.

6.

7. Find the side length of a square whose diagonal measure is 8.

7. ______________

8. The width of a rectangle measures 6 and its length measures 9. Find the length of its diagonal.

8. ______________

9.4 Short Quiz

Name ______________________

Date ______________________

1. If $ABCD$ is a square with side lengths of 5, find the length of the diagonal, AC.

1. 

2. If $ABCD$ is a rectangle, find the value of AB.

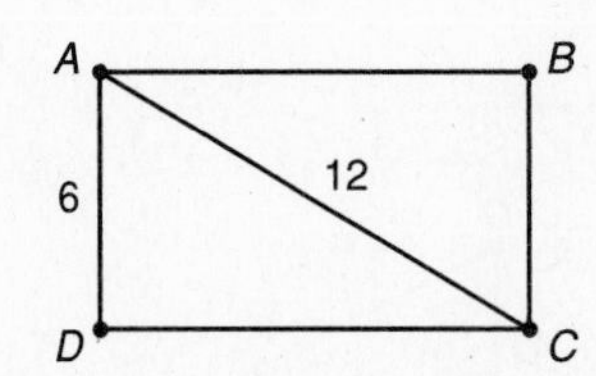

2. ______________

3. Choose the sets that could be the side lengths of a right triangle.

a. 4, 5, 6 **b.** 6, 8, 10 **c.** 5, 7, 10 **d.** 5, 12, 13

3. ______________

4. A triangle has side lengths 9, 13, and 16. Decide whether it is a right, acute, or obtuse triangle. Explain.

4.

5. Find the values of x and y.

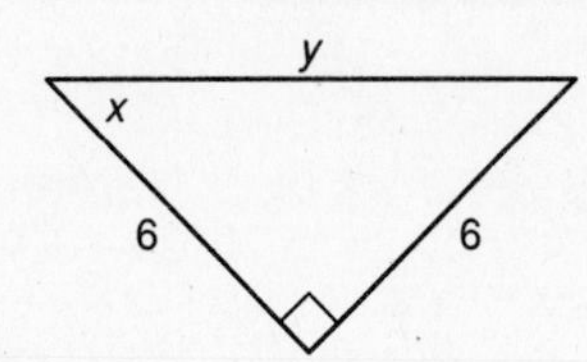

5. ______________

6. Find the values of x and y.

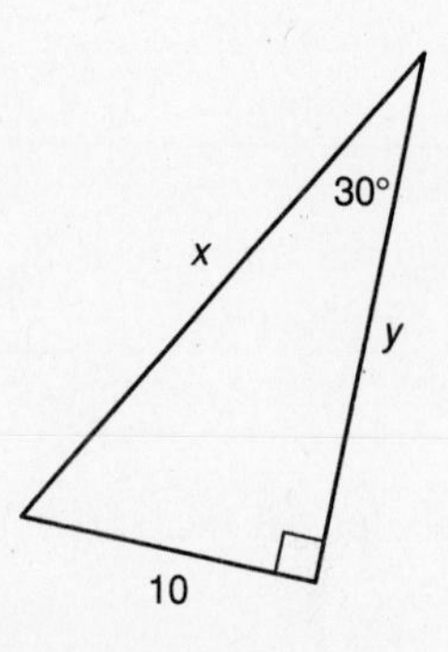

6. ______________

9.6 Short Quiz

Name ______________________

Date ______________________

1. Match the trigonometric ratio with its name.

   i. sine ii. cosine iii. tangent

   a. $\frac{\text{opp.}}{\text{adj.}}$ b. $\frac{\text{opp.}}{\text{hyp.}}$ c. $\frac{\text{adj.}}{\text{hyp.}}$

1. a. ______________ b. ______________ c. ______________

2. Find tan B.

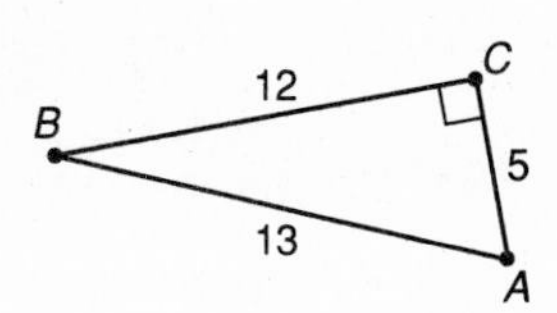

2. ______________

3. Find sin Q. (*Hint:* Use the Pythagorean Theorem.)

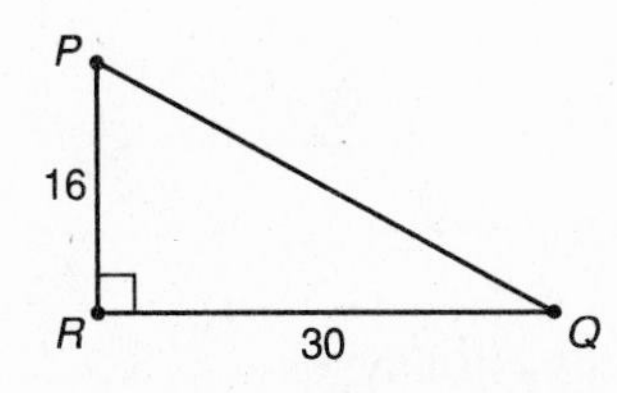

3. ______________

4. Given that tan A = 0.325, use a calculator to approximate the measure of acute angle A. Round your answer to two decimal places.

4. ______________

5. A man 6 feet tall stands x feet from a point at ground level. The angle from ground level to the top of the man's head is 14°. Find the value of x to the nearest foot.

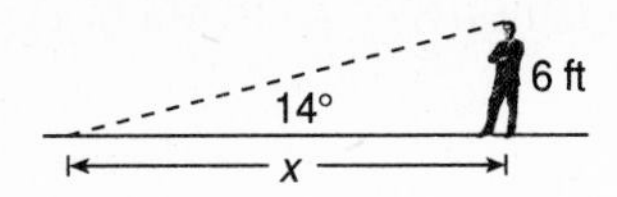

5. ______________

6. Find the measure of $\angle A$ to two decimal places.

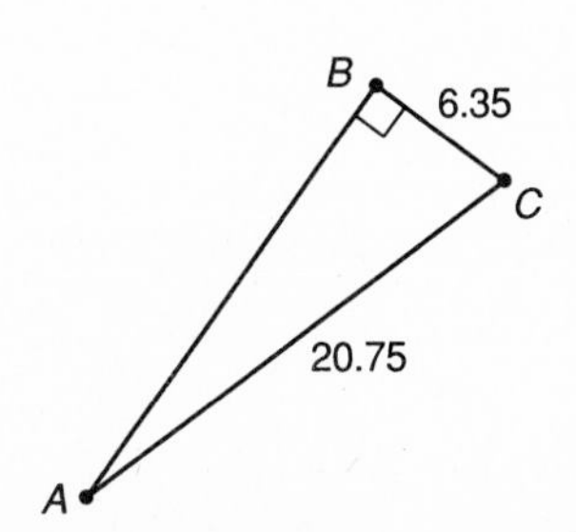

6. ______________

Chapter 9 Test

Form A

(Page 1 of 3 pages)

Name ______________________

Date ______________________

In Problems 1–4, refer to the figure below.

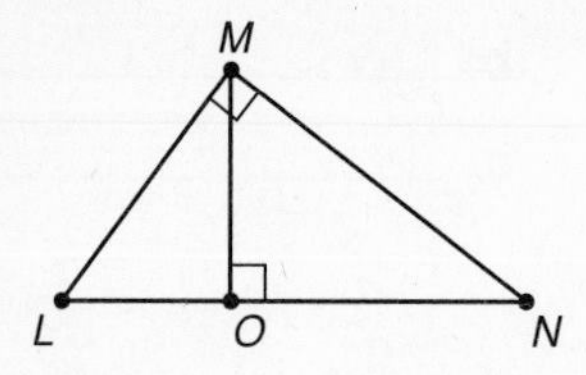

1. Name a triangle similar to $\triangle LOM$. **1.** ____________

2. Complete the proportion. **2.** ____________

$$\frac{OL}{ML} = \frac{\boxed{?}}{LN}$$

3. MO is the geometric mean of which segment lengths? **3.** ____________

4. *Given:* $LO = 3$ and $ON = 6$. Find the value MO. **4.** ____________

5. Which is greater, the geometric mean or the arithmetic mean of 4 and 12? Explain. **5.**

6. Choose the sets that are possible side lengths of a right triangle. **6.** ____________
   a. 1, 1, 2 **b.** 1, 1, $\sqrt{2}$ **c.** 3, 4, 7 **d.** 3, 4, 5

7. A baseball "diamond" is a square with a side length of 90 feet. How far is the throw from third base to first base? (Round your answer to one decimal place.) **7.** ____________

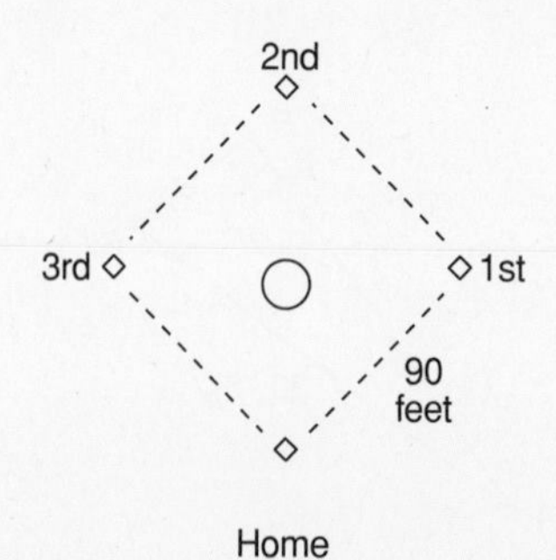

Name ______________________

Date ______________________

8. Which of the following sets are Pythagorean triples? **8.** ______________

a. $\sqrt{3}$, $\sqrt{4}$, $\sqrt{5}$ **b.** 12, 16, 20 **c.** $\frac{1}{3}$, $\frac{1}{4}$, $\frac{1}{5}$ **d.** 3^2, 4^2, 5^2

9. A triangle has side lengths of 6, 9, and 11. Decide whether it is an acute, right or obtuse triangle. Explain. **9.**

10. If $EFGH$ is a rectangle, what value must FH have? **10.** ______________

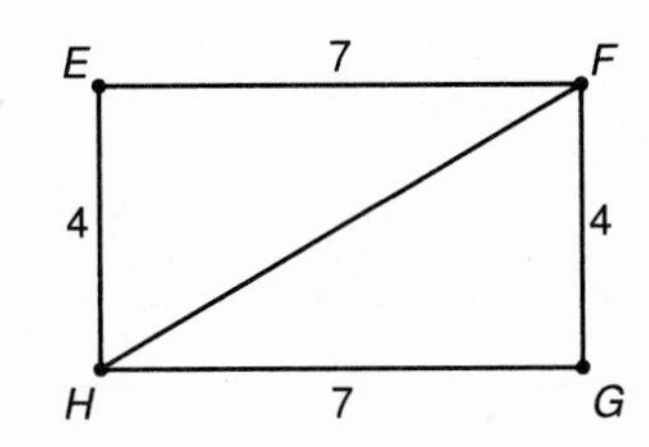

11. Find the altitude of an isosceles triangle whose base is 10 and whose congruent sides are 9. **11.** ______________

12. What is the length of the diagonal of a square whose side lengths are $7\sqrt{2}$? **12.** ______________

13. What is the length of an altitude of an equilateral triangle whose side length is $8\sqrt{3}$? **13.** ______________

14. *Given:* Rectangle $PQRS$ with $QR = 8$ and $SQ = 10$. What is the value of y? **14.** ______________

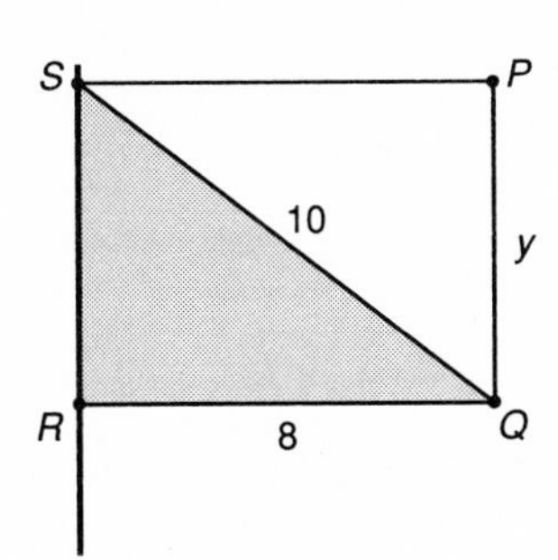

Name ______________________

Date ______________________

15. Write the trigonometric ratio.

a. $\sin A$ **b.** $\tan B$ **c.** $\cos A$

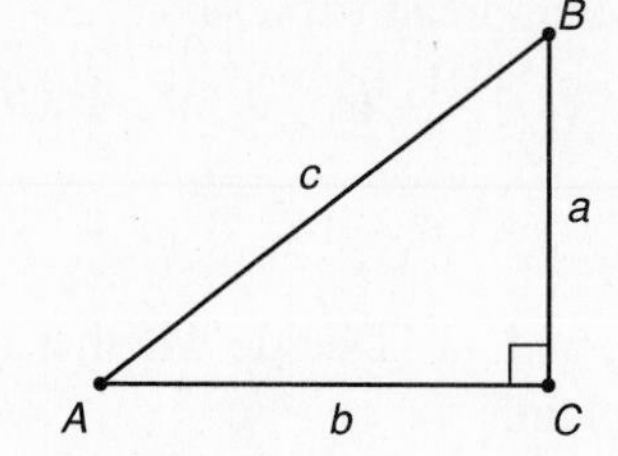

15. a. ______________

b. ______________

c. ______________

16. Find $\tan G$.

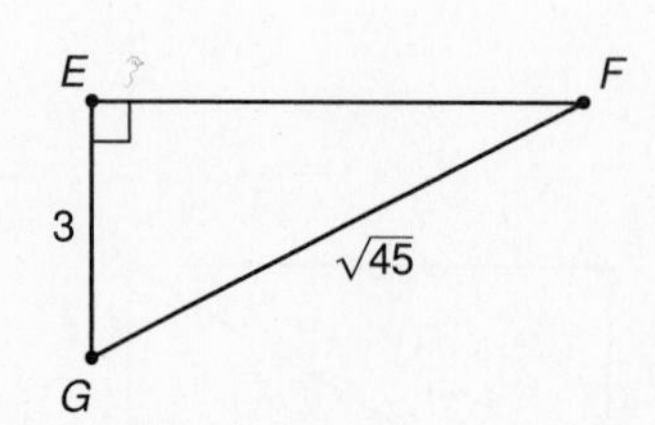

16. ______________

17. Use a calculator to find the value of $\cos 41^\circ$ to four decimal places.

17. ______________

18. Assume that $\angle A$ is an acute angle. If $\sin A = 0.994$, use a calculator to find the measure of $\angle A$ to two decimal places.

18. ______________

19. When solving a right triangle, if you are given both acute angles is that sufficient information? Explain.

19.

20. From a point 135 feet from the base of a tree, the angle from ground level to the top of the tree is 49°. Find the height of the tree to the nearest foot.

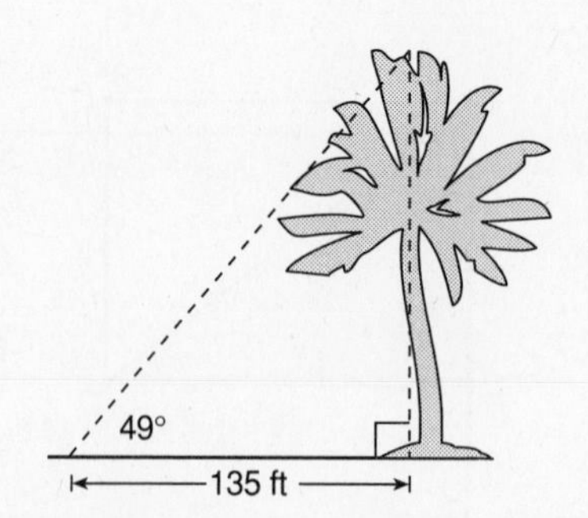

20. ______________

Chapter 9 Test

Form B

Name ______________________

Date ______________________

1. All right triangles are $\boxed{?}$. **1.** ______

a. equilateral or scalene **b.** acute or obtuse
c. isosceles or scalene **d.** similar

2. In the figure, an altitude is drawn to the hypotenuse of a right triangle. Which of the following is *not* true? **2.** ______

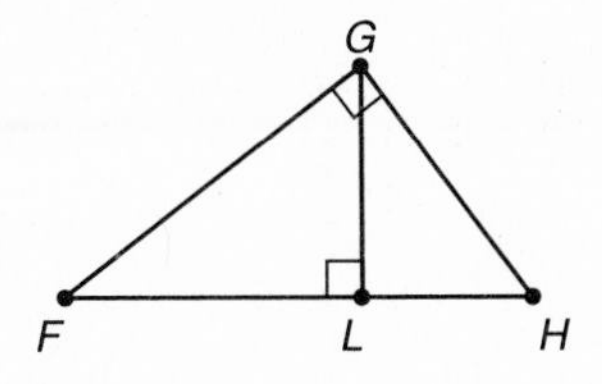

a. $\triangle FLG \sim \triangle GHL$
b. $\triangle GLH \sim \triangle FLG$
c. $\triangle FGH \sim \triangle FLG$
d. $\triangle GLH \sim \triangle FGH$

3. The closest approximation of the geometric mean of 5.8 and 20.4 is $\boxed{?}$. **3.** ______

a. 11 **b.** 10.88 **c.** $2\sqrt{29}$ **d.** 10

4. In $\triangle ABC$, CD is the geometric mean of $\boxed{?}$. **4.** ______

a. BD and DA **b.** BD and BC
c. AD and AC **d.** BC and AC

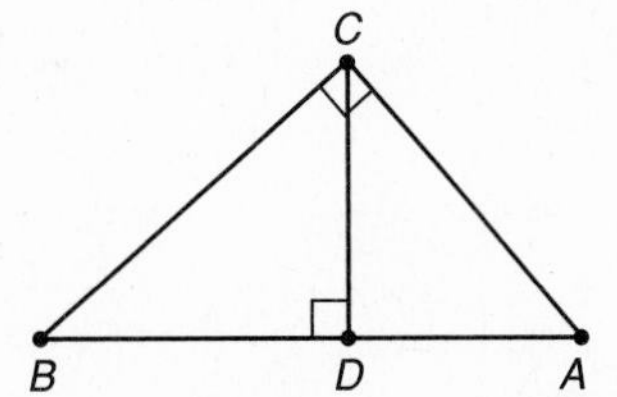

5. *Given:* $PS = 8$, $SR = 16$ **5.** ______

The value of QS is $\boxed{?}$.

a. 12 **b.** $8\sqrt{2}$
c. $8\sqrt{5}$ **c.** $8\sqrt{3}$

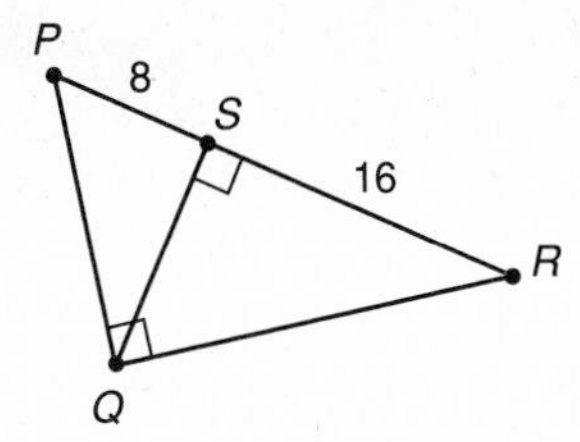

6. $\triangle ABC$ is a right triangle. **6.** ______

$AB = \boxed{?}$.

a. 117 **b.** $3\sqrt{5}$
c. $3\sqrt{13}$ **d.** $3\sqrt{6}$

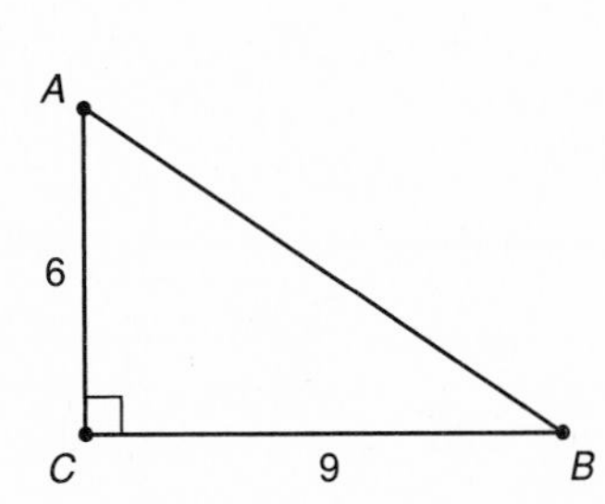

7. A set of Pythagorean triples is $\boxed{?}$. **7.** ______

a. 1, 1, 2 **b.** 6, 9, 12 **c.** 3, 5, 9 **d.** 5, 12, 13

Name ____________________

Date ____________________

8. A 25.5 foot ladder rests against the side of a house at a point 24.1 feet above the ground. The foot of the ladder is x feet from the house. Find the value of x.

a. 1.96 ft **b.** 7.0 ft **c.** 10.1 ft **d.** 8.3 ft

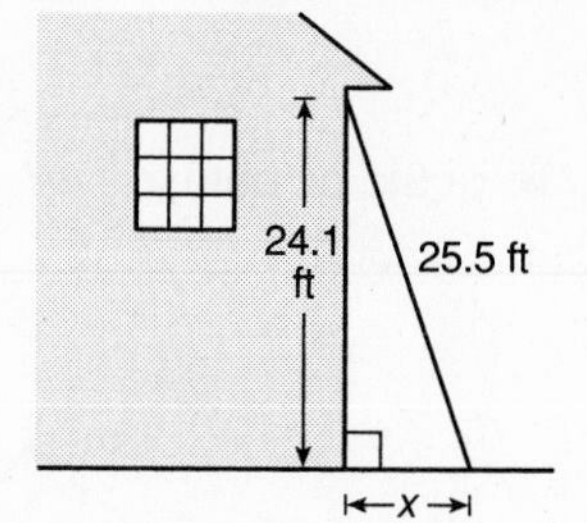

8. ________

9. For the triangle shown, the Pythagorean Theorem states that [?].

a. $f^2 - g^2 = e^2$ **b.** $e = f + g$
c. $e^2 = f^2 + g^2$ **d.** $e^2 + f^2 = g^2$

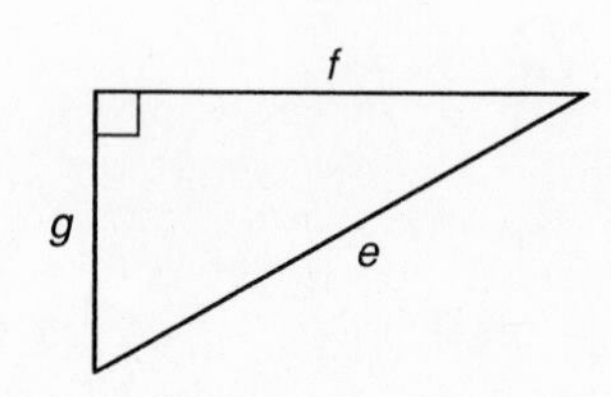

9. ________

10. Which triangle is *not* congruent to the others?

a.

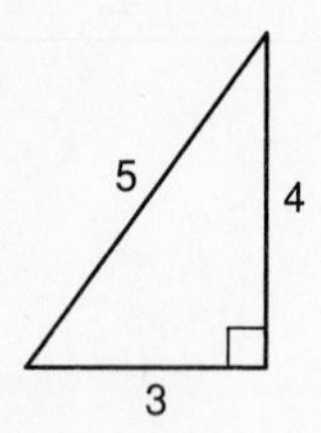

b.

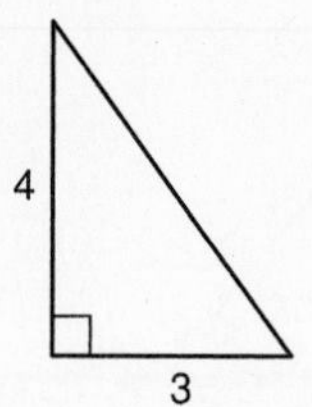

c.

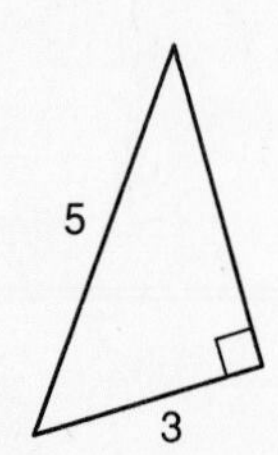

d.

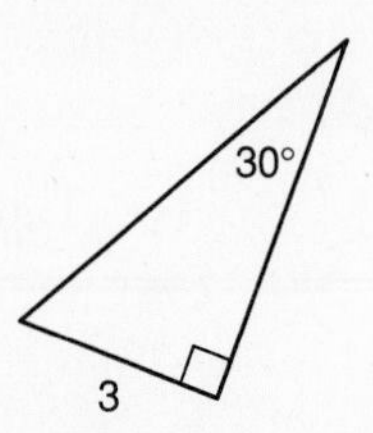

11. If the side lengths of a triangle are 7, 6, and 9, the triangle [?].

a. is an acute triangle **b.** is a right triangle
c. cannot be formed **d.** is an obtuse triangle

12. In a 45°–45°–90° triangle, the ratio of the length of the hypotenuse to the length of a side is [?].

a. $\sqrt{3}:1$ **b.** $\sqrt{2}:1$ **c.** $1:1$ **d.** $2:1$

Name ____________________

Date ____________________

13. In a 30°-60°-90° triangle, the ratio of the length of the hypotenuse to the length of the shorter side is [?].

a. $\sqrt{3}:1$ **b.** $\sqrt{2}:1$ **c.** $2:\sqrt{3}$ **d.** $2:1$

13. ________

14. An equilateral triangle has side lengths of 10. The length of its altitude is [?].

a. $5\sqrt{10}$ **b.** 5 **c.** $5\sqrt{3}$ **d.** $10\sqrt{5}$

14. ________

15. The cosine of $\angle A$ is the ratio [?].

a. $\frac{\text{opp.}}{\text{adj.}}$ **b.** $\frac{\text{opp.}}{\text{hyp.}}$

c. $\frac{\text{adj.}}{\text{hyp.}}$ **d.** $\frac{\text{hyp.}}{\text{adj.}}$

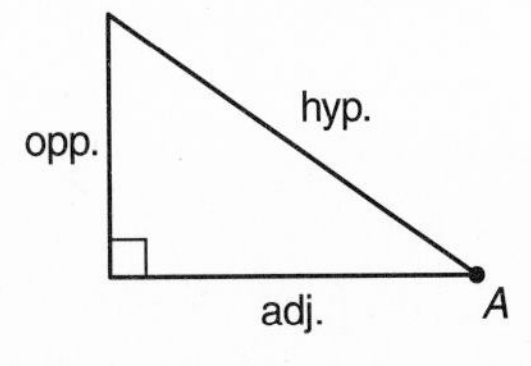

15. ________

16. The tangent of $\angle B$ is [?].

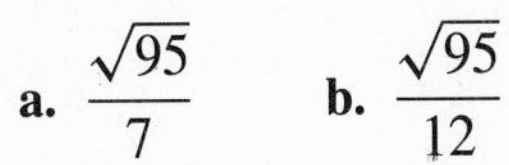

a. $\frac{\sqrt{95}}{7}$ **b.** $\frac{\sqrt{95}}{12}$

c. $\frac{12}{7}$ **d.** $7\sqrt{95}$

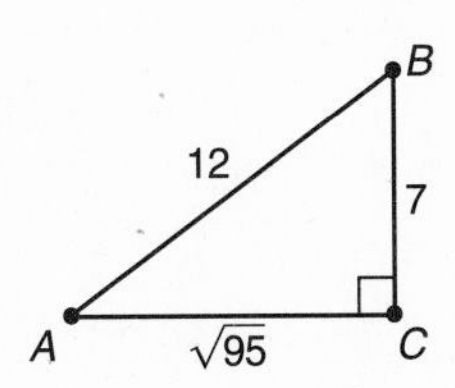

16. ________

17. Use your calculator to determine cos 23°.

a. ≈ 0.390 **b.** ≈ 0.921 **c.** ≈ 1.07 **d.** ≈ 0.424

17. ________

18. Assume that $\angle A$ is an acute angle and $\tan A = 1.230$. The measure of $\angle A$ is [?].

a. $\approx 50.9°$ **b.** $\approx 39.1°$ **c.** $\approx 7.01°$ **d.** $\approx 129.9°$

18. ________

19. Which of the following is *not* enough information to solve a right triangle?

a. Two angles

b. Two sides

c. One side length and one acute angle measure

d. One side length and one trigonometric ratio

19. ________

20. Two legs of a right triangle have lengths 15 and 8. The measure of the smaller acute angle is [?].

a. $\approx 32.2°$ **b.** $\approx 28.1°$ **c.** $\approx 17°$ **d.** $\approx 61.9°$

20. ________

Chapter 9 Test

Form C

(Page 1 of 3 pages)

Name ____________________

Date ____________________

In Problems 1–4, refer to the figure below.

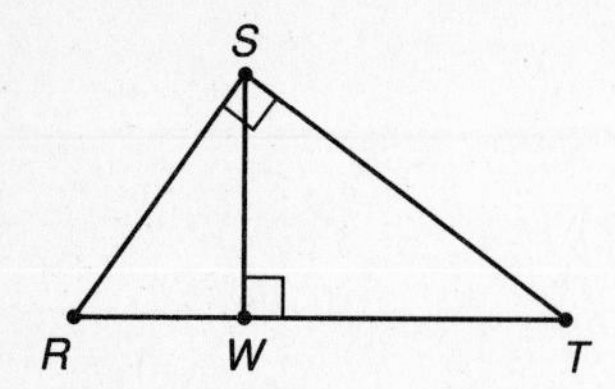

1. Name a triangle similar to $\triangle WST$. **1.** ____________

2. Complete the proportion: $\frac{RS}{SW} = \frac{RT}{\boxed{?}}$. **2.** ____________

3. RS is the geometric mean of the lengths of which segments? **3.** ____________

4. *Given*: $RW = 3$ and $ST = \frac{20}{3}$. Find RT. **4.** ____________

5. Which is greater, the arithmetic mean or the geometric mean of 6 and 27? **5.** ____________

6. Choose the sets which are possible sides of a right triangle. **6.** ____________

a. 4, 9, 13 **b.** $\sqrt{2}$, $\sqrt{2}$, 2 **c.** 8, 15, 17 **d.** 1, 1, 2

7. A baseball "diamond" is a square of side length 90 feet. How far is the throw, to one decimal place, from home plate to second base? **7.** ____________

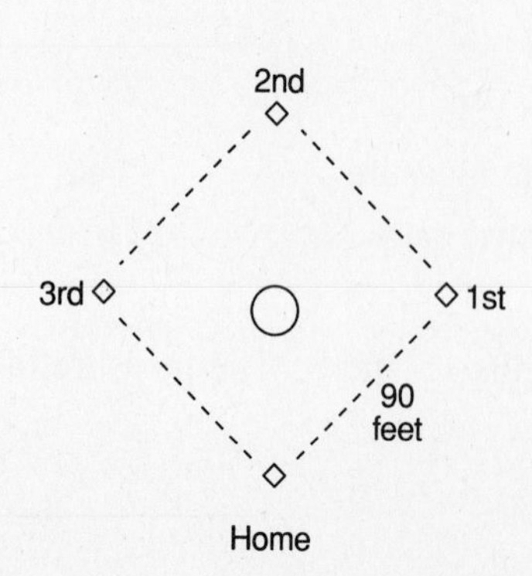

Name ____________________

Date ____________________

8. If a, b, and c are sides of a right triangle, which of the following are also sides of a right triangle?

a. The square root of each length $(\sqrt{a}, \sqrt{b}, \sqrt{c})$

b. Twice the length of each side $(2a, 2b, 2c)$

c. Four more than each length $(a + 4, b + 4, c + 4)$

d. The square of each length

8. ____________________

9. A triangle has side lengths 7, 9, and 11. Decide whether it is an acute, right, or obtuse triangle.

9. ____________________

10. If $PQRS$ is a rhombus, but not a square, what do we know about the length of PR?

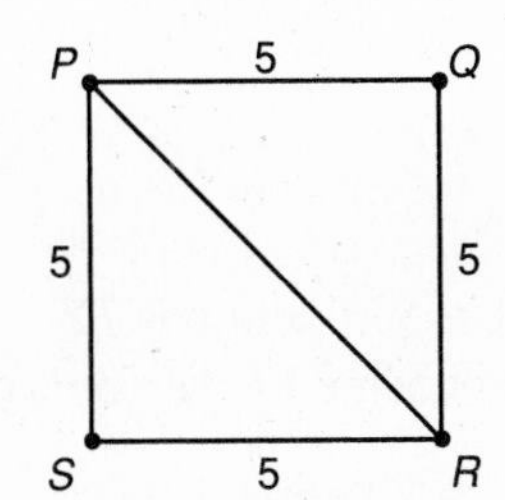

10.

11. The cross section of a V-thread on a screw is an equilateral triangle. The distance p between successive threads is known as the *pitch* of the thread, and the distance d is the *depth* of the thread. If $p = \frac{1}{8}$ inch, what is the value of d?

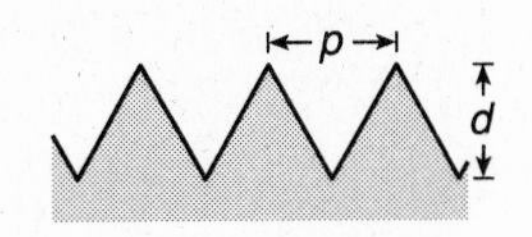

11. ____________________

12. The length of the diagonal of a square is 22. What is the length of each side?

12. ____________________

13. The altitude of an equilateral triangle is 6. What is the length of each side?

13. ____________________

14. $\triangle ABC$ is isosceles. What must the ratio of w to y be to make the triangle equilateral?

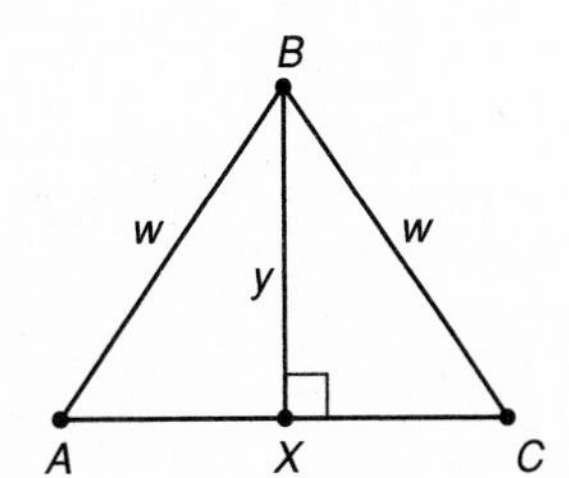

14. ____________________

Name ______________________

Date ______________________

15. Write the trigonometric ratio.
a. $\cos B$ **b.** $\tan A$ **c.** $\sin B$

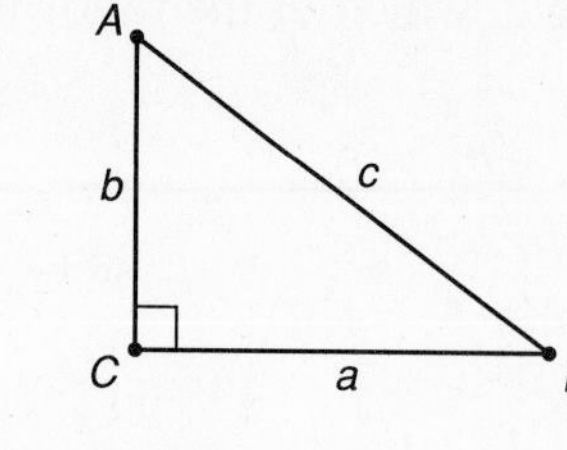

15. a. ______________
b. ______________
c. ______________

16. Find $\tan S$.

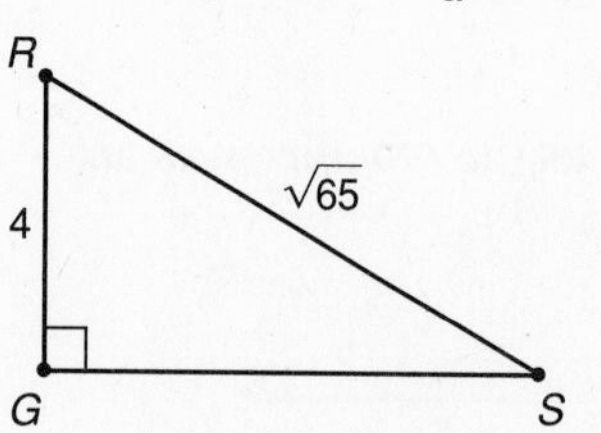

16. ______________

17. Use a calculator to find $\cos 17°$, $\cos 37°$, $\cos 57°$, and $\cos 77°$. As the angle increases, what happens to the cosine of the angle? Explain.

17.

18. Assume that $\angle A$ is an acute angle. If $\sin A = 0.9540$, find the $\tan A$ to four decimal places. (Use your calculator.)

18. ______________

19. If you are solving a right triangle and are given one side length (a leg or a hypotenuse), what additional information do you need?

19. ______________

20. An antenna is atop the roof of a 100-foot building, 10 feet from the edge, as shown in the figure. From a point 50 feet from the base of the building the angle from ground level to the top of the antenna is 66°. Find x, the length of the antenna, to the nearest foot.

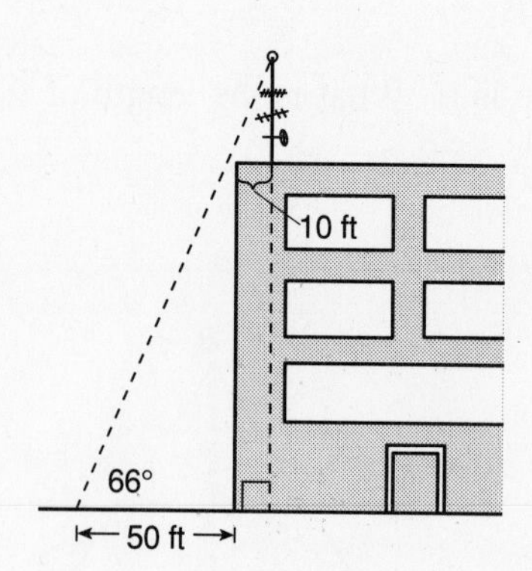

20. ______________

Cumulative Test

Chapters 7–9

(Page 1 of 8 pages)

Name ______________________

Date ______________________

1. Explain what is meant by a transformation that is an isometry.

1.

2. Explain whether or not this transformation is rigid.

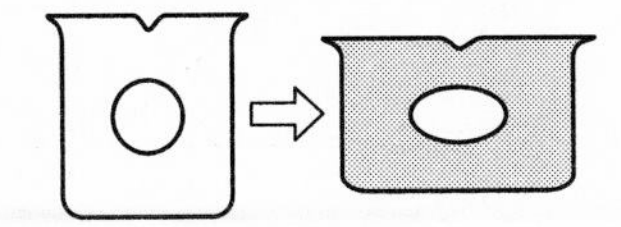

2.

In Problems 3 and 4, name the type of transformation. (Preimages are unshaded; images are shaded.)

3.

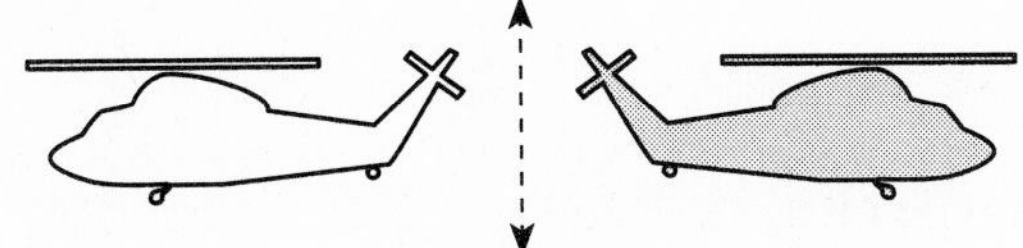

3. ______________________

4.

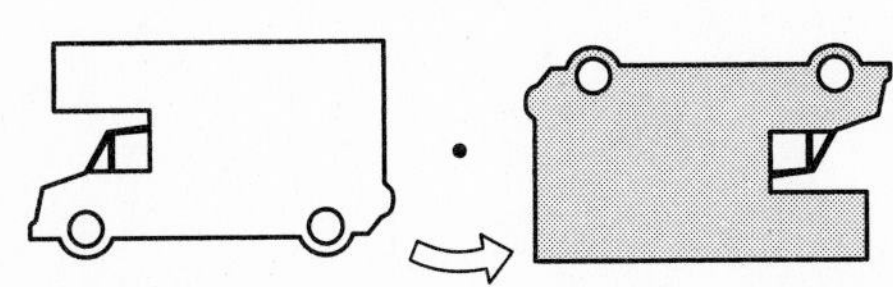

4. ______________________

5. The points in a coordinate plane are reflected in the x-axis. In general, every point (x, y) is mapped onto what point?

5. ______________________

6. $\overline{CD}$ is translated by the motion rule $(x, y) \rightarrow (x + 2, y - 2)$. Find the coordinates of the endpoints of the image $\overline{C'D'}$.

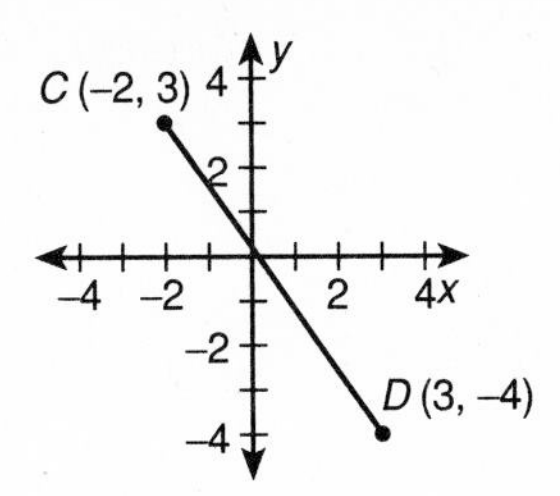

6. ______________________

7. Draw the reflection of $\overline{VW}$ in line ℓ.

7. *Use figure at left.*

Name ____________________

Date ____________________

In Problems 8 and 9, how many lines of symmetry does each octagon have?

8. 

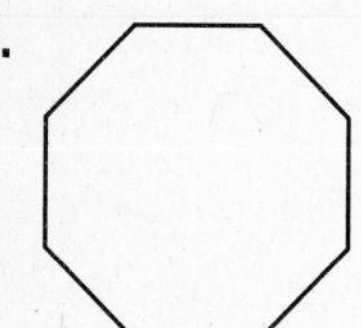

Regular octagon

8. ____________________

9.

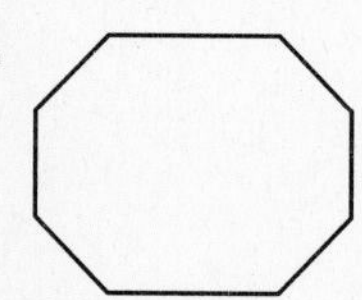

Equiangular octagon

9. ____________________

10. Draw a triangle with exactly one line of symmetry. What type of triangle is it?

10. ____________________

11. $\overline{RS}$ is rotated 180° clockwise about the origin. Find the coordinates of the endpoints of the image $\overline{R'S'}$.

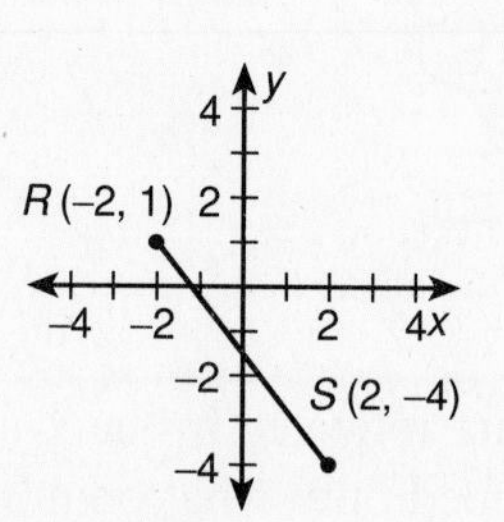

11. ____________________

12. Does the clock face shown have rotational symmetry? If so, list any angles of rotation, 180° or less, that can map it onto itself.

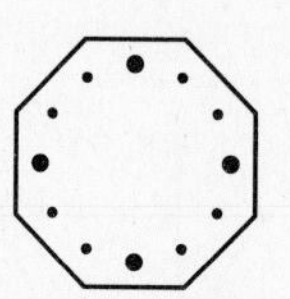

12. ____________________

13. Lines ℓ and m intersect at O and form a 30° angle. Q is reflected in ℓ, followed by a reflection in m. Describe the location of the image Q'.

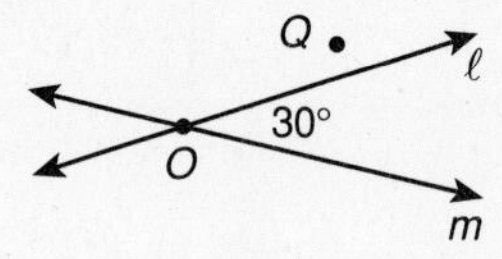

13. ____________________

14. In the figure, $\ell \parallel m$ and the distance between the lines is d. Figure P is reflected in ℓ onto image P' which is then reflected in m onto image P''. The composition of the two reflections about parallel lines results in what transformation?

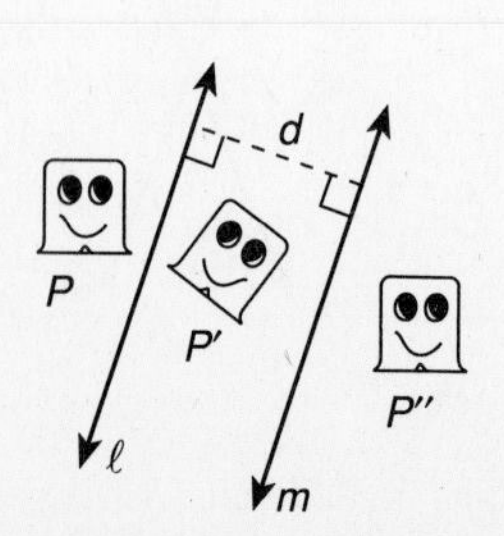

14. ____________________

15. Find the translation vector $\overrightarrow{v}$ that maps $\triangle ABC$ onto $\triangle A'B'C'$.

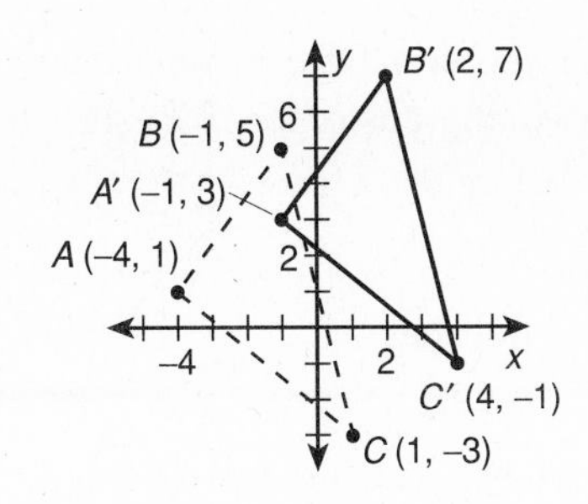

15. ______

16. A translation vector is $\overrightarrow{u} = \langle -7, 1\rangle$. If the coordinates of A', the image of point A, are $(-6, -3)$, find the coordinates of point A.

16. ______

17. Point A is translated by the vector $\overrightarrow{v} = \langle 0, -4\rangle$, and then reflected in the line $x = 1$. If the coordinates of A are $(-3, 7)$, find the coordinates of its image A'.

17. ______

18. The point $A(4, 2)$ is translated by the vector $\overrightarrow{v} = \langle -7, -3\rangle$ and then reflected in the x-axis. Find the coordinates of its image, A'.

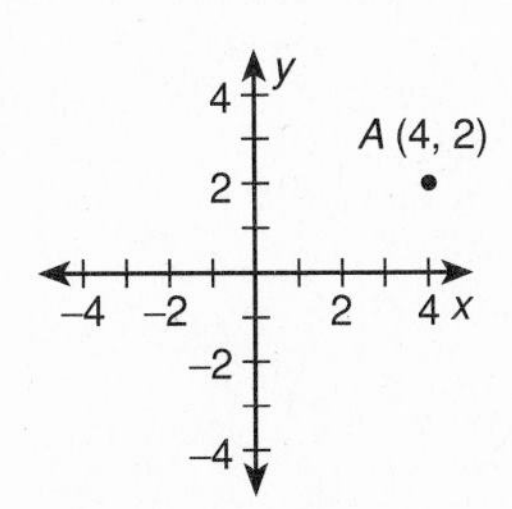

18. ______

19. Classify the frieze pattern.

19. ______

20. Rewrite the fraction so that the numerator and denominator have the same units. Then simplify.

$$\frac{108 \text{ in.}}{5 \text{ yd}}$$

20. ______

21. Solve the proportion.

$$\frac{5}{x-2} = \frac{7}{x+1}$$

21. ______

22. A salesclerk made sales to 18% of the potential customers she contacted. If she made 81 sales, how many customers had she contacted? **22.** ____________

23. If $\frac{a}{b} = \frac{c}{d}$, decide whether it is true or not that **23.**

$$\frac{a+b}{b} = \frac{c+d}{d}.$$

Explain your reasoning.

24. Find the geometric mean of 6 and 30. **24.** ____________

25. A garage ramp has a slope of $\frac{1}{20}$. If its total rise is 32.3 feet, how long is its run? **25.** ____________

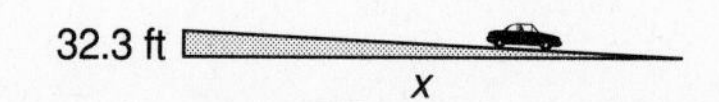

26. Calculate the slope of the line. Does it matter which points are used? **26.** ____________

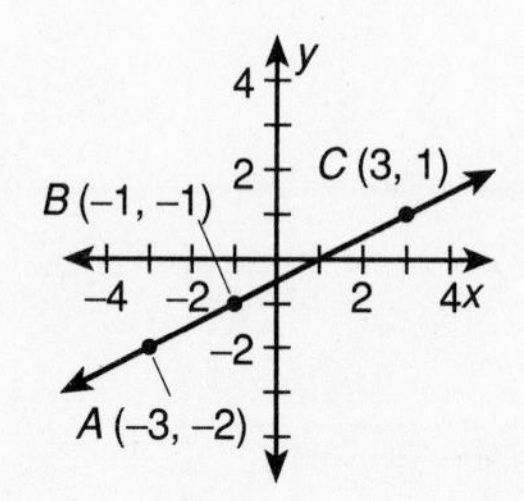

27. If two hexagons are similar, must they be regular hexagons? Explain. **27.**

28. Which triangle is not similar to any of the others? **28.** ____________

a.
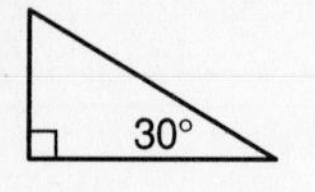

b.
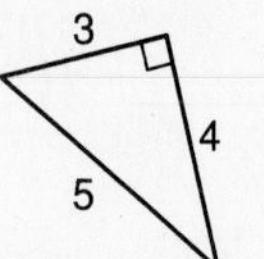

c.
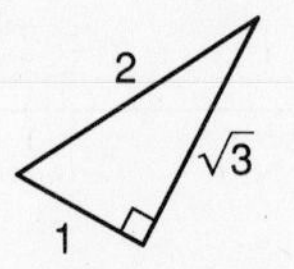

d.
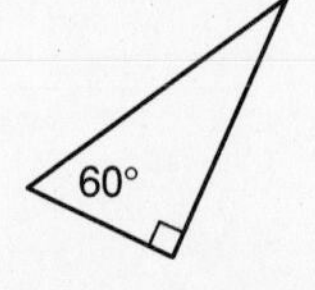

Name ______________________

Date ______________________

29. Sketch, if possible, two isosceles trapezoids that are not similar.

29.

30. What value of x will make the two triangles similar?

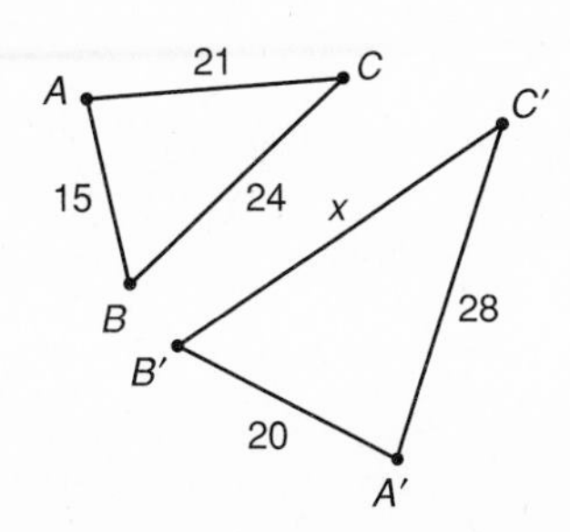

30. ______________________

31. *Given:* $\frac{PQ}{P'Q'} = \frac{QR}{Q'R'} = \frac{RP}{R'P'}$

State the postulate or theorem that can be used to prove that the two triangles are similar.

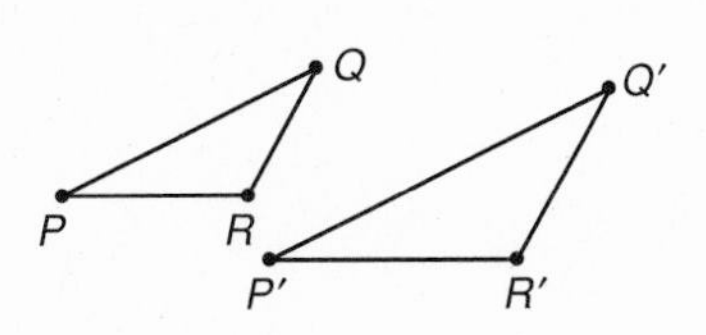

31. ______________________

32. Given that $\frac{DG}{FH} = \frac{GE}{HE}$, state a relationship between $\overline{GH}$ and $\overline{DF}$.

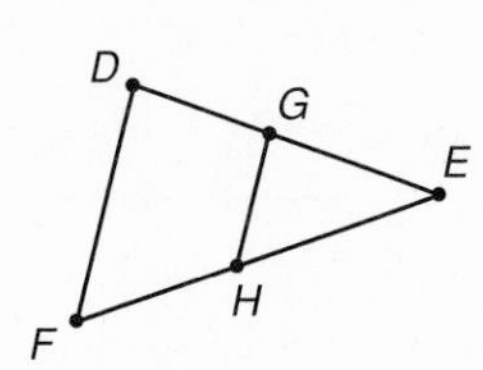

32. ______________________

33. Find the value of x. (Round your answer to one decimal place.)

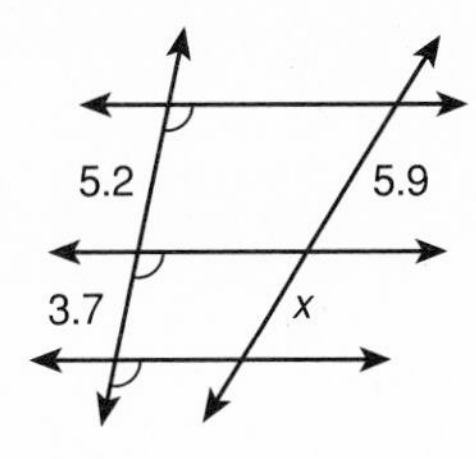

33. ______________________

34. Which two rectangles are similar?

a. 8 in. × 10 in. **b.** 5 in. × 7 in.

c. 2 ft × $2\frac{1}{2}$ ft **d.** $\frac{11}{12}$ ft × $\frac{7}{6}$ ft

34. ______________________

Name ______________________

Date ______________________

35. The scale factor of the dilation shown is 2.2. Find the length of x. (Round the result to one decimal place.)

35. ______________

In Problems 36–39, refer to the figure below.

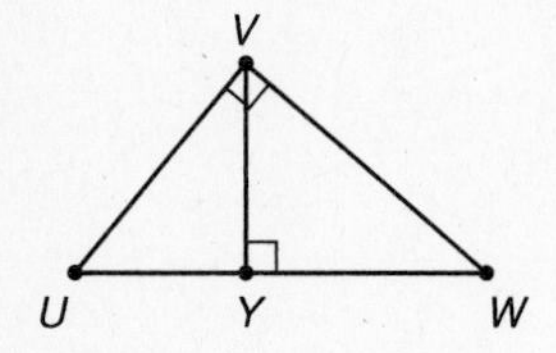

36. Name a triangle similar to $\triangle VYW$.

36. ______________

37. VY is the geometric mean of which segment lengths?

37. ______________

38. Complete the proportion.

$$\frac{UY}{UV} = \frac{UV}{\boxed{?}}$$

38. ______________

39. *Given:* $UY = 4$ and $YW = 5$.
Find the value of UV.

39. ______________

40. 8 is the geometric mean of 2 and what number?

40. ______________

41. If the sides of a triangle are 5, 7, and x, what possible values of x would make it a right triangle? Make a sketch of your solution(s).

41. ______________

Name ______________________

Date ______________________

42. The so-called "golden rectangle" has side lengths in the ratio of approximately 1 : 1.618. If the shorter side of the rectangle is 10 inches, find the diagonal length. (Round the result to one decimal place.)

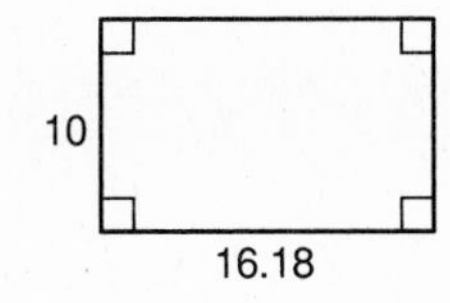

42. ______________________

43. In Problem 42, if a square is removed from the left side of the golden rectangle, the rectangle shown is formed. Find the ratio of length to width for the new rectangle. What does the result suggest?

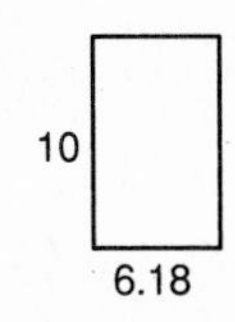

43.

44. The side lengths of three triangles are given. Classify each triangle as right, acute, or obtuse.

a. 2, 3, 4 **b.** 3, 4, 5 **c.** 4, 5, 6

44. a. ______________________

b. ______________________

c. ______________________

45. What is the side length of an equilateral triangle whose altitude has a length of $6\sqrt{3}$?

45. ______________________

46. Find the altitude to the base of an isosceles triangle whose base is 12 and whose congruent sides are 9.

46. ______________________

47. Are all rectangles with a diagonal of 10 centimeters similar? Explain.

47. ______________________

48. Write the trigonometric ratios.
a. $\cos Q$ **b.** $\tan P$ **c.** $\sin P$

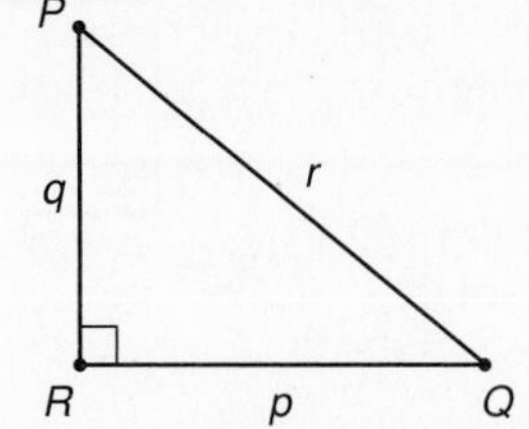

48. a. ______________________
b. ______________________
c. ______________________

49. Find $\cos N$.

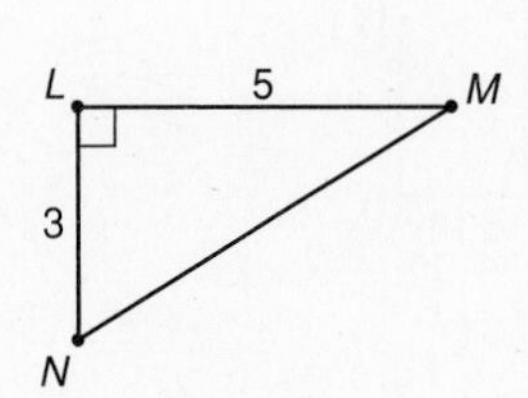

49. ______________________

50. Use a calculator to find the value of $\tan 53^\circ$ to four decimal places.

50. ______________________

51. Assume $\angle A$ is acute. If $\sin A = 0.240$, use your calculator to find the measure of $\angle A$ to two decimal places.

51. ______________________

52. Assume $\angle B$ is acute. Use your calculator to find $\cos B$ to three decimal places given that $\tan B = 0.352$.

52. ______________________

53. From an observation platform 180 feet above the water's edge, the line of sight to a sailboat makes an 83.5° angle with the vertical. Find x, the distance from the water's edge to the boat, to the nearest foot.

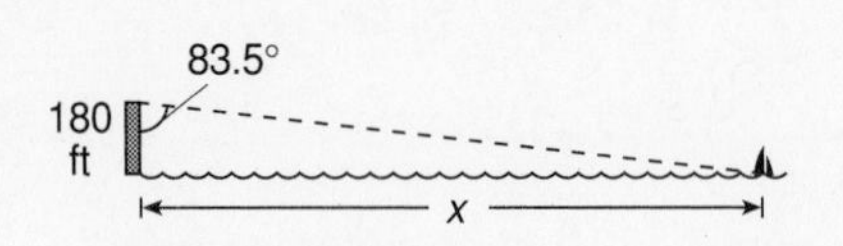

53. ______________________

10.2 Short Quiz

Name ______________________

Date ______________________

In Problems 1–6, use the figure below and the given information.

Given: P is the center of the circle.
$m\angle PAD = 90°$

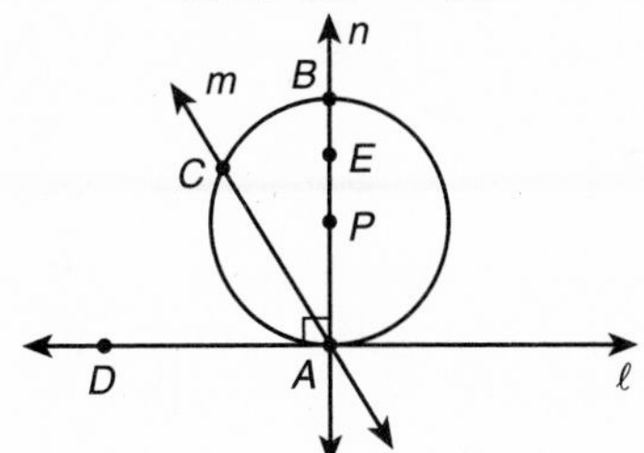

1. Name a chord that is not a diameter. **1.** ______________

2. Name an interior point other than the center of the circle. **2.** ______________

3. Name a radius of the circle. **3.** ______________

4. Name an exterior point of the circle. **4.** ______________

5. Name a tangent line to the circle. **5.** ______________

6. Name a secant to the circle that does not contain a diameter. **6.** ______________

In Problems 7–10, use the figure below.

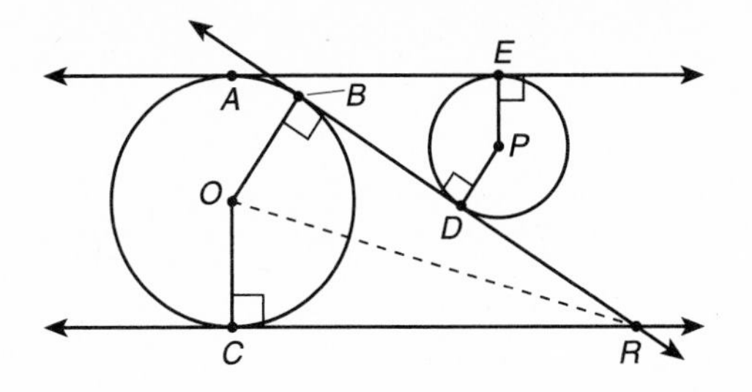

7. Name a common internal tangent to $\odot O$ and $\odot P$. **7.** ______________

8. Name a segment congruent to $\overline{BR}$. **8.** ______________

9. Name a common external tangent to $\odot O$ and $\odot P$. **9.** ______________

10. Name an angle congruent to $\angle BRO$. **10.** ______________

Mid-Chapter 10 Test Form A

(Use after Lesson 10.3)

Name ______________________

Date ______________________

1. The diameter of a circle is 12.74 inches. Find the radius.

1. ______________

2. *Given:* $\overline{AC}$ is a diameter of $\odot O$;
$AC = 8$;
$\overleftrightarrow{BA}$ is tangent to $\odot O$ at A;
$OB = 10$.
Find AB.

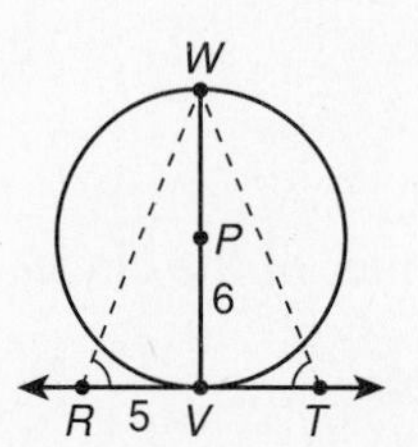

2. ______________

3. *Given:* $\odot P$ with $PV = 6$;
$\angle WRV \cong \angle WTV$;
$RV = 5$.
Find WT.

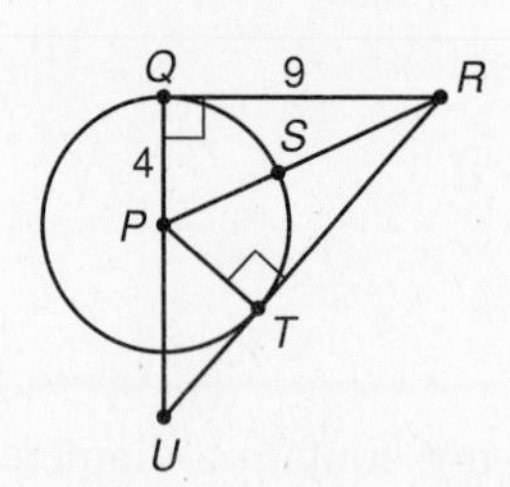

3. ______________

4. *Given:* $\odot P$ with $PQ = 4$, $QR = 9$
Find SR.

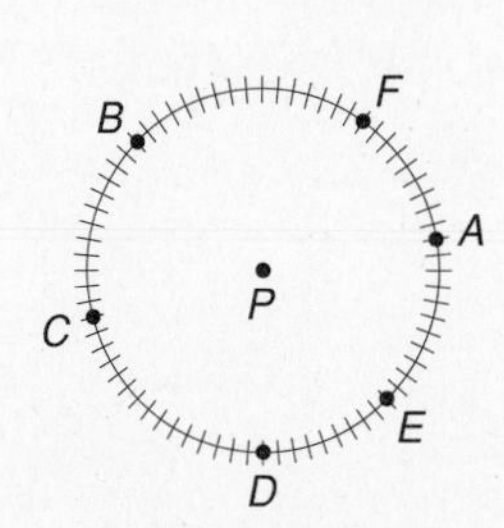

4. ______________

5. *Given:* $\odot P$
Using a protractor, find the measures of the following.
a. $\angle BPA$ **b.** $\angle DPA$ **c.** $\widehat{CB}$

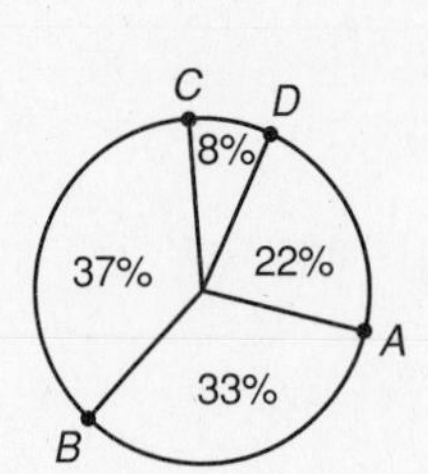

5. a. ______________
b. ______________
c. ______________

6. Use the circle graph to determine the arc measures (to the nearest degree) of $\widehat{AB}$, $\widehat{BC}$, $\widehat{CD}$, and $\widehat{DA}$.

6. ______________

Mid-Chapter 10 Test Form B

(Use after Lesson 10.3)

Name ____________________

Date ____________________

1. The diameter of a circle is 18.78 inches. Find the radius.

1. 

2. *Given:* $\overline{RV}$ is a diameter of $\odot P$;
$PV = 7$;
$\overleftrightarrow{RT}$ is tangent to $\odot P$ at R;
$TP = 11$.
Find RT.

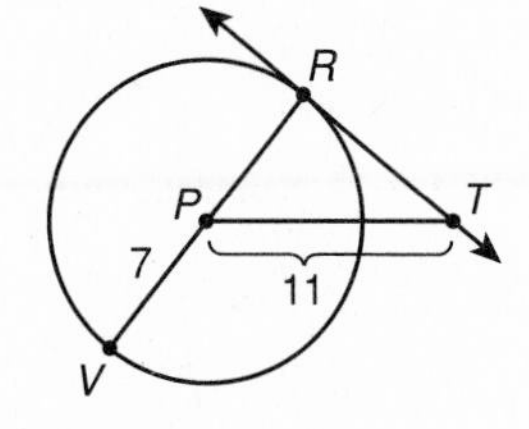

2. ____________________

3. *Given:* $\overline{PA}$ is tangent to $\odot Q$ at A;
$\overline{QA} = 3$; $\overline{QB} = 1$.
Find AP.

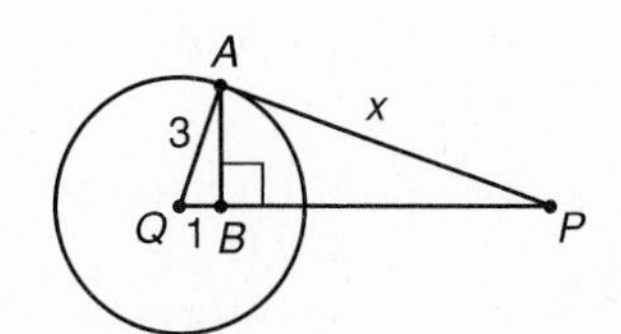

3. ____________________

4. *Given:* $\odot P$ with $PK = 5$,
$GH = 15$
Find LH.

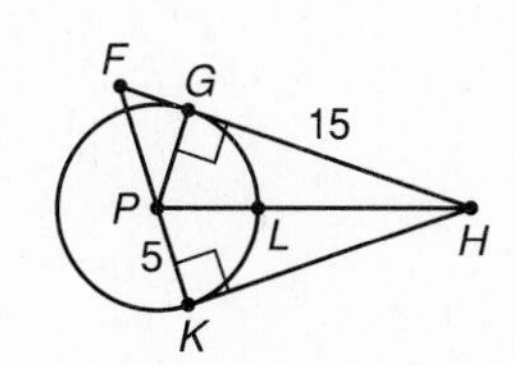

4. ____________________

5. *Given*: $\odot P$
Using a protractor, find the measures of the following.
a. $\overparen{CBA}$ **b.** $\overparen{FA}$ **c.** $\overparen{CDE}$

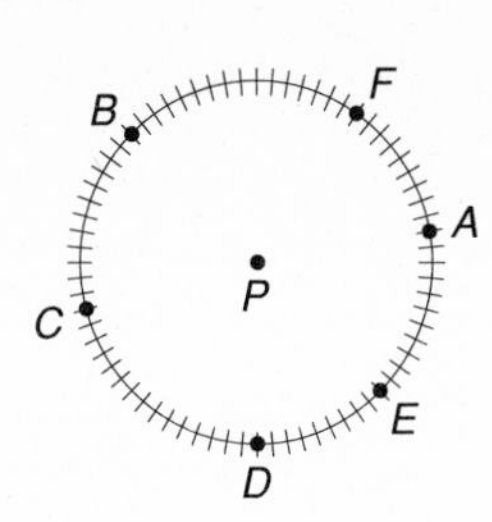

5. a. 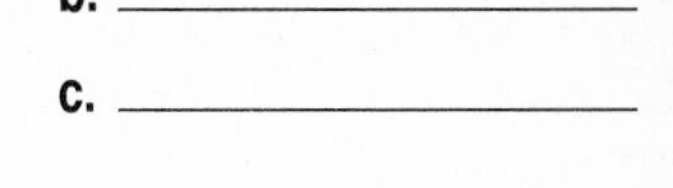
b. ____________________
c. ____________________

6. Use the circle graph to determine the arc measures (to the nearest degree) of $\overparen{AB}$, $\overparen{BC}$, $\overparen{CD}$, $\overparen{DA}$.

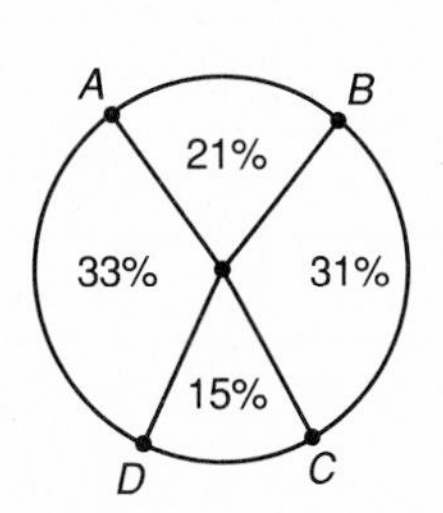

6. ____________________

10.4 Short Quiz

Name ______________________

Date ______________________

1. *Given:* $\odot P$ as shown, with
   $m\overset{\frown}{BC} = 46°$ and
   $m\overset{\frown}{AD} = 45°$.
   Find $m\overset{\frown}{AC}$.

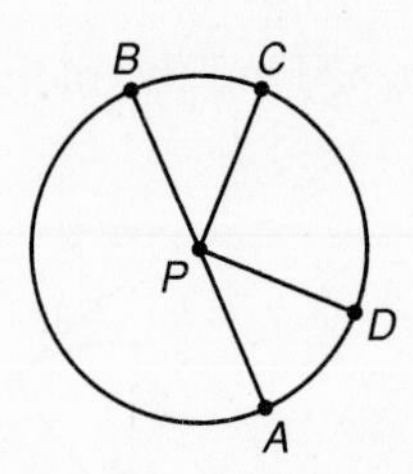

1. ______________________

2. In a recent survey, 900 students were asked to name their majors (Business, Science, Liberal Arts, and other). The results are shown in the circle graph where
   $m\overset{\frown}{AB} = 75.6°$; $m\overset{\frown}{BC} = 115.2°$;
   $m\overset{\frown}{CD} = 32.4°$; $m\overset{\frown}{DA} = 136.8°$.
   Find the number of students in each category.

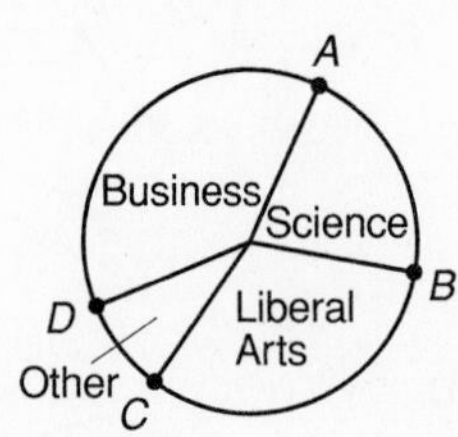

2. ______________________

In Problems 3–5, use the figure at the right.

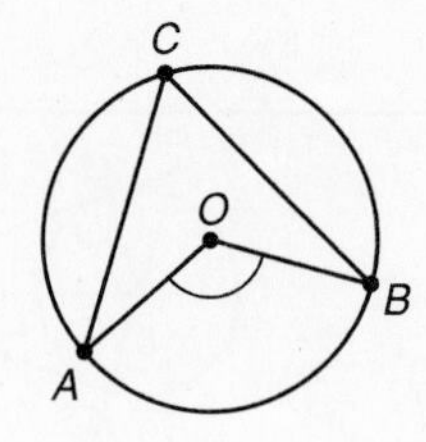

3. Name a central angle.

3. ______________________

4. Name a minor arc.

4. ______________________

5. Name a major arc.

5. ______________________

6. Which two segments (not radii) of $\odot P$ are congruent?

T
R
S
L
P
Q

6. ______________________

7. *Given:* $\odot Q$ with $RU = 7$.
   Find VW.

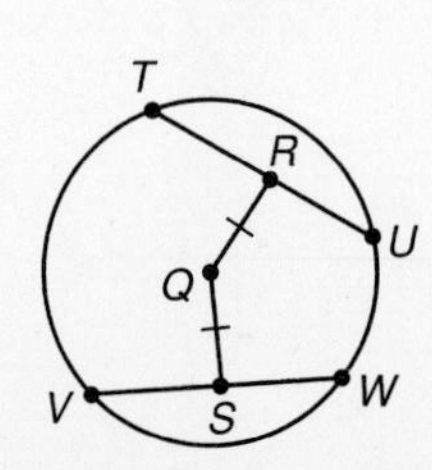

7. ______________________

10.6 Short Quiz

Name ______________________

Date ______________________

1. Find $m\angle ACB$ in $\odot Q$ as shown.

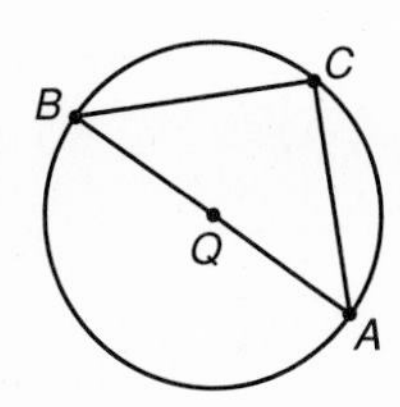

1. ______________________

2. Find $m\angle ACD$.

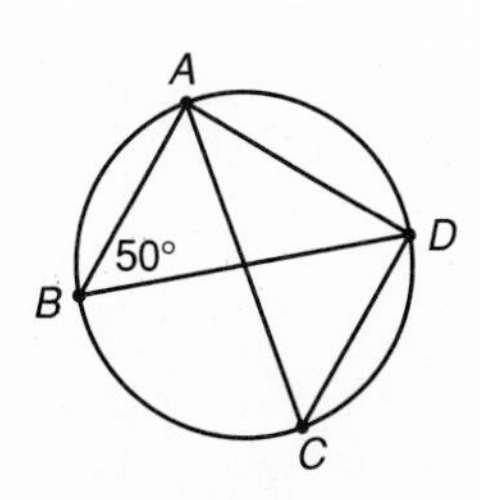

2. ______________________

3. *Given:* $\odot Q$ with $m\widehat{AB} = 48°$
Find $m\angle DCA$.

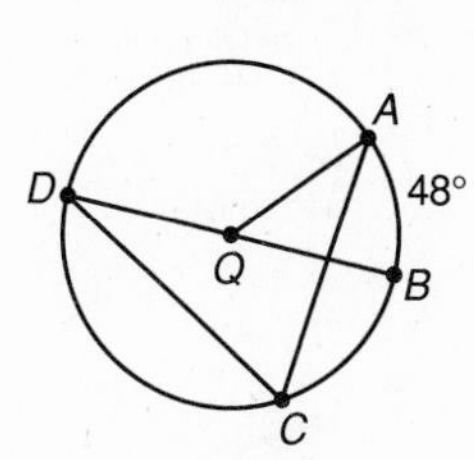

3. ______________________

4. Find $m\angle 1 + m\angle 2$. Explain.

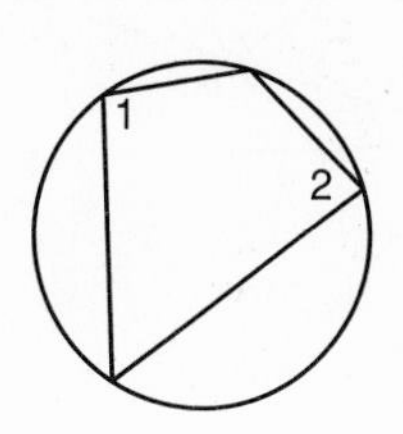

4.

5. *Given:* $m\widehat{AB} = 130°$;
ℓ is tangent to $\odot P$ at A.
Find $m\angle 1$.

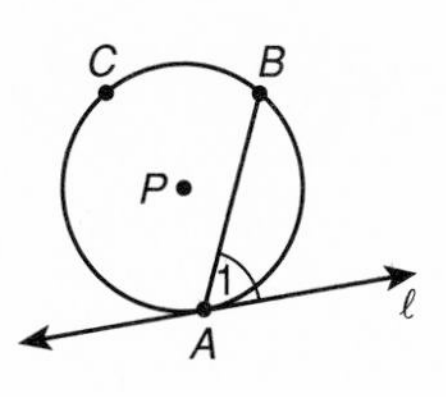

5. ______________________

6. *Given:* $m\widehat{AD} = 132°$;
$m\widehat{BC} = 44°$
Find $m\angle x$.

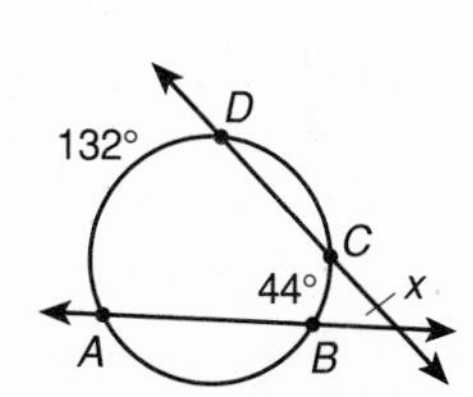

6. ______________________

Chapter 10 Test **Form A**

(Page 1 of 3 pages)

Name ____________________

Date ____________________

1. Define a chord of a circle.

1. ____________________

2. Draw a common external tangent to $\odot P$ and $\odot Q$.

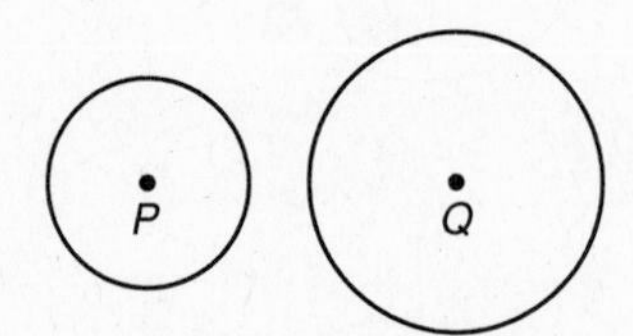

2. *Use figure at left.*

3. Define a secant line of a circle and illustrate the definition on the circle at the right.

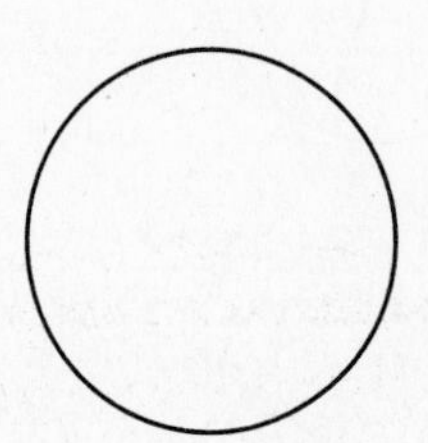

3. ____________________ *Use figure at left.*

4. *Given:* $\overleftrightarrow{OA}$ is tangent to $\odot Q$ at A. List any right angles in the figure. Explain.

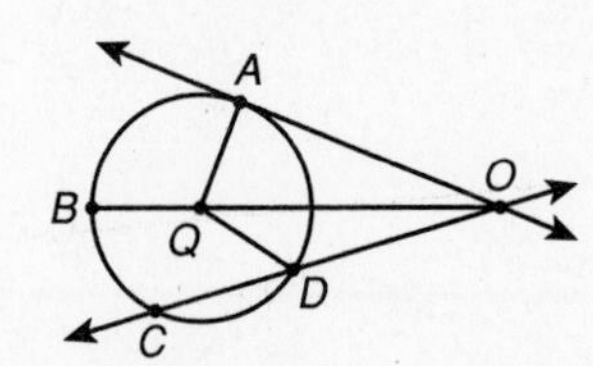

4. ____________________

5. *Given:* $\overline{ST}$ is tangent to $\odot R$ at S. Find RT.

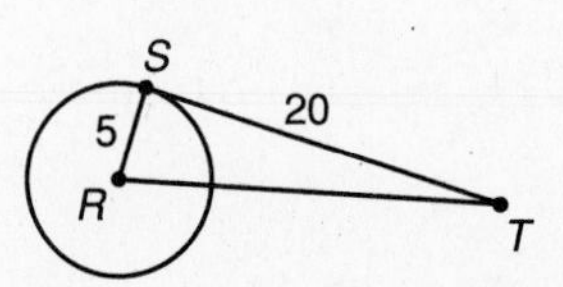

5. ____________________

6. A circle is inscribed in a polygon. What can you say about the sides of the polygon with respect to the circle?

6. ____________________

7. *Given*: P is the center of the circle as shown.

Find $m\widehat{DBC}$.

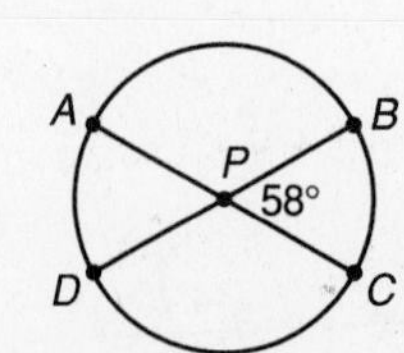

7. ____________________

Name ______________________

Date ______________________

8. On a given day, a local blood bank had a stock of blood types (O, A, B, and AB) as indicated by the circle graph. If they had 1000 units in all, how many units of each type did they have?

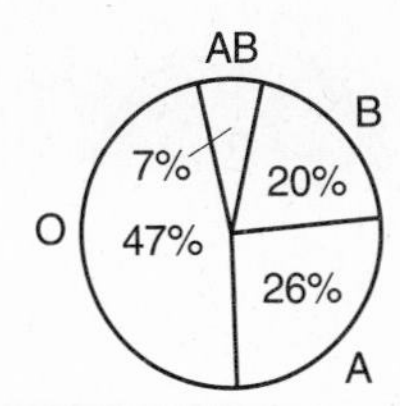

8. ______________________

9. Find $m\overset{\frown}{QW}$ in $\odot P$, as shown.

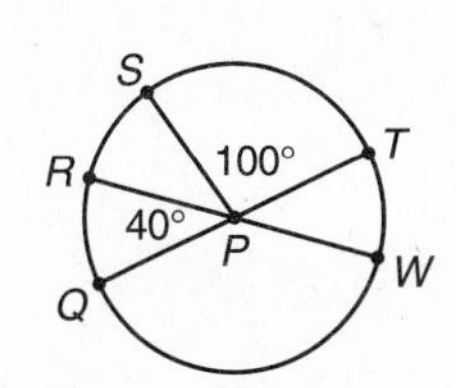

9. ______________________

10. *Given:* $\odot O$ with diameters $\overline{AD}$ and $\overline{BC}$.
Decide whether or not $\overline{AB} \cong \overline{CD}$.
Explain your reasoning.

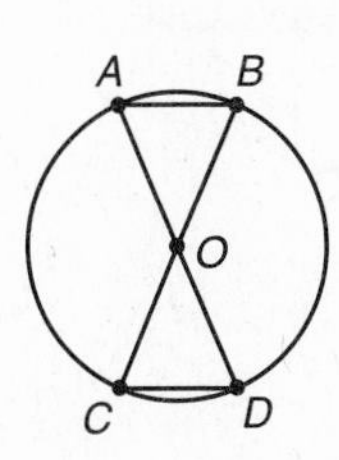

10.

11. Find RS in $\odot C$, as shown.
Explain your reasoning.

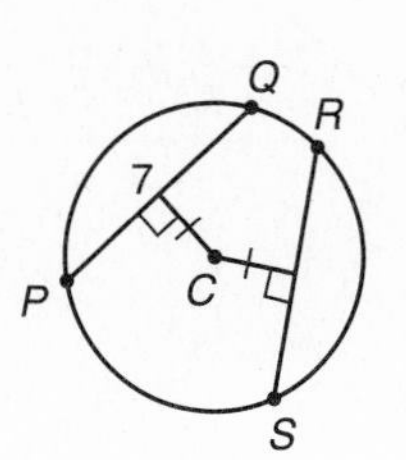

11.

12. *Given:* $\odot P$ and $\overline{PT} \perp$ to chord $\overline{RS}$ at T.
Decide whether or not $RT = TS$.
Explain your reasoning.

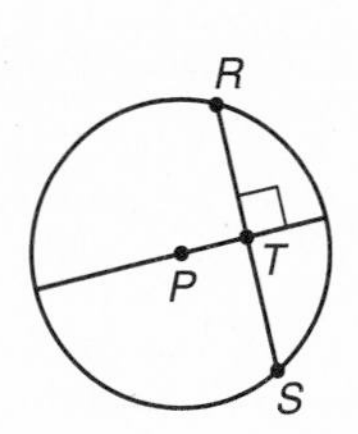

12.

13. *Given:* $\odot G$ with intersecting chords $\overline{AB}$ and $\overline{CD}$ that meet at P.
Can triangles $\triangle ADP$ and $\triangle CBP$ be congruent? Can they be similar? Or are they neither congruent nor similar? Explain.

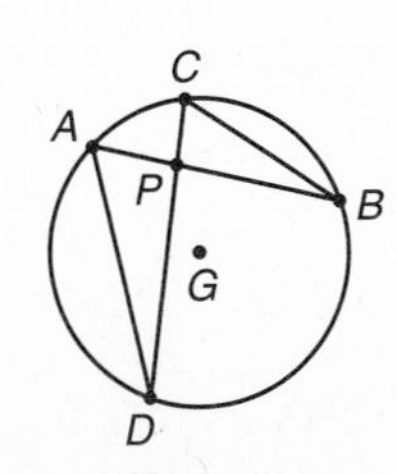

13.

Name ____________________

Date ____________________

14. Find $m\overset{\frown}{BC}$ and $m\angle BDC$.

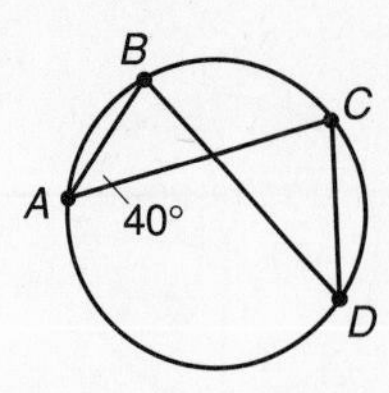

14. ____________________

15. What must be the measures of $\angle C$ and $\angle D$ so that a circle may be circumscribed about $ABCD$?

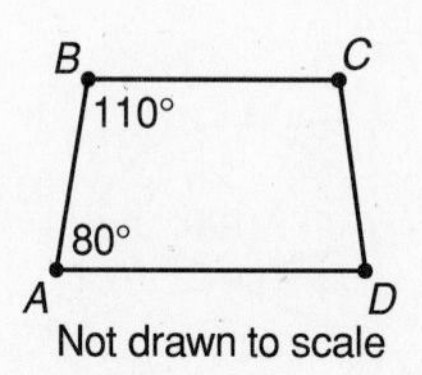

15. ____________________

16. Find the measure of $\angle 1$.

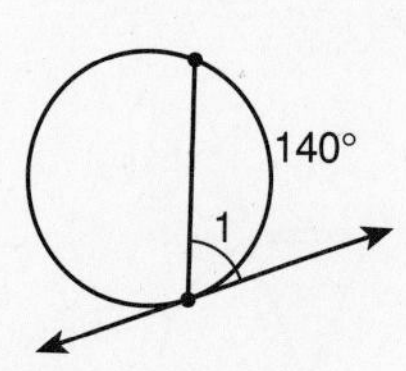

16. ____________________

17. Find the measure of $\angle 1$.

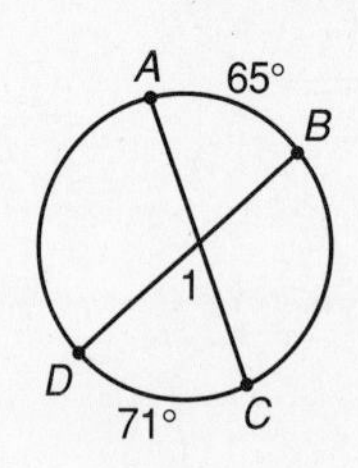

17. ____________________

18. Find the measure of $\angle 1$.

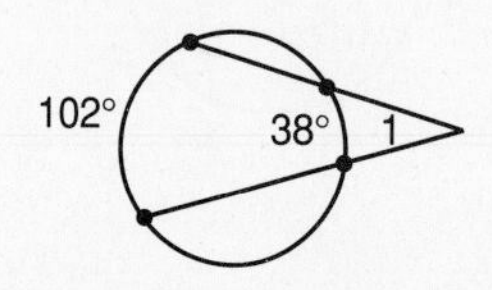

18. ____________________

19. Write the standard equation of the circle whose center is $(-3, 5)$ and whose radius is $\frac{5}{2}$.

19. ____________________

20. Sketch the graph of the equation $(x - 2)^2 + (y + 1)^2 = 13$. Label the coordinates of the center and the y-intercepts.

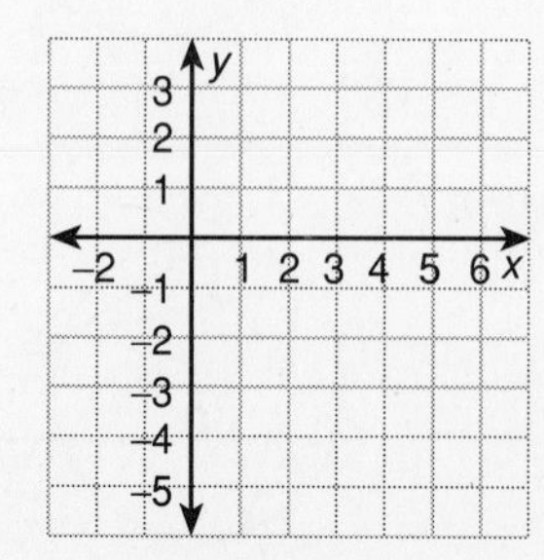

20. *Use grid at left.*

Name ____________________

Date ____________________

1. A circle is the set of all points in a plane that [?]. 1. ______
   a. have a center b. lie within a given radius
   c. are equidistant from a given point d. have a diameter

2. A segment whose endpoints are on a circle is a [?]. 2. ______
   a. tangent b. chord c. radius d. secant

3. The center of a circle lies on [?]. 3. ______
   a. the circle b. every chord
   c. a tangent line d. every diameter

4. If a circle has a diameter of 12, then it has [?]. 4. ______
   a. a radius of 24 b. a radius of 4
   c. a diameter of 6 d. a radius of 6

5. A line which intersects a circle at exactly one point is called [?]. 5. ______
   a. a tangent of the circle b. a secant of the circle
   c. the point of tangency d. a chord

6. The segment that joins the centers of two circles [?]. 6. ______
   a. is a chord of both circles
   b. is intersected by a common internal tangent, if there is one
   c. is intersected by a common external tangent
   d. is the diameter of both circles

7. Two circles are concentric if [?]. 7. ______
   a. they have exactly one point of intersection
   b. they have congruent radii
   c. they have no points of intersection
   d. they have the same center

Name ______________________

Date ______________________

In Problems 8 and 9, use the figure below and the given information.

Given: $\overline{SR}$ is tangent to $\odot Q$ at R.

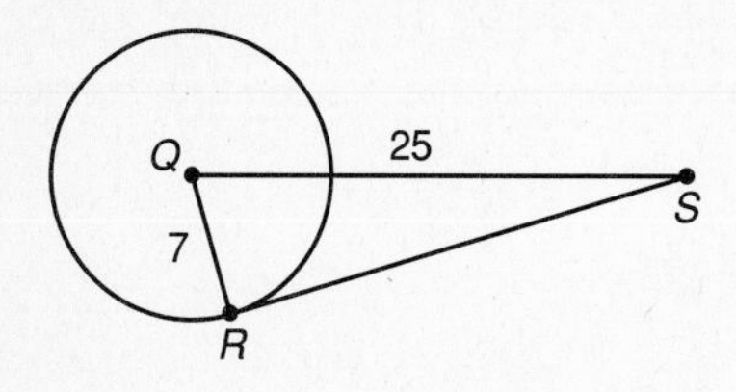

8. Choose the true statement. **8.** ________

a. $m\angle RQS = 90°$ **b.** $\overline{SQ} \cong \overline{SR}$

c. $m\angle SRQ = 90°$ **d.** $m\angle RQS + m\angle QSR = 180°$

9. $SR = \boxed{?}$ **9.** ________

a. 24 **b.** $\sqrt{764}$ **c.** $\sqrt{674}$ **d.** $\sqrt{567}$

10. $\overline{RP}$ is tangent to $\odot G$ at P. **10.** ________
$\overline{RQ}$ is tangent to $\odot G$ at Q.
Choose the statement that is *not* true.

a. $\overline{PR} \cong \overline{QR}$

b. $\angle PGR \cong \angle QGR$

c. $\angle GRP \cong \angle GRQ$

d. $\angle GPR$ is obtuse.

11. A circle is *inscribed* in a polygon if $\boxed{?}$. **11.** ________

a. every point of the circle is inside the polygon

b. each side of the polygon is tangent to the circle

c. each vertex of the polygon is on the circle

d. each side of the polygon is a chord of the circle

12. A minor arc of $\odot P$ is $\boxed{?}$. **12.** ________

a. $\overparen{EHF}$ **b.** $\overparen{FEH}$

c. $\overparen{FGH}$ **d.** $\overparen{EFG}$

13. The measure of $\overparen{ADB}$ is $\boxed{?}$. **13.** ________

a. 223° **b.** 137°

c. 131° **d.** 229°

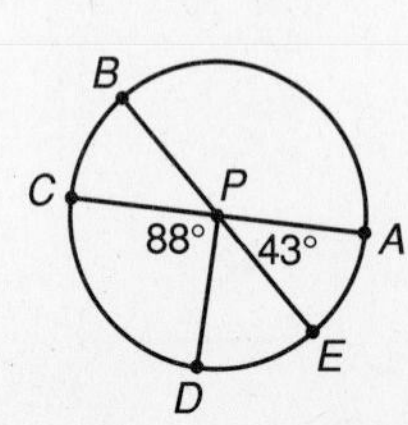

Chapter 10 Test Form B

Name ______________________

Date ______________________

14. Refer to the figure and choose the true statement.

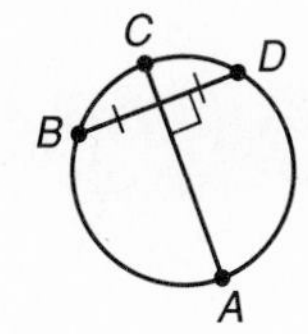

a. $\widehat{BAD}$ is a minor arc.

b. $\overline{AC}$ is a diameter.

c. $\widehat{BCD}$ is a major arc.

d. $\widehat{CD} \cong \widehat{BA}$.

14. ________

15. *Given:* P is the center of both circles.
$RW = 5$;
$\overline{UT}$ is tangent to the smaller circle at V.
$\overline{RS}$ is tangent to the smaller circle at W.

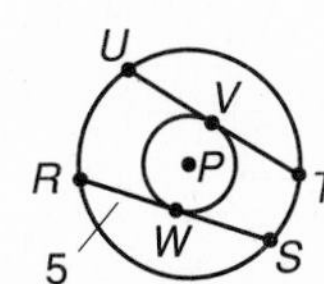
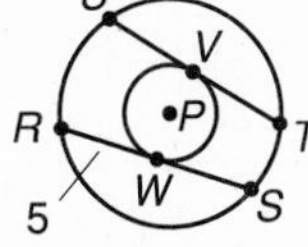

Then, $UT = \boxed{?}$.

a. $5\sqrt{2}$ b. $5\sqrt{5}$ c. 10 d. 5

15. ________

16. *Given:* $\odot Q$ and $m\angle ABC = 62°$.

Then, $m\widehat{AC} = \boxed{?}$.

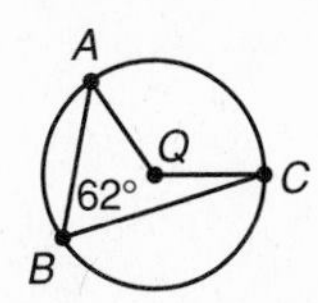

a. 124° b. 236° c. 62° d. 248°

16. ________

17. A circle can be circumscribed about which quadrilateral?

a. 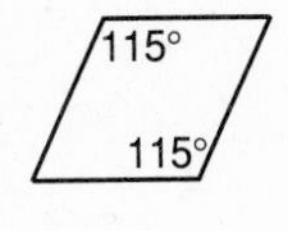

b. 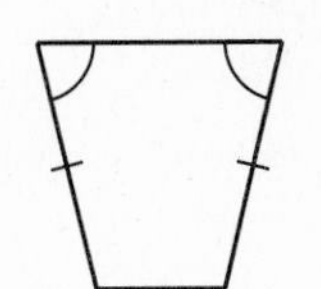

c. 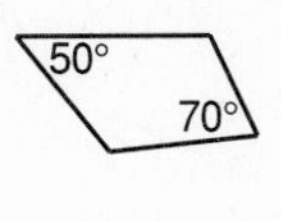

d. 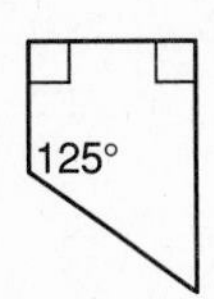

17. ________

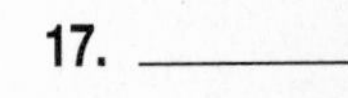

18. *Given:* $m\widehat{SQ} = 106°$, $m\widehat{PR} = 120°$

Then, $m\angle x = \boxed{?}$.

a. 226° b. 67° c. 134° d. 113°

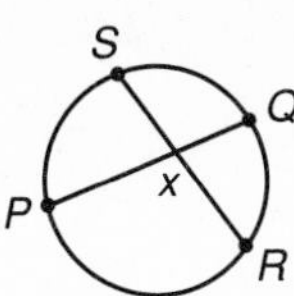

Not drawn to scale

18. ________

19. *Given:* $m\widehat{AB} = 82°$, $m\widehat{CD} = 30°$

Then, $m\angle DOC = \boxed{?}$.

a. 112° b. 56° c. 26° d. 52°

Not drawn to scale

19. ________

20. A standard equation of a circle whose center is (−4, 3) and whose radius is 7 is $\boxed{?}$.

a. $(x - 4)^2 + (y + 3)^2 = 49$ b. $(x + 4) + (y - 3) = 7$

c. $(x - 4)^2 + (y + 3)^2 = 7$ d. $(x + 4)^2 + (y - 3)^2 = 49$

20. ________

Chapter 10 Test Form C

Name ____________

Date ____________

1. Define a *diameter* of a circle. **1.**

2. Draw a common internal tangent to $\odot R$ and $\odot S$.

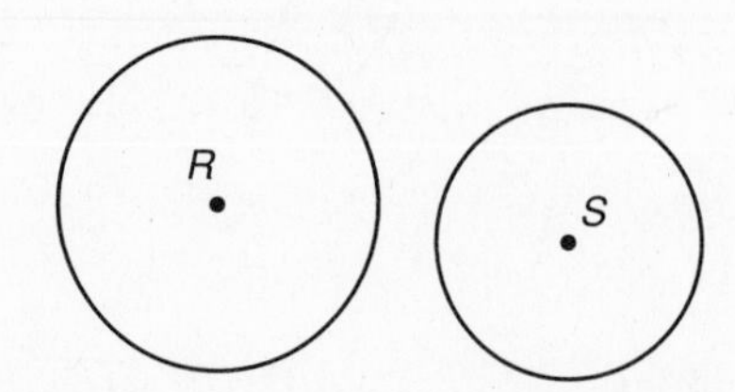

2. *Use figure at left.*

3. Define a tangent line to a circle. Draw a sketch to illustrate the definition. **3.**

4. *Given*: $\overleftrightarrow{OA}$ and $\overleftrightarrow{OC}$ are tangent to $\odot Q$ at A and C, respectively.
List any right angles in the given figure.

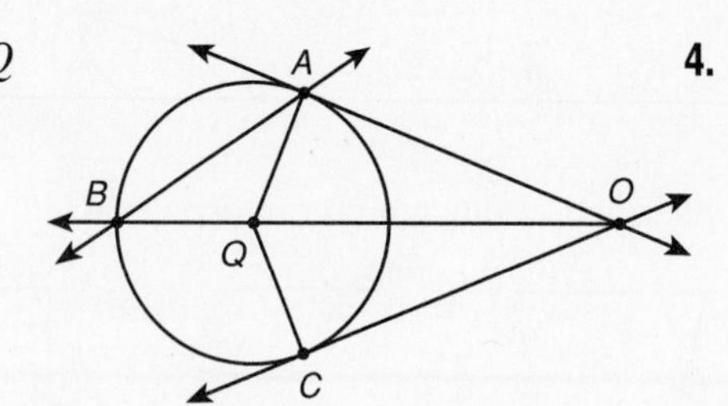

4. ____________

5. *Given*: $RP = 22$, $RA = 6$
$\overline{PQ}$ is tangent to $\odot R$ at Q.
Find PQ.

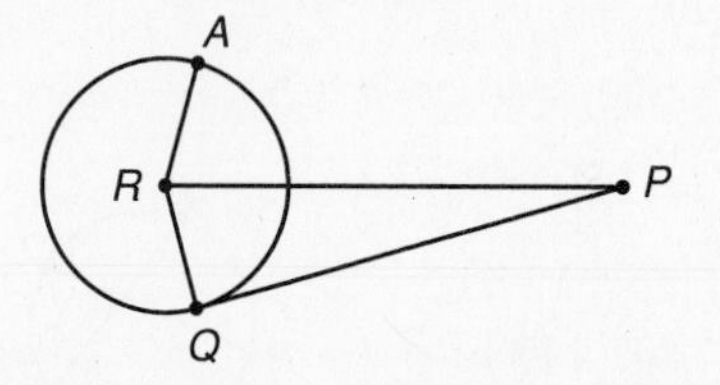

5. ____________

6. Define what is meant by a circle circumscribed about a polygon. **6.**

7. The points $P(-1, 2)$, $Q(3, 2)$, and $R(3, -4)$ lie on a circle. Write the equation of the circle.

7. ____________

Name ______________________

Date ______________________

8. On a given day, a local blood bank had a stock of blood types (O, A, B, and AB) as indicated by the circle graph. Determine the arc measures for each category to the nearest degree.

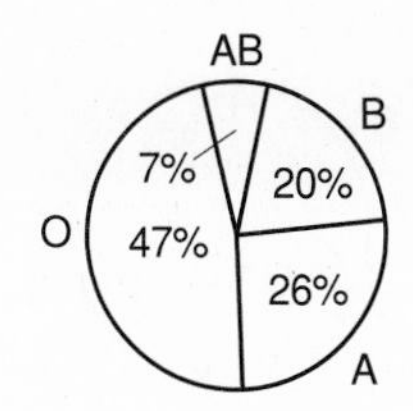

8.

9. Find $m\widehat{CEB}$ and $m\widehat{EA}$ in $\odot O$.

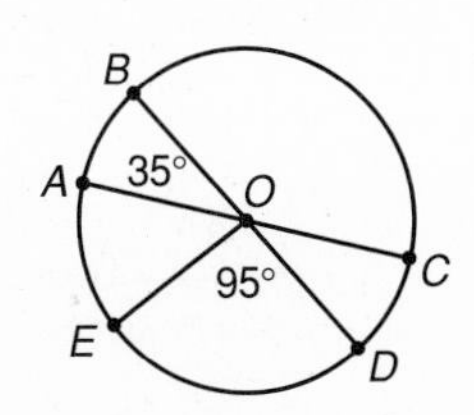

9. ______________________

10. *Given*: $\odot Q$ and diameters $\overline{PV}$ and $\overline{RT}$. Decide whether or not $\overline{PT} \cong \overline{RV}$. Explain your reasoning.

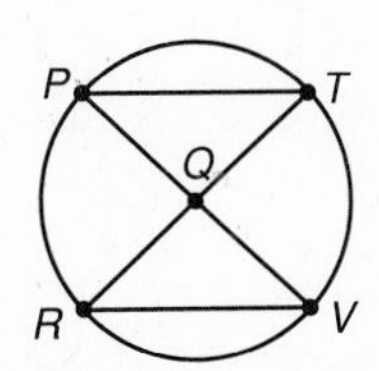

10.

11. *Given*: Figure as shown. Explain why $ABCD$ is a parallelogram.

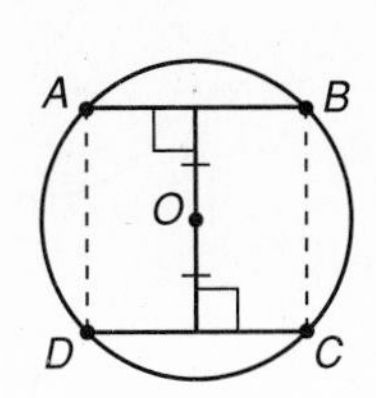

11.

12. *Given*: $\overline{PR}$ and $\overline{QS}$ are intersecting chords and $\overline{PQ} \cong \overline{RS}$.

Prove: $\overline{PR} \cong \overline{QS}$

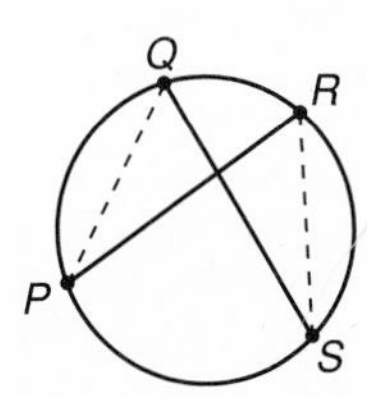

12. ______________________

Name ______________________

Date ______________________

13. Find $m\widehat{ABD}$ and $m\angle ACD$.

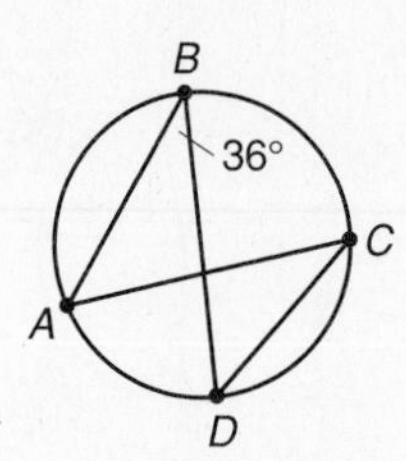

13. ______________

14. What must be the measures of $\angle B$ and $\angle C$ so that a circle can be circumscribed about $ABCD$.

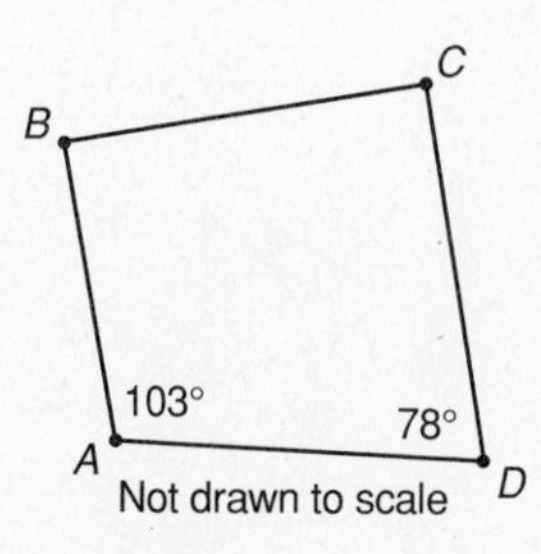

14. ______________

15. Find the measure of $\angle 1$.

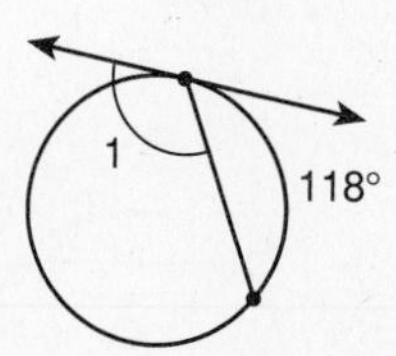

15. ______________

16. Write an equation that can be used to solve for x. Then solve the equation for x.

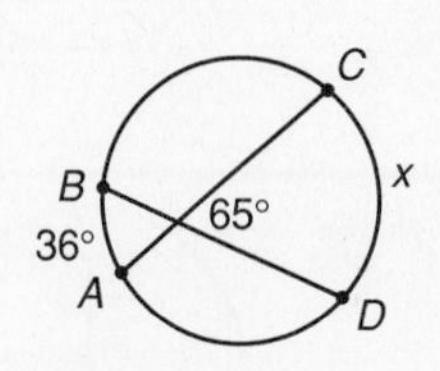

16.

17. Find the measure of $\angle 1$.

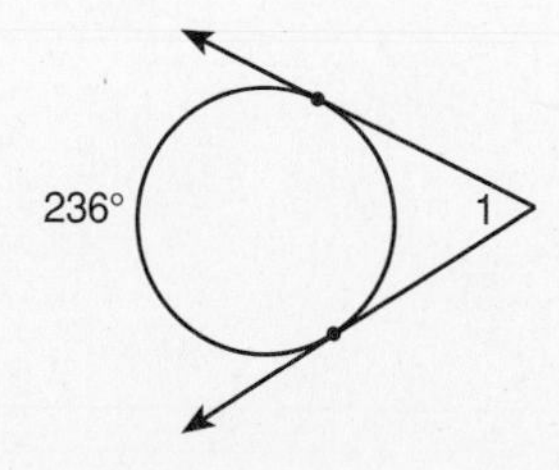

17. ______________

18. Sketch the graph of $(x-4)^2 + (y+3)^2 = 25$. Label the coordinates of the center and the intercepts.

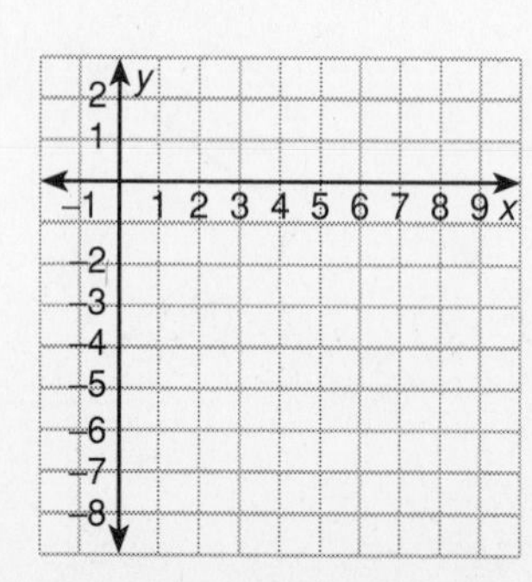

18.

11.2 Short Quiz

Name ______________________

Date ______________________

1. Find the perimeter of this triangle.

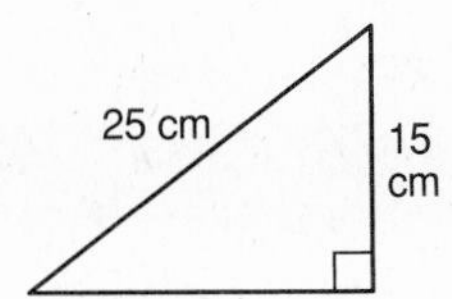

1. ______________________

2. Find the area of the triangle in Problem 1.

2. ______________________

3. Find the perimeter and area of the shaded region.

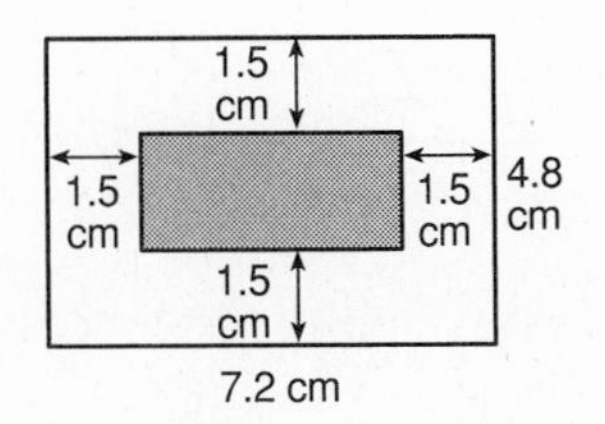

3. ______________________

4. Find the area of the parallelogram.

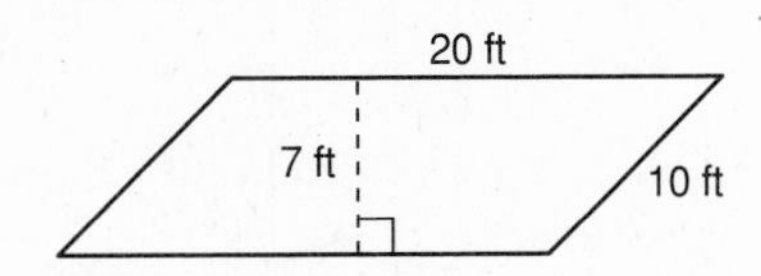

4. ______________________

5. Find the area of the shaded region.

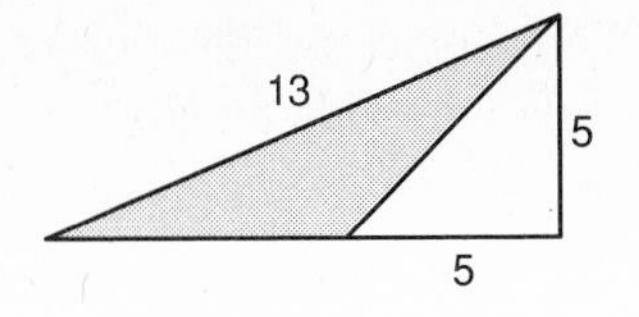

5. ______________________

6. Find the area of the parallelogram.

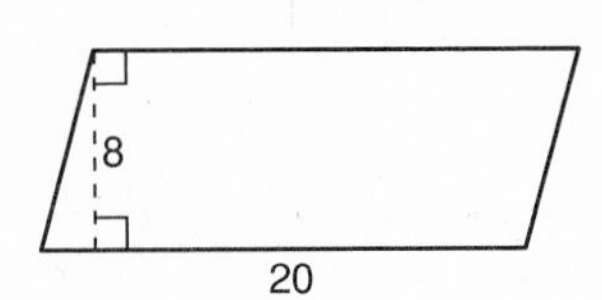

6. ______________________

Mid-Chapter 11 Test Form A

(Use after Lesson 11.3)

Name ____________________

Date ____________________

1. Find the perimeter and area of the polygon.

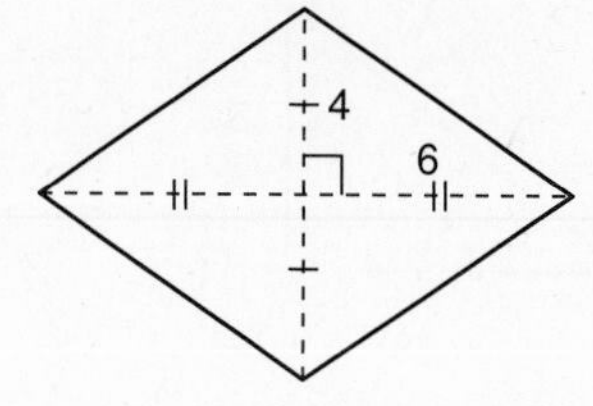

1. ____________________

2. Find the total perimeter (both inner and outer) and the area of the shaded region of the picture frame.

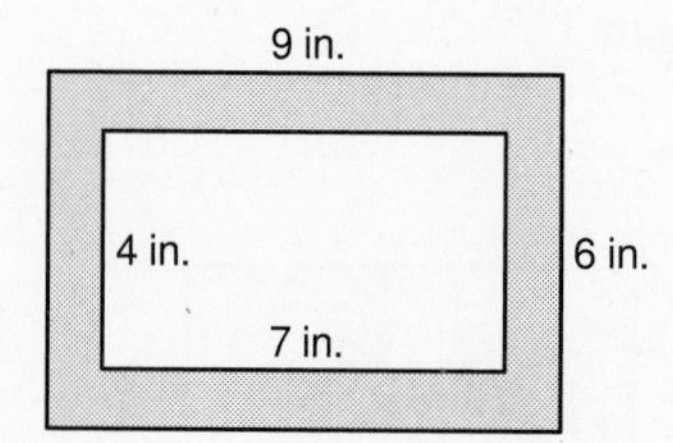

2. ____________________

3. Find the area of the $\square ABCD$.

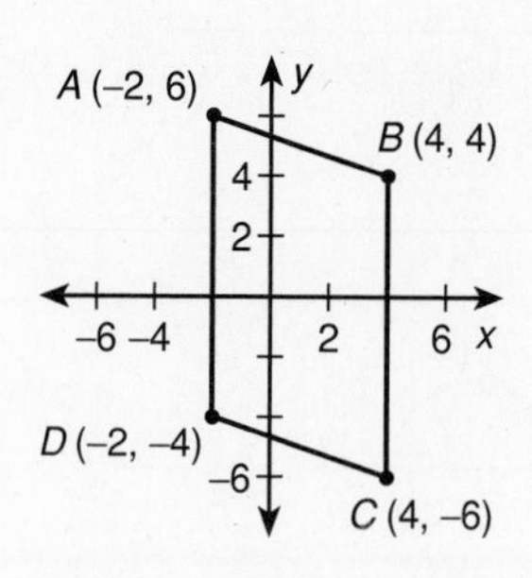

3. ____________________

4. The area of a triangle is given by $A = \sqrt{s(s-a)(s-b)(s-c)}$, where a, b, and c are side lengths of the triangle and s is one-half of its perimeter. Find the area of a triangle with side lengths of 7, 9, and 12.

4. ____________________

5. Find the area of this trapezoid.

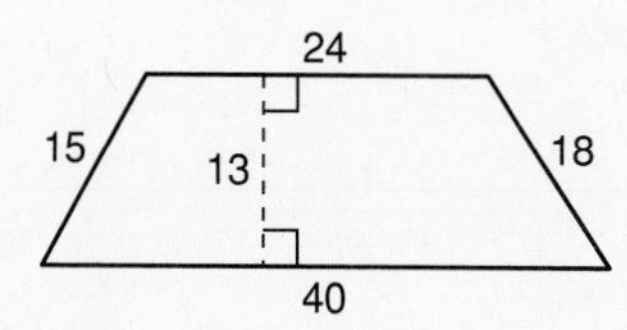

5. ____________________

6. The area of this quadrilateral is 460 square units. Find the value of x.

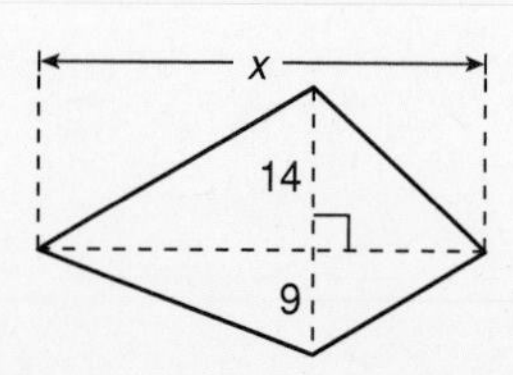

6. ____________________

Mid-Chapter 11 Test Form B

(Use after Lesson 11.3)

Name ______________________

Date ______________________

1. Find the perimeter and area of the polygon.

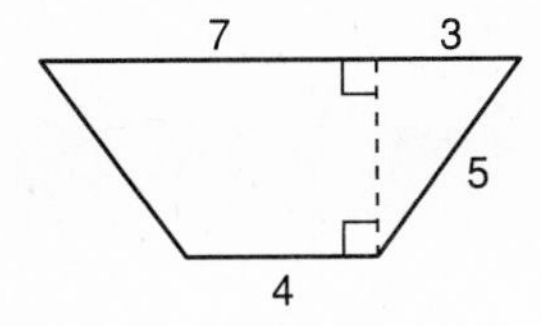

1. ______________________

2. Find the total perimeter (inner and outer) and the area of the shaded region of this picture frame.

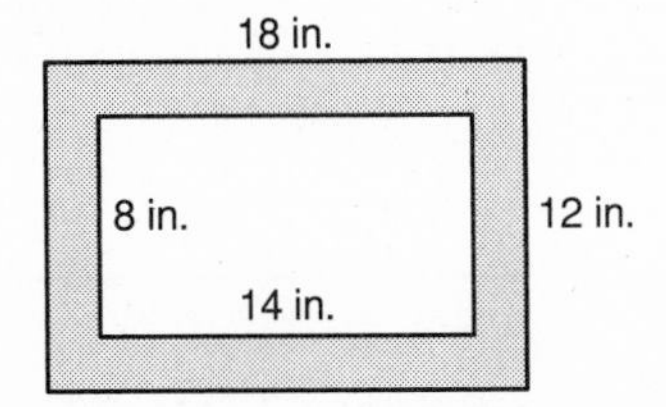

2. ______________________

3. Find the area of $\triangle ABC$.

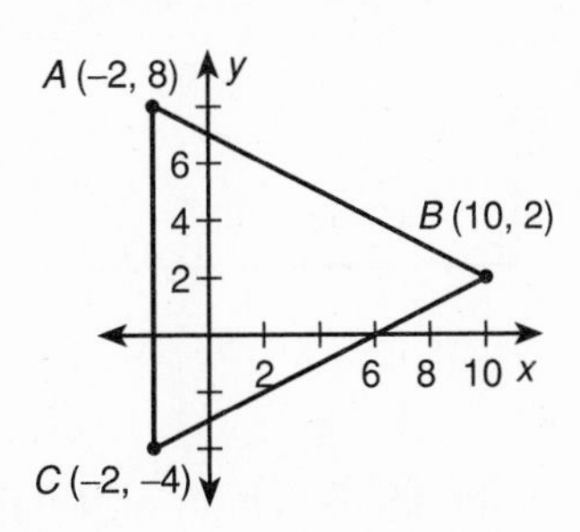

3. ______________________

4. The area of a triangle is given by $A = \sqrt{s(s-a)(s-b)(s-c)}$, where a, b, and c are side lengths of the triangle and s is one-half of its perimeter. Find the area of a triangle with side lengths of 6, 9, and 11.

4. ______________________

5. Find the area of this trapezoid.

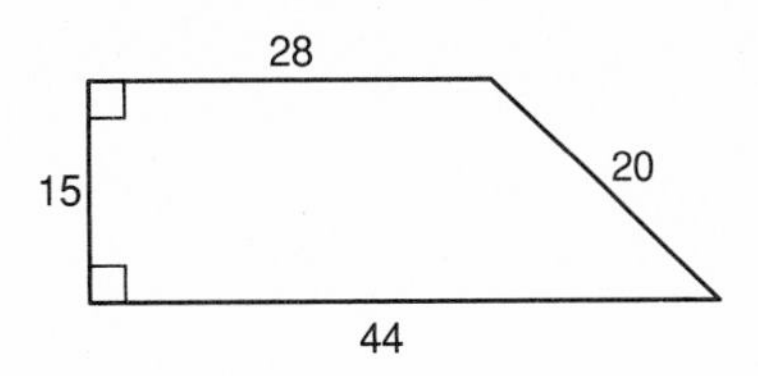

5. ______________________

6. The area of this quadrilateral is 60 square units. Find the value of x.

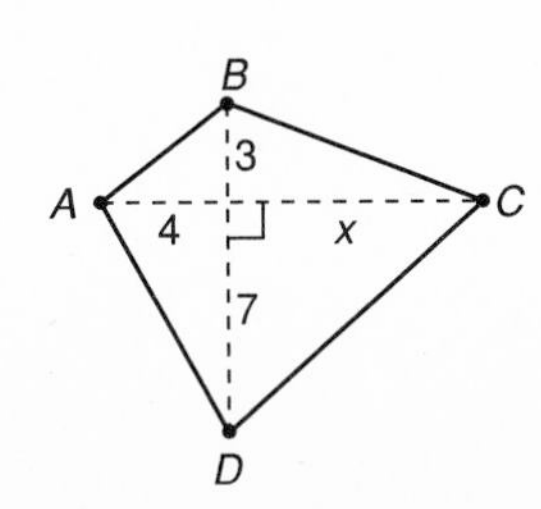

6. ______________________

11.4 Short Quiz

Name ______________________

Date ______________________

1. Find the area of this quadrilateral.

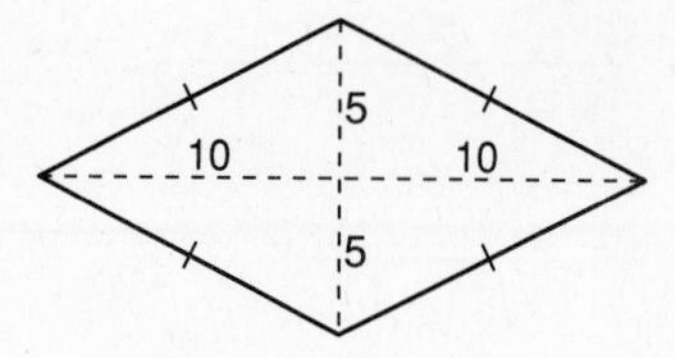

1. ______________

2. The quadrilateral has an area of 378 square units. Find the value of x.

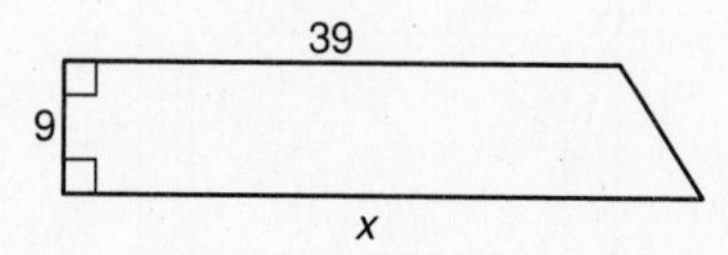

2. ______________

3. Find the area of the rectangle.

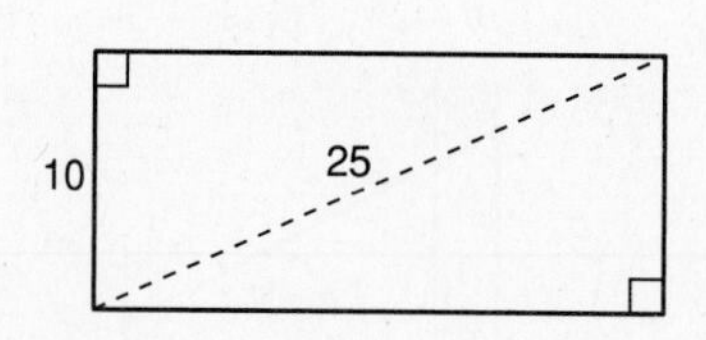

3. ______________

4. Find the area of an equilateral triangle with a side length of 2 units.

4. ______________

5. A regular hexagon has a side length of 2 units. Find its area.

5. ______________

6. Find the apothem of a regular octagon of side length 2 units.

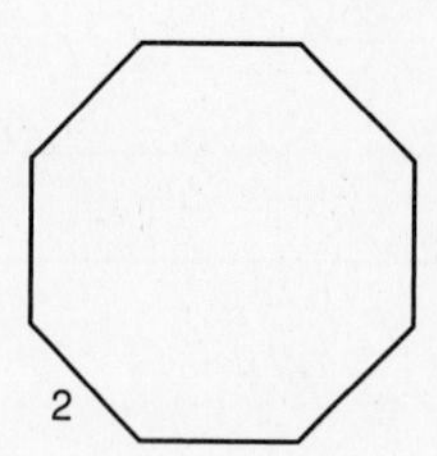

6. ______________

11.6 Short Quiz

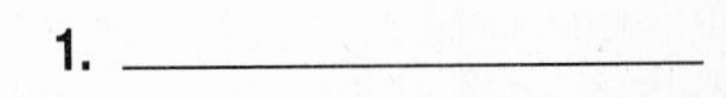

Name ______________________

Date ______________________

1. A circle has a circumference of 28 inches. Find the value of its diameter to two decimal places.

1. ______________

2. Find the value of the arc length, x, to two decimal places.

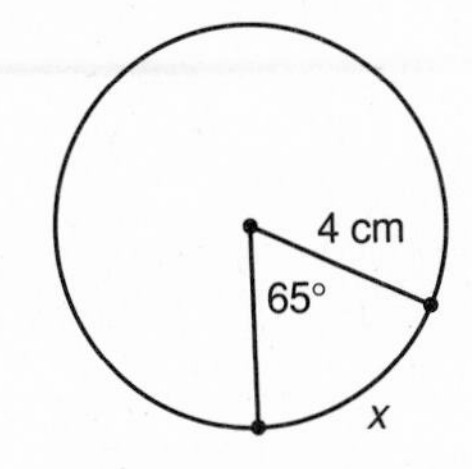

2. ______________

3. An automobile has tires with 22-inch diameters. If the wheels revolve 20 times, how far does the car move? (Round the answer to the nearest foot.)

3. ______________

4. Find the area of a circle whose diameter is 5.75 inches. (Round the answer to two decimal places.)

4. ______________

5. Find the area of the shaded sector to two decimal places.

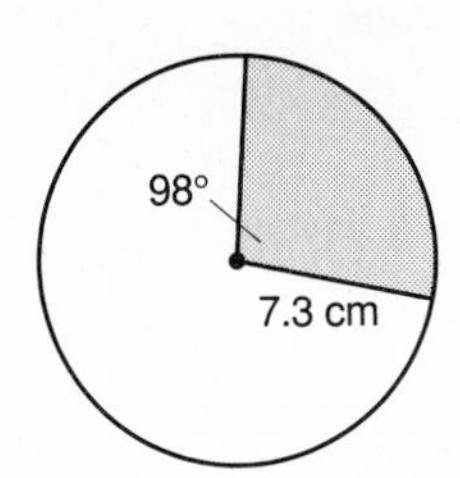

5. ______________

6. A regular pentagon is inscribed in a circle of radius 2. Use a calculator to find the area of the shaded segment assuming $a = 1.618$ and $b = 2.35$.

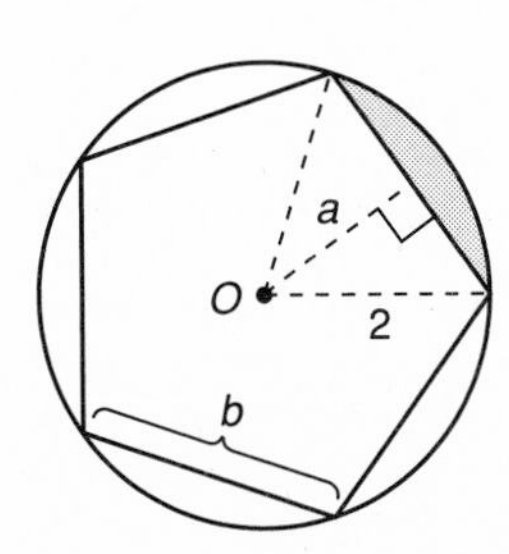

6. ______________

Chapter 11 Test Form A

Name ______________________

(Page 1 of 3 pages) Date ______________________

In Problems 1 and 2, use the figure at the right. Each square in the grid is 1 square unit in area.

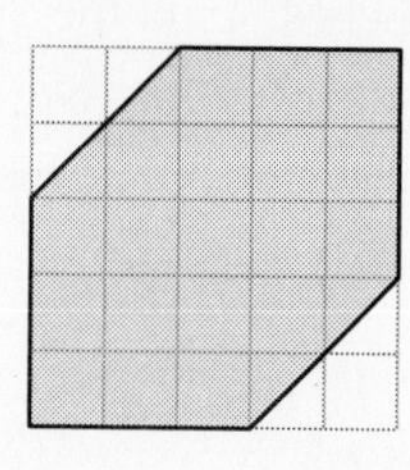

1. Find the perimeter of the polygon. 1. ____________

2. Find the area of the polygon. 2. ____________

3. Give an example to show that two rectangles with the same area need not have the same perimeter. 3.

4. Must two squares with the same perimeter be congruent? Explain. 4.

5. A target is in the shape of a square whose side length is 20 yards. It is placed within a rectangular playing field 100 yards in length and 40 yards wide. Assume a sky diver landing on the field is equally likely to touch down anywhere in the field. What is the probability that the first point of touchdown is inside the square? 5. ____________

In Problems 6–8, refer to the parallelogram $ABCD$ at the right.

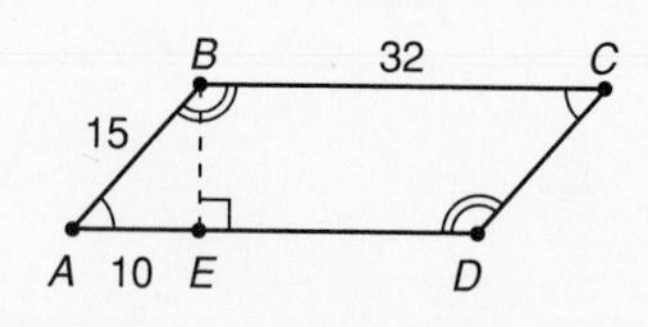

6. Find the height, BE. 6. ____________

7. Find the area of $\triangle ABE$. 7. ____________

8. Find the area of parallelogram $ABCD$. 8. ____________

Name ______________________

Date ______________________

9. Find the area of $\triangle ABC$.

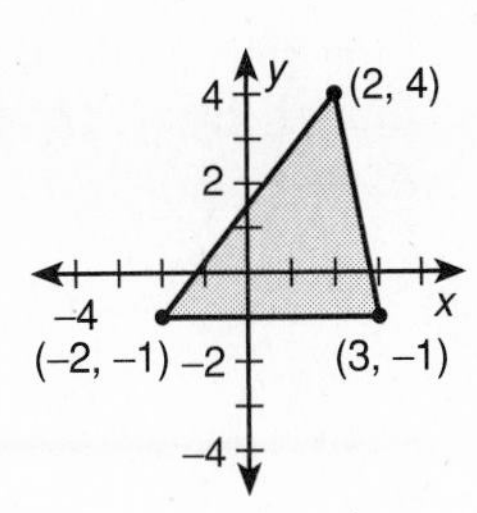

9. ______________________

10. Find the area of trapezoid $ABCD$.

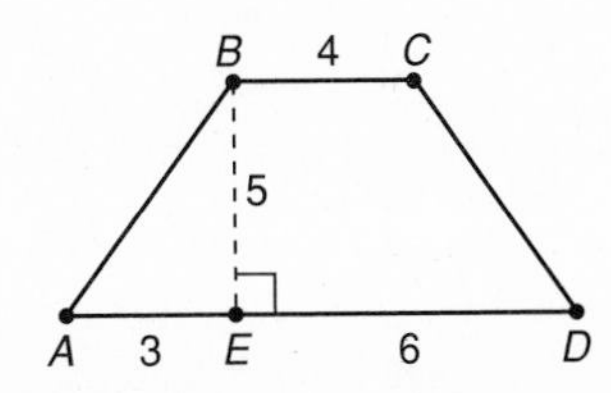

10. ______________________

11. Find the area of quadrilateral $ABCD$.

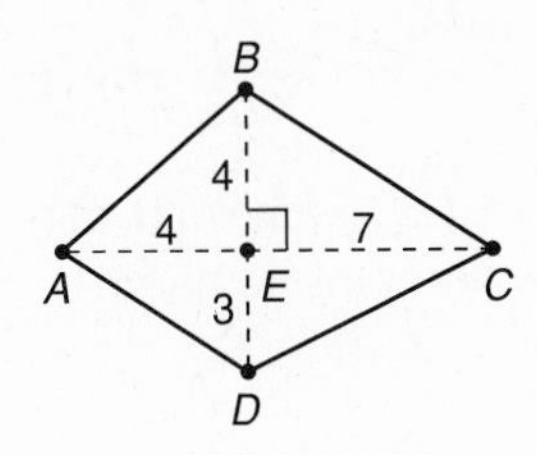

11. ______________________

12. Find the area of the quadrilateral $ABCD$.

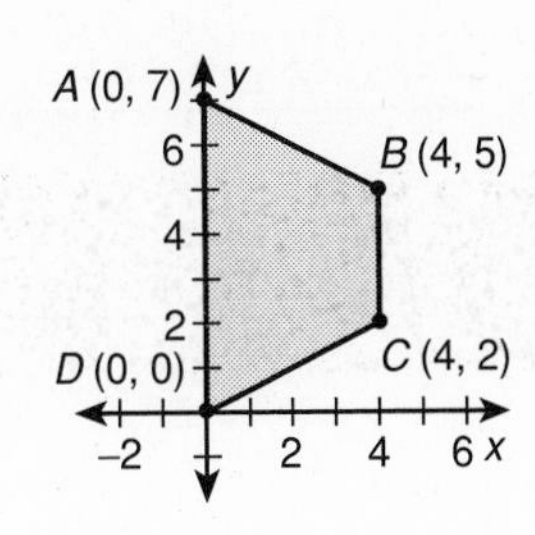

12. ______________________

13. An equilateral triangle has a side length of $2\sqrt{3}$. Find its area.

13. ______________________

14. A square is inscribed in a circle of radius $3\sqrt{2}$. Find the area of the square.

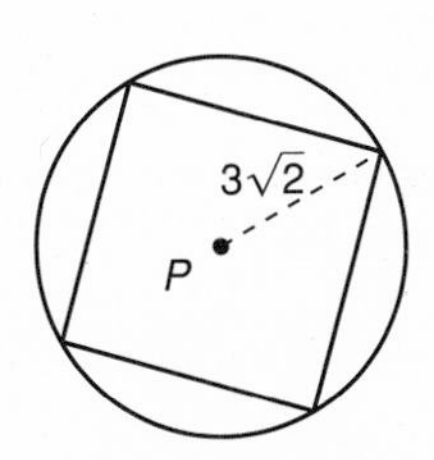

14. ______________________

Chapter 11 Test Form A

Name ____________________

Date ____________________

15. The base of a gazebo is a regular 20-gon with 6-foot sides. Find its apothem, a, to the nearest tenth of a foot.

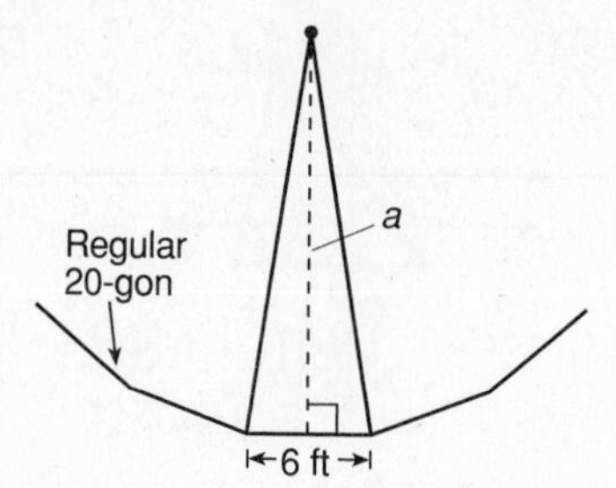

15. ____________________

16. Find the arc length of $\widehat{AB}$ to two decimal places.

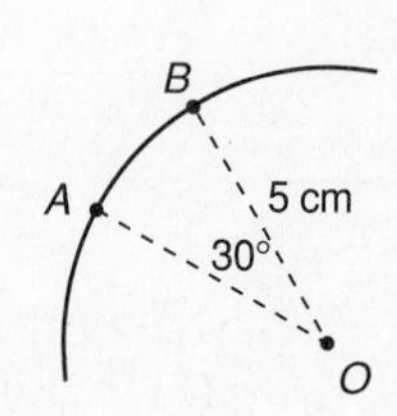

16. ____________________

17. The tires of an automobile have a diameter of 22 inches. If the wheels revolve ten times, how far does the automobile move? (Round the result to nearest tenth of a foot.)

17. ____________________

18. Find the area of the shaded region. (Assume that the ends of the figure are semicircles.)

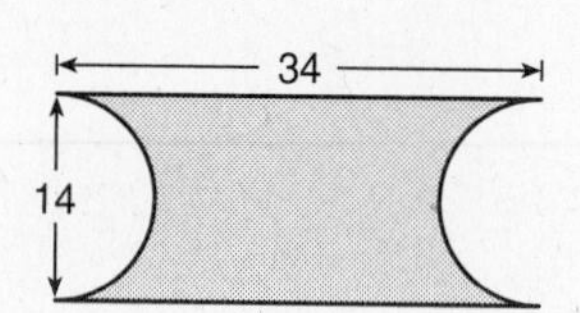

18. ____________________

19. *Given:* $\odot$ O with $\overline{OC} \perp \overline{AB}$.

If $OA = 7$ and $OC = 2$, find the arc length of $\widehat{AB}$ to two decimal places. Explain your reasoning.

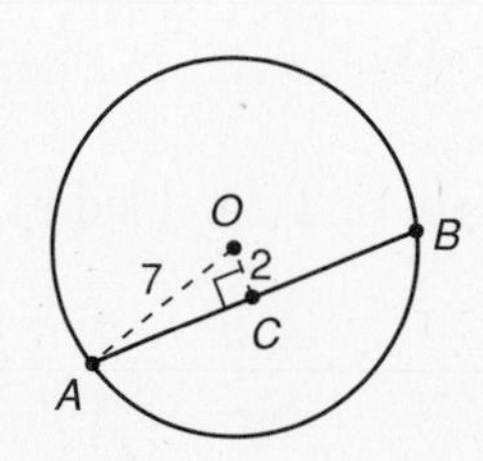

19.

20. $\triangle ABC$ and $\triangle A'B'C'$ are similar triangles with

$$\frac{A'B'}{AB} = \frac{5}{4}.$$

If the area of $\triangle ABC$ is 80 square units, find the area of $\triangle A'B'C'$.

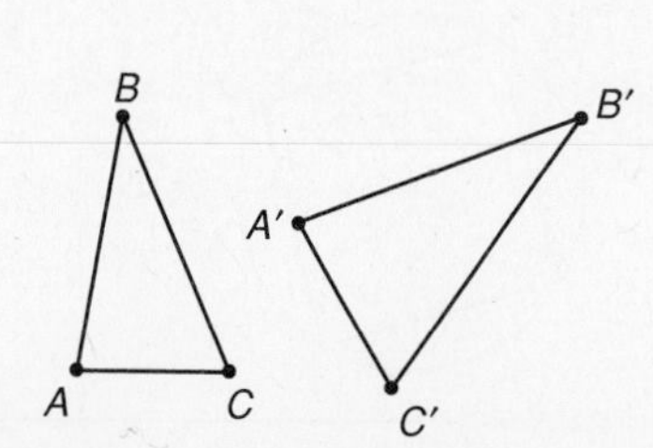

20. ____________________

Chapter 11 Test

Form B (Page 1 of 3 pages)

Name ______________

Date ______________

1. The perimeter of an equilateral triangle with a side length of 4 is [?]. 1. ______

   a. $4\sqrt{3}$ units **b.** 12 units **c.** 8 units **d.** $(6+2\sqrt{3})$ units

2. The area of an equilateral triangle whose side length is 4 is [?]. 2. ______

   a. $6+2\sqrt{3}$ units^2 **b.** 12 units^2 **c.** $4\sqrt{3}$ units^2 **d.** 8 units^2

3. The diagonal of a square is 6 units. Its perimeter is [?]. 3. ______

   a. $12\sqrt{2}$ units **b.** 18 units **c.** 12 units **d.** 24 units

4. A rectangle has length ℓ and width w. An expression for its area is [?]. 4. ______

   a. $\sqrt{\ell^2+w^2}$ **b.** $\ell \cdot w$ **c.** $2(\ell+w)$ **d.** $\sqrt{\ell w}$

5. A rectangular garden, 42 feet by 20 feet, is surrounded by a walkway of uniform width. If the total area of the garden and walkway is 1248 square feet, the width of the walkway is [?]. 5. ______

   a. 2.5 ft **b.** 2 ft
   c. 3.29 ft **d.** 3 ft

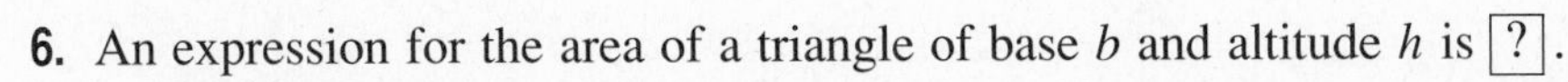

6. An expression for the area of a triangle of base b and altitude h is [?]. 6. ______

   a. $2bh$ **b.** $\frac{1}{2}(a+b+c)$ **c.** $\frac{1}{2}bh$ **d.** $\frac{1}{2}(b+h)$

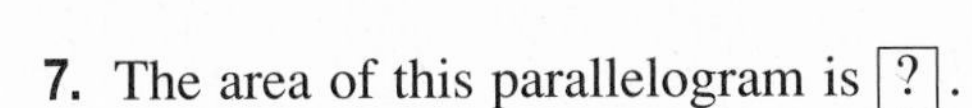

7. The area of this parallelogram is [?]. 7. ______

   a. 800 units^2 **b.** 680 units^2
   c. $40\sqrt{111}$ units^2 **d.** 340 units^2

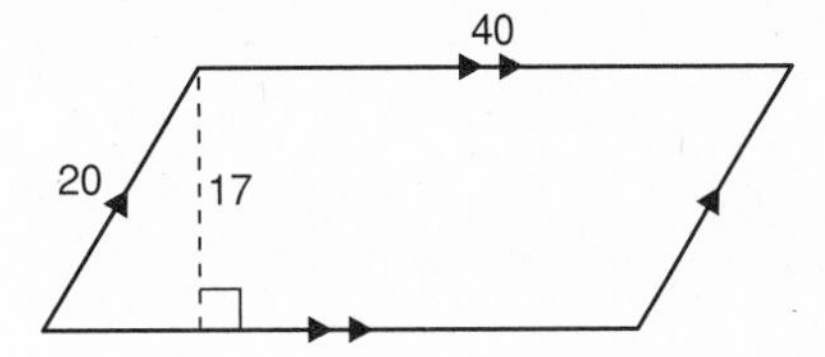

8. An expression for the area of this trapezoid is [?]. 8. ______

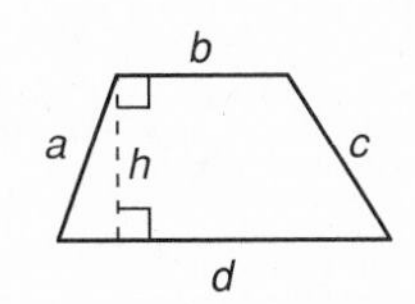

   a. $\frac{a+b+c+d}{4} \cdot h$ **b.** $\frac{(a+c)(b+d)}{2}$
   c. $\frac{h(b+d)}{2}$ **d.** $\frac{bd}{2} \cdot h$

Chapter 11 Test Form B

(Page 2 of 3 pages)

Name ____________________

Date ____________________

9. An expression for the area of quadrilateral $ABCD$ is $\boxed{?}$.

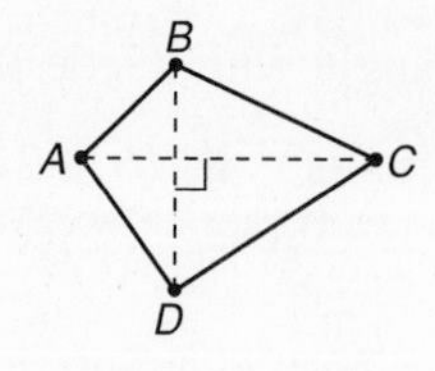

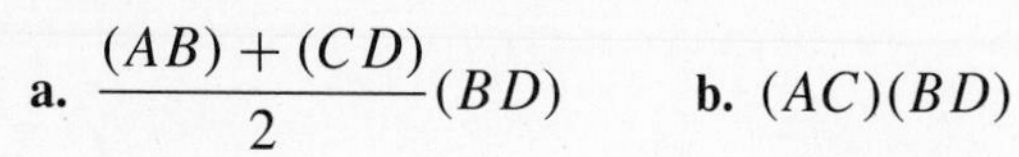

a. $\dfrac{(AB)+(CD)}{2}(BD)$ **b.** $(AC)(BD)$

c. $(AB)(AD)$ **d.** $\dfrac{(BD)(AC)}{2}$

9. ______

10. The area of this quadrilateral is $\boxed{?}$.

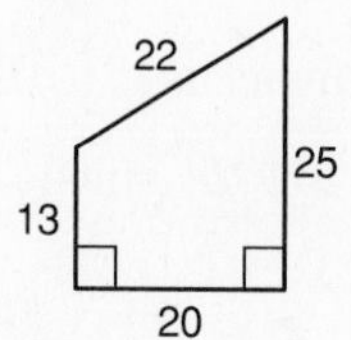

a. 380 units^2 **b.** 16.25 units^2
c. 23.3 units^2 **d.** 58 units^2

10. ______

11. The area of this quadrilateral is $\boxed{?}$.

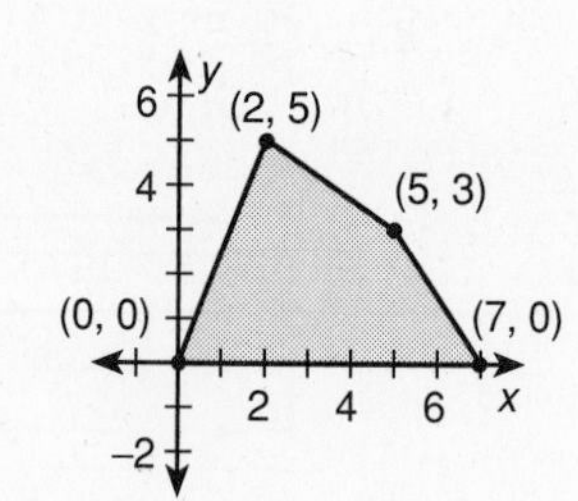

a. 17.5 units^2 **b.** 20 units^2
c. 21 units^2 **d.** $\frac{21}{2}$ units^2

11. ______

12. *Given:* $\triangle PQR$ is equilateral. If the coordinates of P and R are $(-10, 0)$ and $(10, 0)$, then the area of $\triangle PQR$ is approximately $\boxed{?}$.

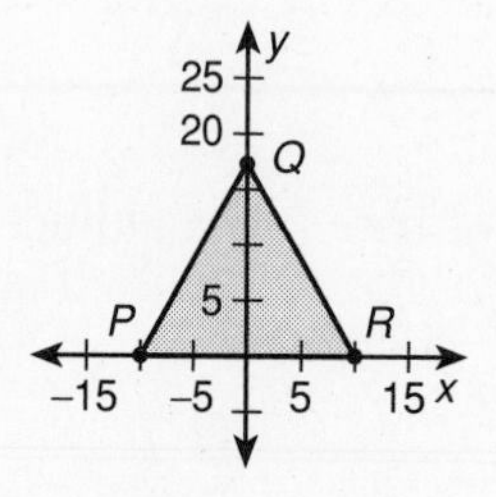

a. 86.6 units^2 **b.** 173.2 units^2
c. 43.3 units^2 **d.** 200 units^2

12. ______

13. A regular hexagon has an apothem of 2 and a side length

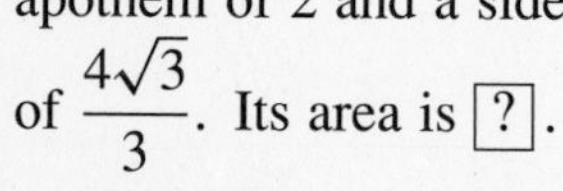

of $\dfrac{4\sqrt{3}}{3}$. Its area is $\boxed{?}$.

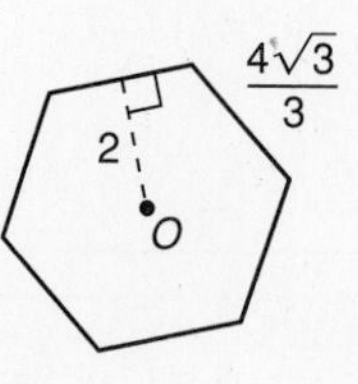

a. $4\sqrt{3}$ units^2 **b.** 4π units^2

c. $8\sqrt{3}$ units^2 **d.** $\dfrac{16\pi}{3}$ units^2

13. ______

14. An expression for the area of a circle with diameter d is $\boxed{?}$.

a. πd **b.** πd^2 **c.** $2\pi r$ **d.** $\dfrac{\pi d^2}{4}$.

14. ______

Name ____________________

Date ____________________

15. An expression for the circumference of a circle with diameter d is $\boxed{?}$. **15.** ______

a. $\dfrac{\pi d^2}{4}$ **b.** πd **c.** πr **d.** $2\pi d$

16. Find the length of a 40°-arc in a circle with radius of 4. **16.** ______

a. $\dfrac{9\pi}{8}$ **b.** 8π **c.** $\dfrac{16\pi}{9}$ **d.** $\dfrac{8\pi}{9}$

17. An automobile has 20-inch diameter wheels. If the wheels revolved three times after the brakes were applied, the stopping distance was approximately $\boxed{?}$. **17.** ______

a. 15.7 ft **b.** 7.85 ft **c.** 26.2 ft **d.** 157 ft

18. Each circle is tangent to the other two. If the diameter of the large circle is 12, the area of the shaded region is $\boxed{?}$. **18.** ______

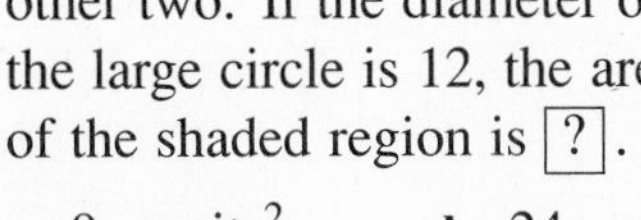

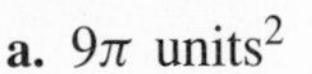

a. 9π units^2 **b.** 24π units^2

c. 18π units^2 **d.** 36π units^2

19. The radius of the circle is $\sqrt{2}$. The distance from the center of the circle to the chord is 1. If the measure of $\overset{\frown}{AB}$ is 90°, the area of the shaded segment is $\boxed{?}$. **19.** ______

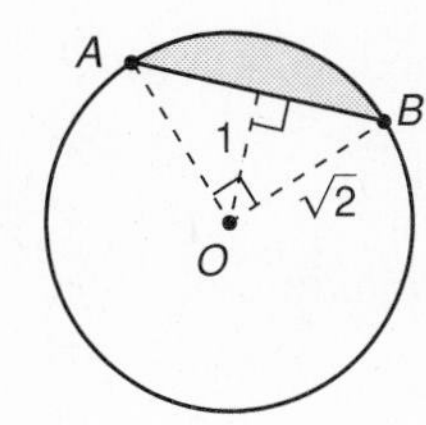

a. $\left(\dfrac{\pi}{2} - 1\right)$ units^2 **b.** $(2\pi - 1)$ units^2

c. $\dfrac{\pi}{4}$ units^2 **d.** $\dfrac{2}{3}$ units^2

20. The ratio of the side lengths of two regular hexagons is 4 to 9. If the area of the smaller hexagon is 16 units^2, then the area of the larger hexagon is $\boxed{?}$. **20.** ______

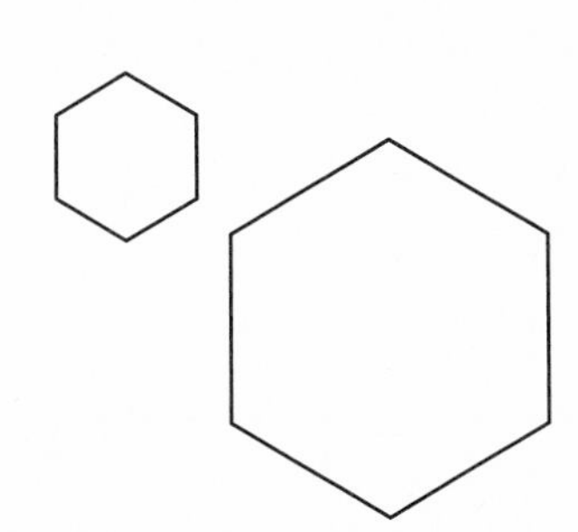

a. 36 units^2 **b.** 81 units^2

c. $\dfrac{64}{9}$ units^2 **d.** $\dfrac{256}{81}$ units^2

Chapter 11 Test

Form C

(Page 1 of 3 pages)

Name ______________________

Date ______________________

In Problems 1 and 2, use the figure at the right. Each square in the grid is 1 square unit in area.

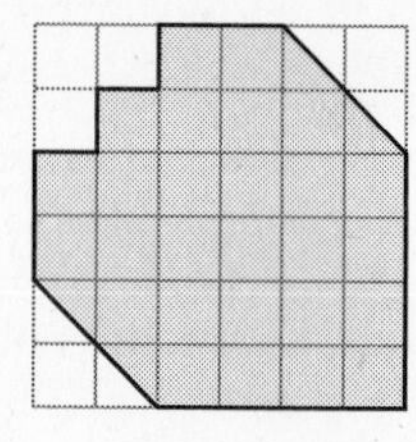

1. Find the perimeter of the polygon. 1. ______________

2. Find the area of the polygon. 2. ______________

In Problems 3–5, use the following.

A rectangular bleacher area is to have a 120-ft frontage and a 60-ft slant width. You must allow 1 yard by 1 yard area for each seated spectator.

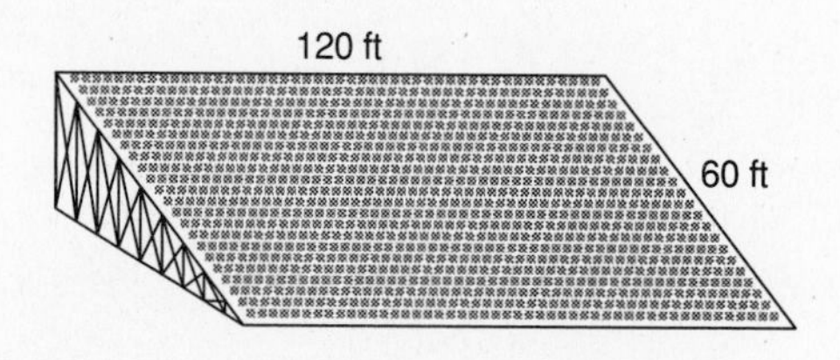

3. How many rows will there be? 3. ______________

4. How many people can you accommodate in a row? 4. ______________

5. How many people can be seated at maximum? 5. ______________

In Problems 6–8, refer to the parallelogram at the right.

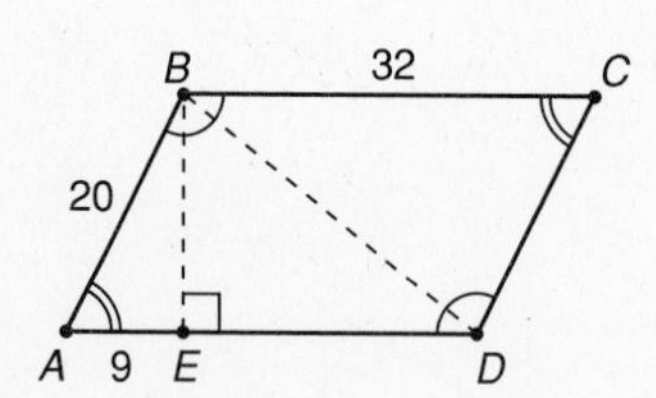

6. Find the area of $\triangle ABE$. 6. ______________

7. Find the area of parallelogram $ABCD$. 7. ______________

8. Find the length of diagonal $\overline{BD}$. 8. ______________

Name ______________________

Date ______________________

9. Find the area of the trapezoid.

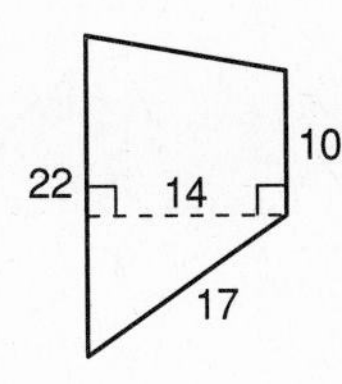

9. ______________

10. *Given*: $AC \perp BD$

If $AC = 18$ and $BD = 15$, find the area of quadrilateral $ABCD$.

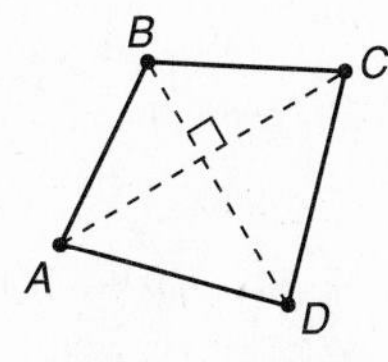

10. ______________

11. Find the area of the quadrilateral.

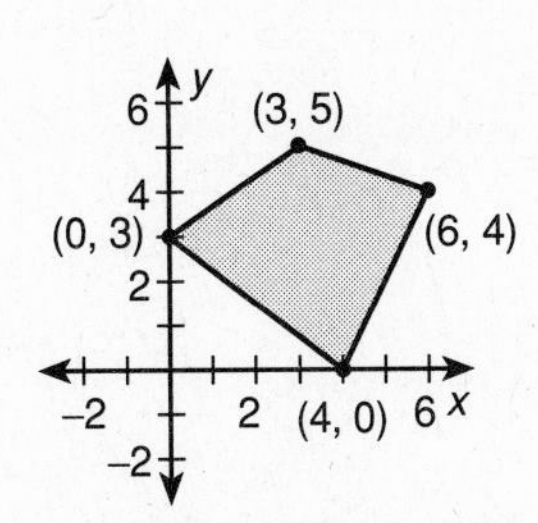

11. ______________

12. Explain why the apothem of a regular polygon is always less than the radius of the circumscribed circle.

12.

13. The floor of a gazebo is a regular 20-gon. If the apothem is 30 feet, find the length of each side to two decimal places.

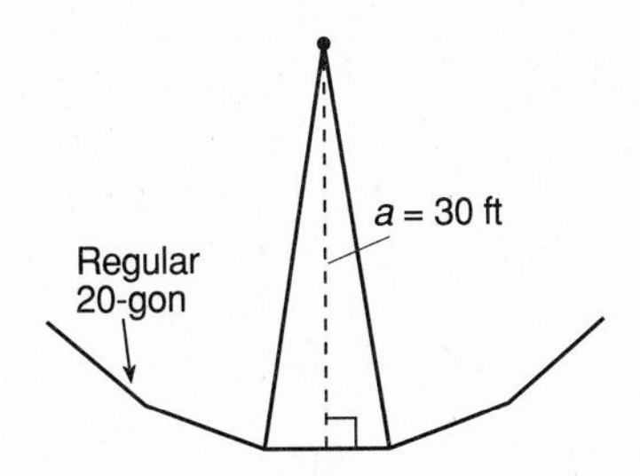

13. ______________

14. The perimeter of an equilateral triangle is 18. Find its area.

14. ______________

Name ____________

Date ____________

15. Circle O has a radius of 7.39. If $\angle AOB$ is 112°, then find the length of $\widehat{AB}$ to two decimal places.

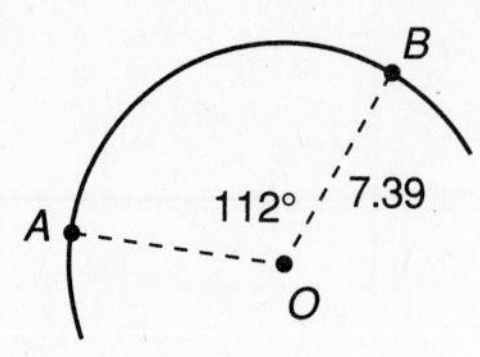

15. ____________

16. A vehicle travels 125.7 feet while its wheels revolve 16 times. Find the diameter of the wheels to the nearest inch.

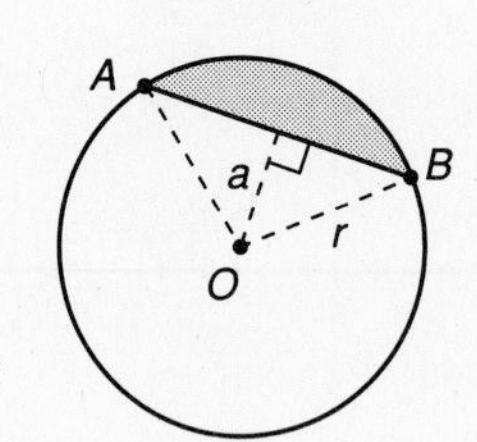

16. ____________

17. *Given*: $m\widehat{AB} = 100°$
$a = 5.14$
$r = 8.00$
Find the area of the shaded segment to two decimal places.

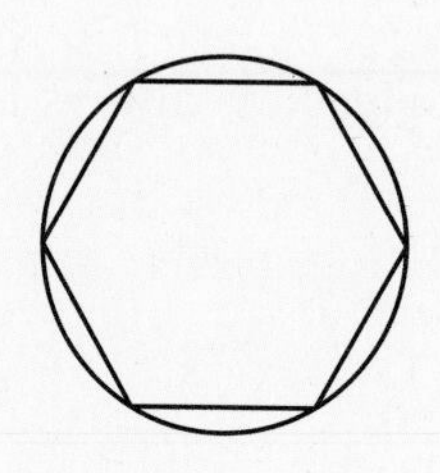

17. ____________

18. Assume a regular polygon (such as a hexagon) is circumscribed by a circle. If the number of sides of the regular polygon increases, then what happens to the area of the segments between the circle and the new polygon? (Illustrate your reasoning by using the figure at the right.)

18.

19. Quadrilaterals $ABCD$ and $A'B'C'D'$ are similar with $\frac{AB}{A'B'} = \frac{5}{2}$. If the area of $ABCD$ is 115 square units, what is the area of $A'B'C'D'$?

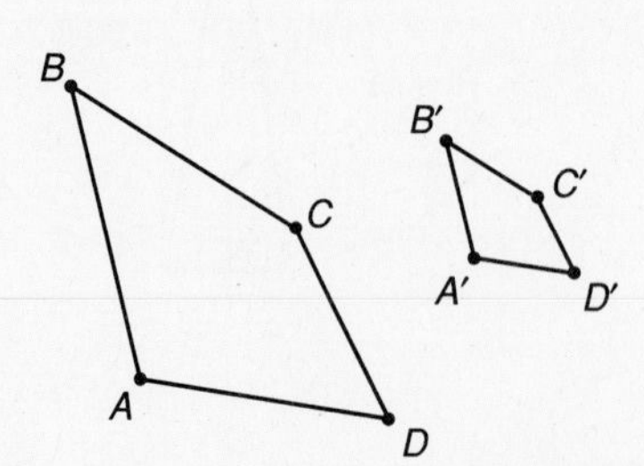

19. ____________

12.2 Short Quiz

Name ______________________

Date ______________________

1. Is the solid shown a polyhedron? Why or why not?

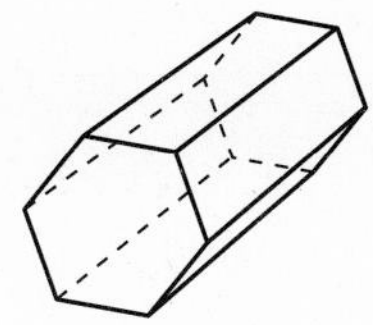

1. ______________________

2. Sketch (if possible) a nonconvex polyhedron.

2.

3. Sketch (if possible) two different polyhedrons with 6 faces.

3.

4. Sketch (if possible) a polyhedron with faces that are triangles and rectangles.

4.

5. Sketch (if possible) a polyhedron with faces that are rectangles and circles.

5.

6. Sketch (if possible) a polyhedron that has 5 faces and 5 vertices.

6.

7. A polyhedron has 15 edges and 10 vertices. How many faces does it have? (Use Euler's formula.)

7. ______________________

8. Find the surface area of this right circular cylinder.

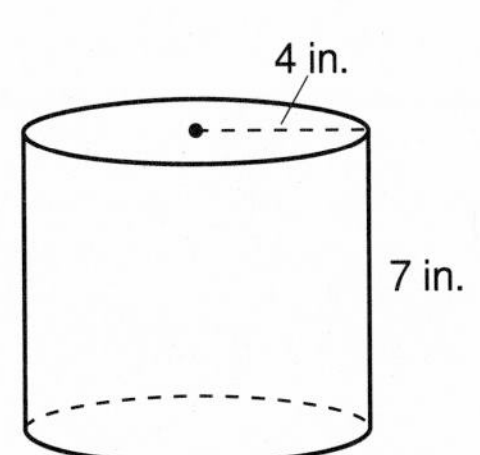

8. ______________________

(Use after Lesson 12.3)

Name ____________

Date ____________

In Problems 1–3, use the solids shown.

A. 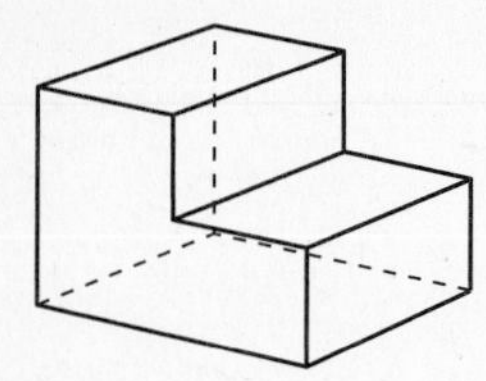 B. 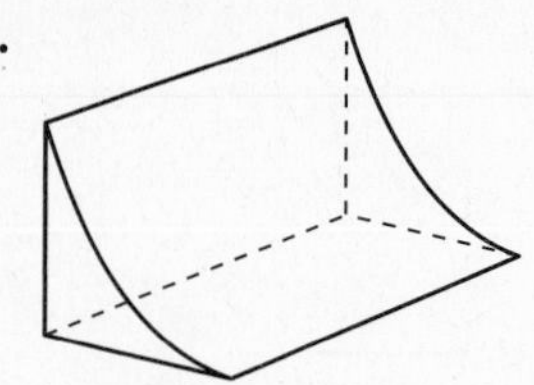 C. 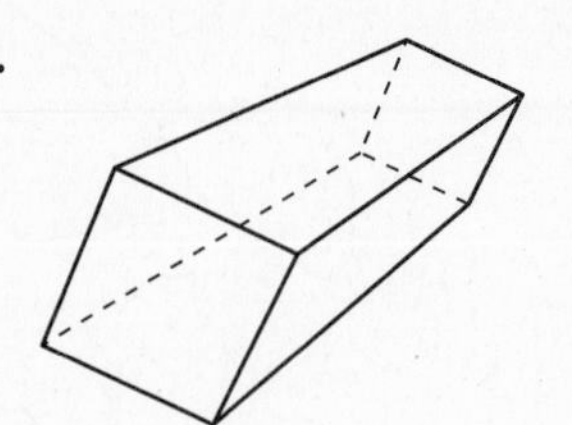 D. 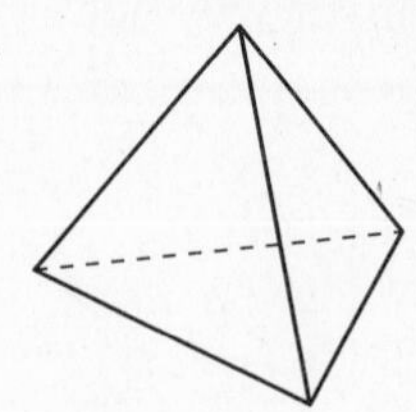

1. One of the solids is not a polyhedron. Which one? Explain. **1.**

2. One of the solids is a regular polyhedron. Which one? Explain. **2.**

3. One of the solids is a polyhedron, but it is not a convex polyhedron. Which one? Explain. **3.**

4. Sketch a polyhedron that has 8 faces and 12 vertices. **4.**

5. Find the lateral area of this right circular cylinder. **5.** ____________

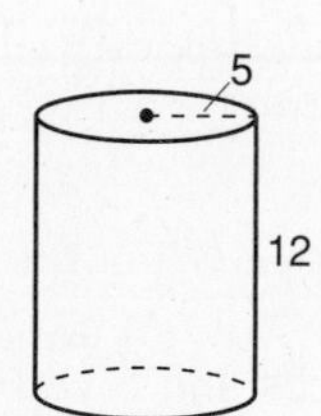

6. Find the total surface area of this prism. **6.** ____________

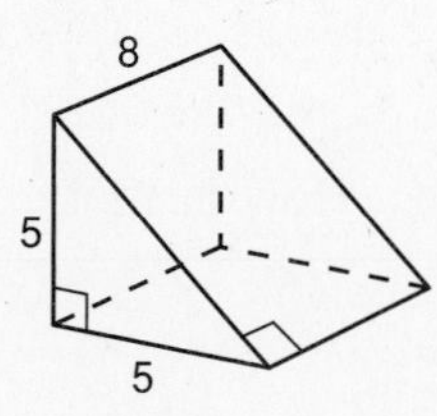

7. Find the total surface area of the right circular cone whose base radius is 5 and whose slant height is 11. **7.** ____________

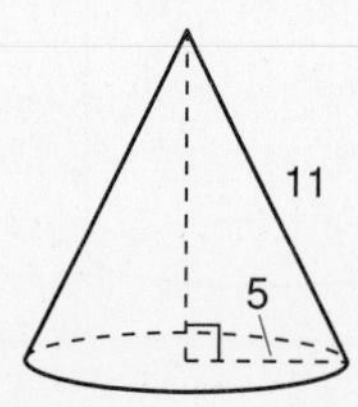

Mid-Chapter 12 Test **Form B**

(Use after Lesson 12.3)

Name ______________________

Date ______________________

In Problems 1–3, use the solids shown.

A.

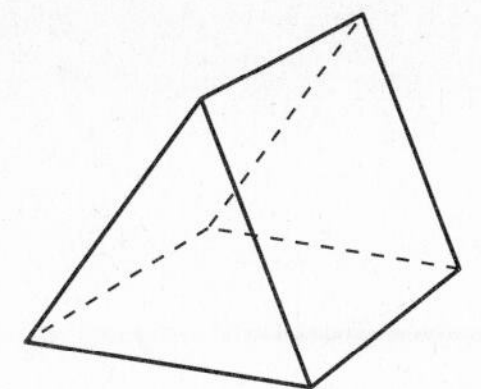

B.

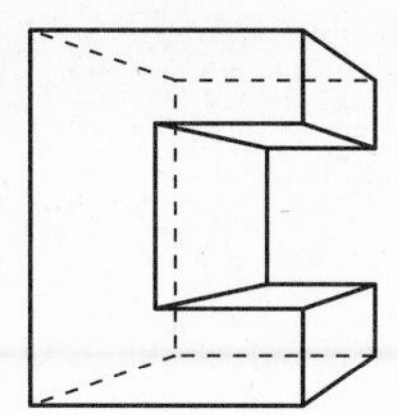

C.

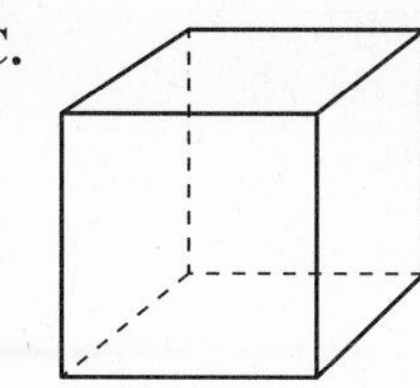

D.

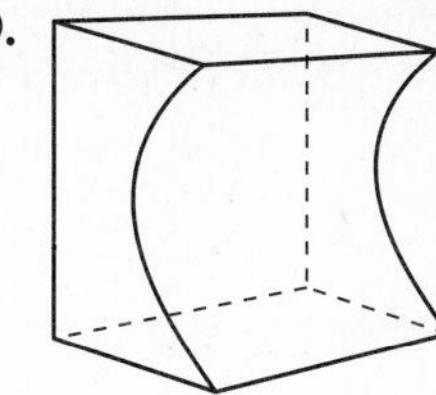

1. One of the solids is a regular polyhedron. Which one? Explain. **1.**

2. One of the solids is not a polyhedron. Which one? Explain. **2.**

3. One of the solids is a polyhedron, but it is not a convex polyhedron. Which one? Explain. **3.**

4. Sketch a polyhedron that has 10 faces and 16 vertices. **4.**

5. Find the lateral surface area of this right circular cylinder.

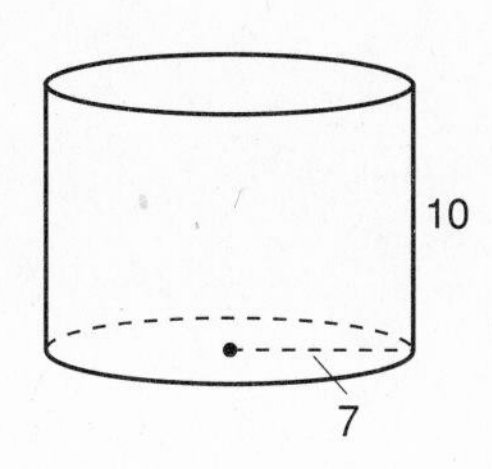

5. ______________________

6. The ratios of the length to the width to the height of a rectangular prism is 1:2:3. What are the actual dimensions of the prism if its surface area is 198 square inches? **6.** ______________________

7. Find the total surface area of the right circular cone whose base radius is 8 and whose slant height is 15.

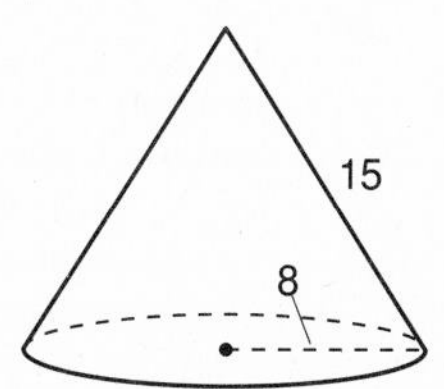

7. ______________________

12.4 Short Quiz

Name ______________________

Date ______________________

1. The regular pyramid has a square base with 13-centimeter side lengths and a slant height of 12 centimeters. Find its surface area.

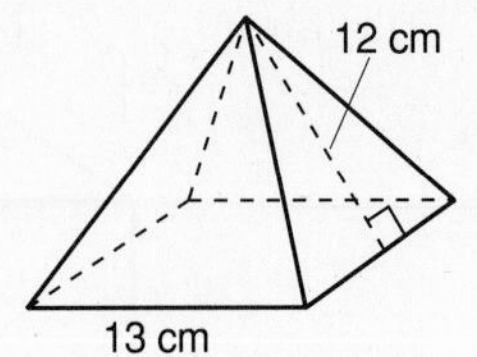

1. ______________

2. Find the surface area of the right cone whose base radius is 3 and whose slant height is 8. (Round the result to one decimal place.)

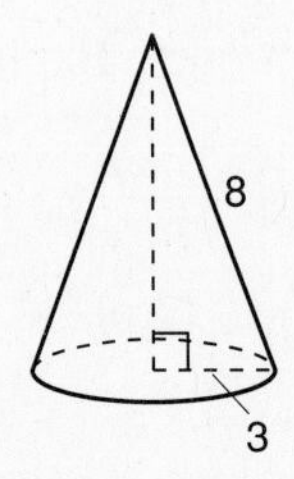

2. ______________

3. Find the surface area of the solid. The cone and the cylinder are right. (Round the result to one decimal place.)

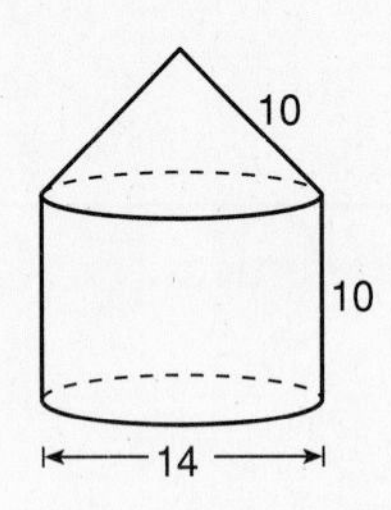

3. ______________

4. Draw and label a rectangular prism that has a volume of 364 cubic inches.

4.

5. A cylinder has a base area of 36π cubic centimeters and a height of 15 centimeters. Find its volume.

5. ______________

6. The volume of a circular cylinder is 972π cubic feet. If the height of the cylinder is 12 feet, find the diameter of its base.

6. ______________

12.6 Short Quiz

Name ______________________

Date ______________________

1. A pile of gravel is in the form of a right circular cone 30 feet in diameter and 21 feet high. Find its volume.

1. ______________________

2. Draw a regular pyramid with a square base whose height is equal to the side length of the base. If the height of the pyramid is 3 yards, find its volume.

2.

3. Find the volume of the solid formed by two right cones. (Round the result to one decimal place.)

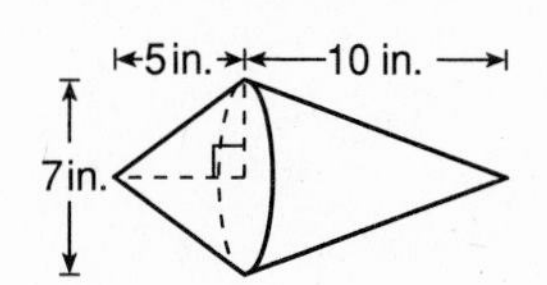

3. ______________________

4. A child's spherical ball has a diameter of 12.3 inches. Find its volume to one decimal place.

4. ______________________

5. Find the surface area of the ball described in Problem 4. (Round the result to one decimal place.)

5. ______________________

6. What is the diameter of a sphere whose surface area is 385 square feet? (Round the result to three decimal places.)

6. ______________________

Chapter 12 Test

Form A (Page 1 of 3 pages)

Name ______________________

Date ______________________

1. Sketch (if possible) a polyhedron with 3 faces. **1.**

2. Sketch (if possible) a polyhedron with 6 faces and 8 vertices. **2.**

3. Sketch (if possible) two different polyhedrons with 6 vertices. **3.**

4. A polyhedron has 6 faces and 7 vertices. How many edges does it have? Explain your answer. **4.**

5. Find the surface area of this right prism. **5.** ______________

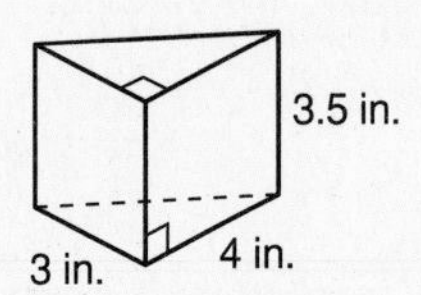

6. Find the surface area of this right cylinder. (Round the result to two decimal places.) **6.** ______________

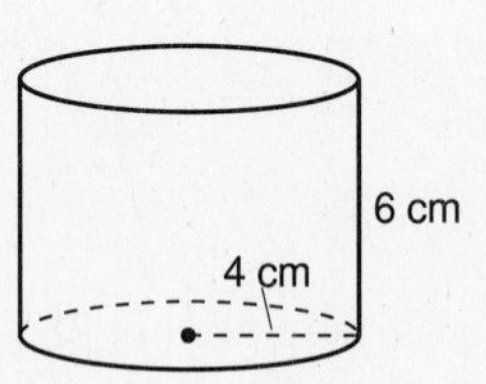

7. The pyramid has a square base and a slant height of 7 ft. Find its surface area. **7.** ______________

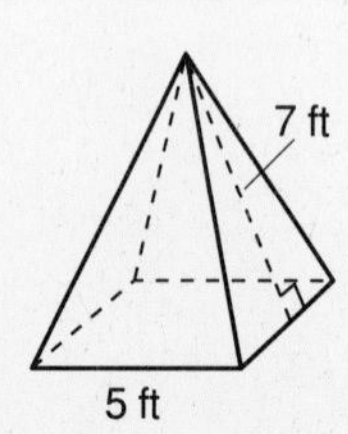

Name ______________________

Date ______________________

8. Find the surface area of this right cone. (Round the result to two decimal places.)

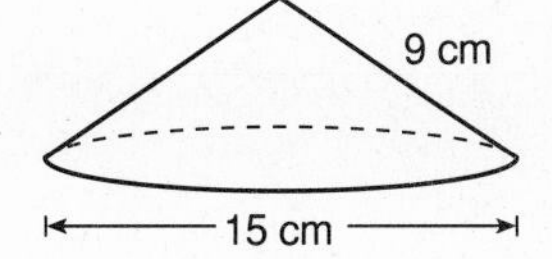

8. ______________________

9. A drinking cup is made in the shape of a right cone, with an open top as shown. If the height of the cup is increased by 1 centimeter, then what happens to its volume?

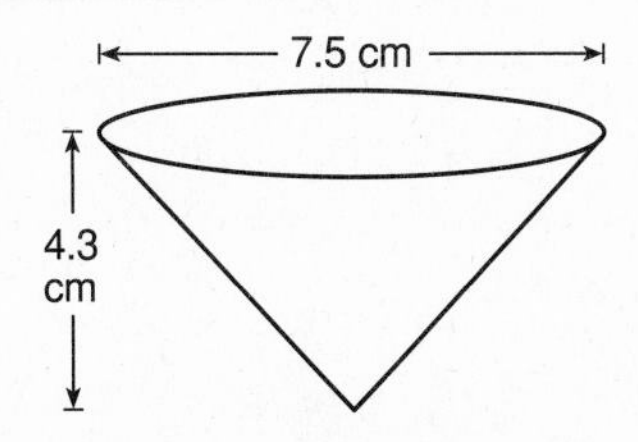

9. ______________________

10. Find the volume of this right prism.

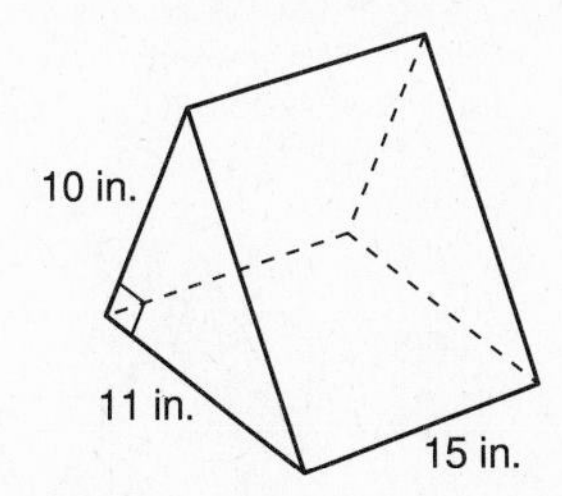

10. ______________________

11. Find the volume of this cylinder. (Round the result to one decimal place.)

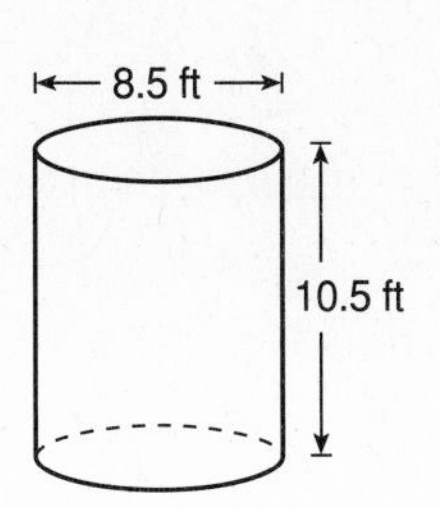

11. ______________________

12. Find the surface area and volume of the cylinder that can be folded from this net shown.

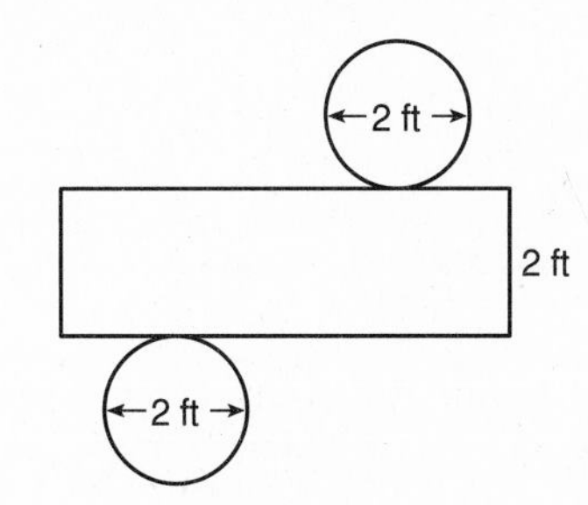

12. ______________________

Name ______________________

Date ______________________

13. This pyramid has a rectangular base. Find its volume to the nearest tenth of a cubic meter.

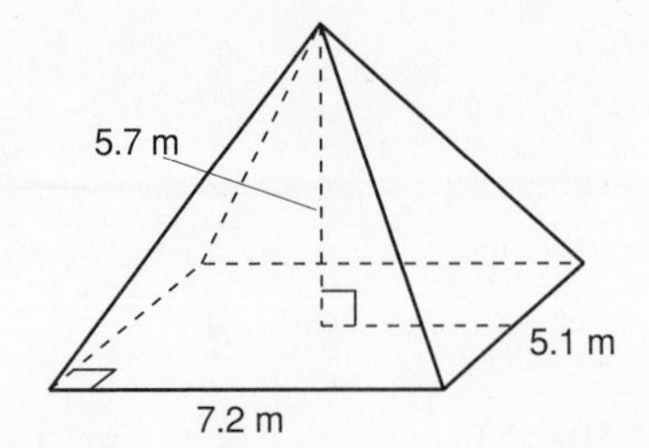

13. ______________________

14. Find the volume of this cone. (Round the result to the nearest tenth of a cubic foot.)

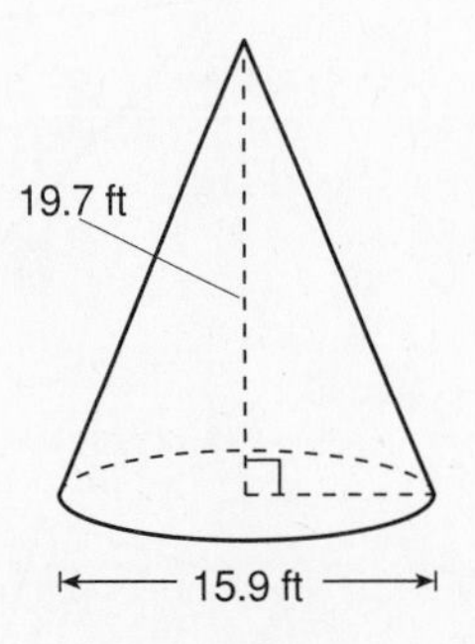

14. ______________________

15. The base of this pyramid is a non-regular septagon with an area of 30.0 square yards. The height of the pyramid is 6.6 yards. Find the volume of the pyramid.

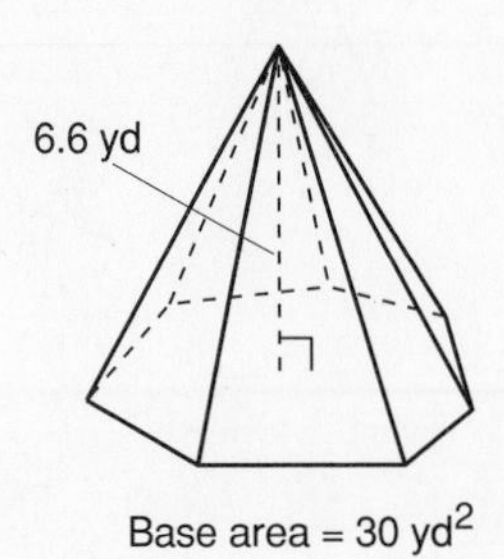

15. ______________________

16. Find the surface area of a sphere whose radius is 17.3 inches to the nearest tenth of a square inch.

16. ______________________

17. For the sphere in Problem 16, find its volume in cubic feet to the nearest tenth of a cubic foot.

17. ______________________

18. What is the area of a great circle of a sphere whose surface area is 44.8 square yards?

18. ______________________

19. Suppose that each dimension of a prism is increased by 10%. By what percent does its volume increase?

19. ______________________

Chapter 12 Test

Form B

(Page 1 of 3 pages)

Name ______________________

Date ______________________

1. The solid shown is [?].

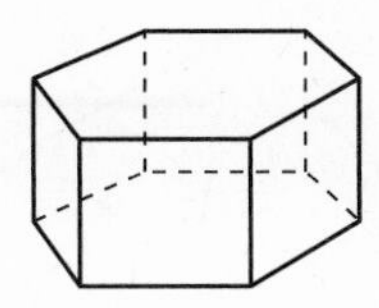

a. a hexahedron
b. a hexagon
c. an octagon
d. an octahedron

1. ______

2. This solid figure is best described as [?].

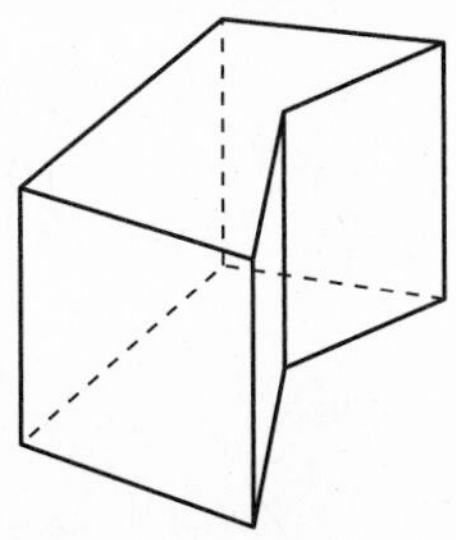

a. a regular polyhedron
b. an nonconvex polyhedron
c. a semiregular polyhedron
d. a polygon

2. ______

3. According to Euler's Theorem, the number of faces (F), vertices (V), and edges (E) of a polyhedron is related by the formula [?].

a. $F + V = E + 2$ b. $F + E = V + 2$
c. $F + V = E - 2$ d. $E + V = F + 2$

3. ______

4. A prism is a polyhedron that has two faces called bases that are always [?].

a. rectangular
b. perpendicular with the lateral edges
c. congruent and parallel
d. isosceles right triangles

4. ______

5. If all the angles in the faces of this polyhedron are right angles, then its surface area is [?].

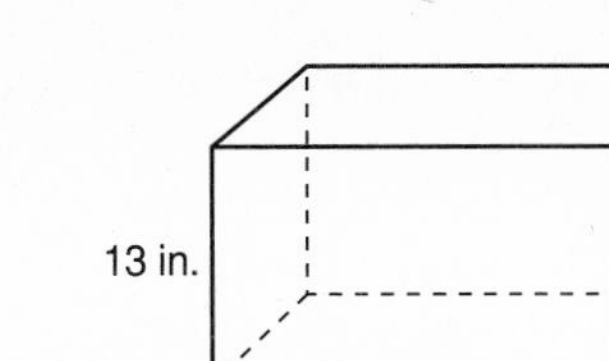

a. 1678 in.^{2} b. 4485 in.^{2}
c. 1794 in.^{2} d. 839 in.^{2}

5. ______

6. The surface area, in square centimeters, of this right cylinder is [?].

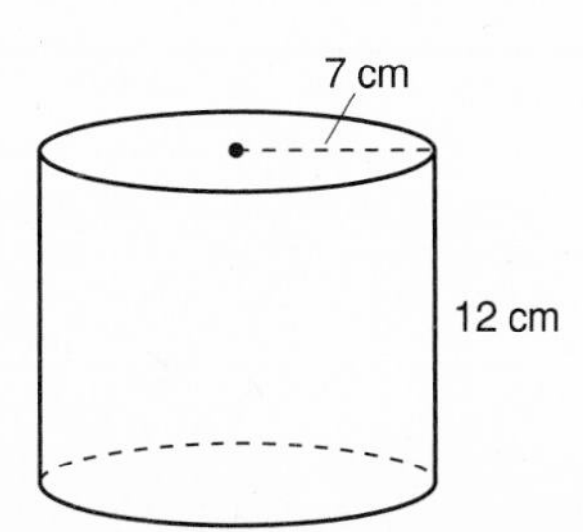

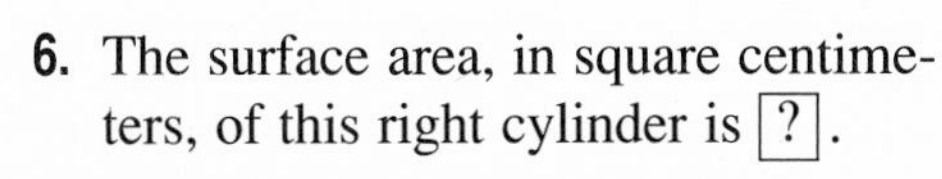

a. $14\pi(12) = 168\pi$
b. $98\pi + (14\pi)12 = 266\pi$
c. $(7^2)\pi + 14\pi(12) = 217\pi$
d. $(7^2\pi)(12) = 588\pi$

6. ______

Name ______________________

Date ______________________

7. The lateral area of this regular pyramid is [?].

   a. 84 ft^2 b. 168 ft^2
   c. 336 ft^2 d. 532 ft^2

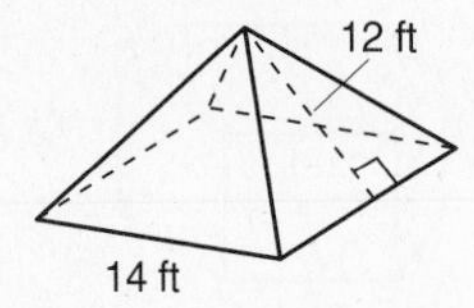

7. ________

8. A regular pyramid has a base area of $6\sqrt{3}$ in.^{2}, a base perimeter of 12 in., and a slant height of $4\sqrt{3}$ in. Its surface area is [?].

   a. $24\sqrt{3}$ in.^{2} b. $54\sqrt{3}$ in.^{2} c. 36 in.^{2} d. $30\sqrt{3}$ in.^{2}

8. 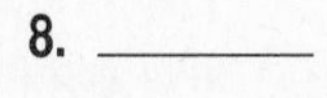

9. The surface area of this right cone is [?].

   a. 36π in.^{2}
   b. 44π in.^{2}
   c. 112π in.^{2}
   d. $16\sqrt{33}\pi$ in.^{2}

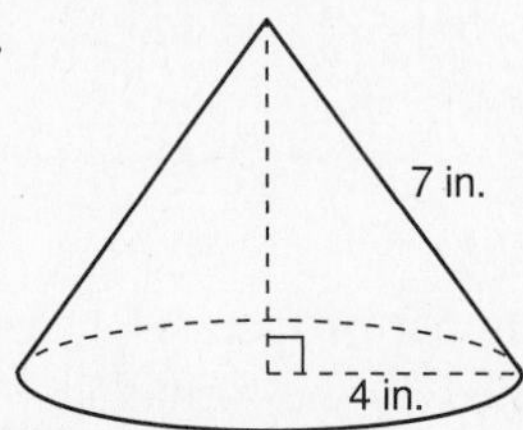

9. ________

10. If two solids have the same height and the same cross-sectional area at every level, then they have the same [?].

   a. surface area b. base perimeter c. volume d. slant height

10. ________

11. The volume of this right prism is [?].

   a. $10\sqrt{13}$ ft^3
   b. $30\sqrt{13}$ ft^3
   c. 120 ft^3
   d. 60 ft^3

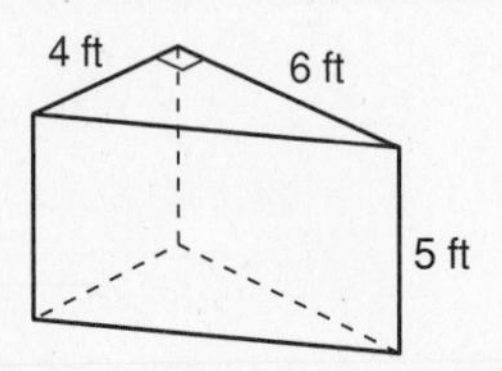

11. ________

12. The volume of this right circular cylinder is about [?].

   a. 1061.9 m^3 b. 326.7 m^3
   c. 1036.9 m^3 d. 265.5 m^3

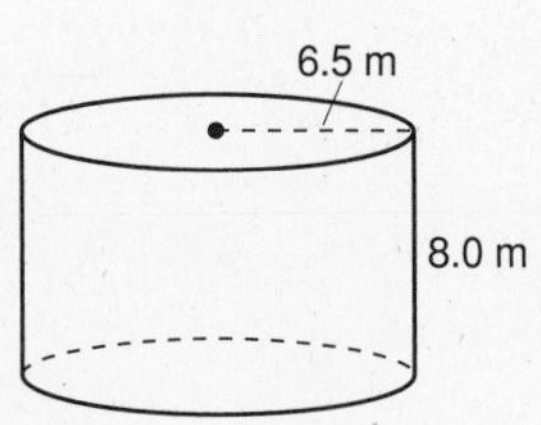

12. ________

13. Describe what happens to the volume of a cone if its radius is doubled while its height is halved. The volume is [?].

   a. unchanged b. increased by a factor of $\frac{1}{3}$
   c. doubled d. not able to be determined

13. ________

Name ______________

Date ______________

14. The volume of this pyramid is [?].

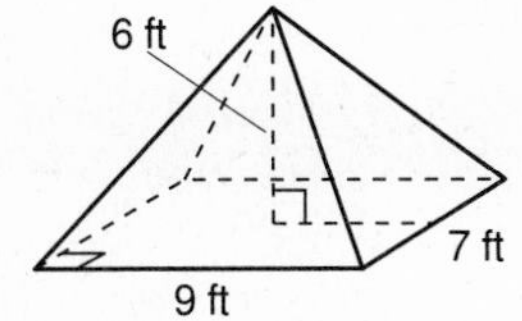

a. 126π ft^3 **b.** 126 ft^3
c. 378 ft^3 **d.** 195π ft^3

14. ______

In Problems 15 and 16, refer to the figure below.

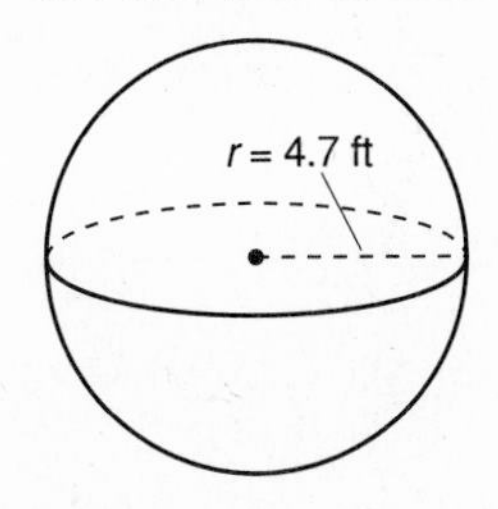

15. The surface area of this sphere is about [?].

a. 69.4 ft^2 **b.** 92.5 ft^2 **c.** 434.9 ft^2 **d.** 277.6 ft^2

15. ______

16. The volume of the sphere is about [?].

a. 277.6 ft^3 **b.** 434.9 ft^3 **c.** 69.4 ft^3 **d.** 92.5 ft^3.

16. ______

17. The top of this cylindrical container has the shape of a hemisphere, as shown. The total volume of the container is [?].

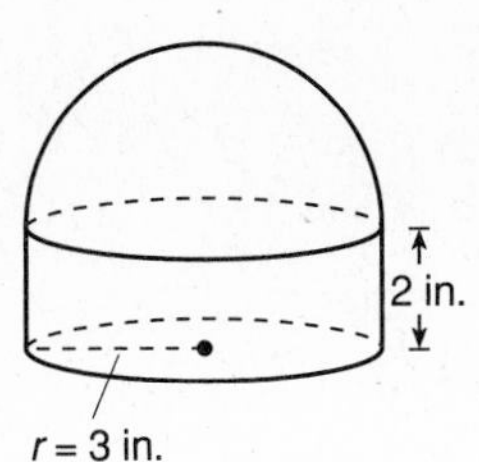

a. 36π in.^{3} **b.** 18π in.^{3}
c. 27π in.^{3} **d.** 36 in.^{3}

17. ______

18. The surface area of a solid is 10 square feet. The dimensions of a similar solid are three times as great as the first. The surface area of the new solid is [?].

a. 40 ft^2 **b.** 270 ft^2 **c.** 80 ft^2 **d.** 90 ft.^{2}.

18. ______

Chapter 12 Test Form C

(Page 1 of 3 pages)

Name ______________________

Date ______________________

1. Draw a polyhedron which is not convex. **1.**

2. Each face of the solid shown is an equilateral triangle. Explain why it is not a regular polyhedron. **2.**

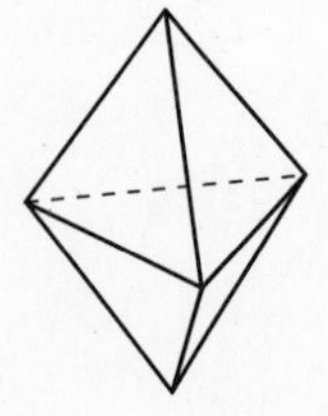

3. A polyhedron has 9 faces and 21 edges. How many vertices does it have? Explain your answer. **3.**

4. A right prism has bases which are equilateral triangles of side length 4 cm. Its height is 5 cm. Find its surface area. **4.** ______________

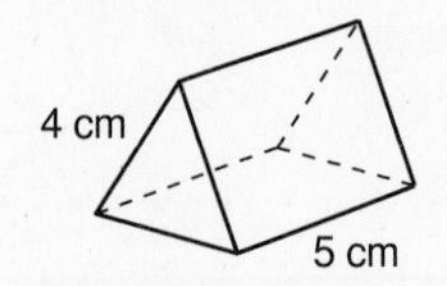

5. The figure shown is a cylindrical solid with a circular cylindrical hole drilled out of the center. Find the surface area of the resulting solid. **5.** ______________

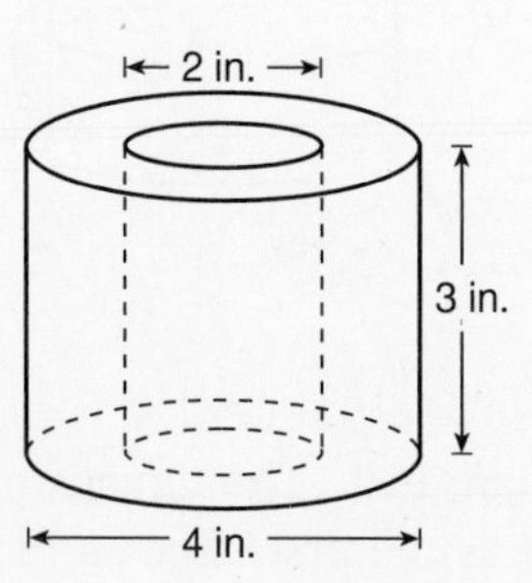

6. A pyramid is in the form of a regular tetrahedron of edge length 2 in. Find its surface area. **6.** ______________

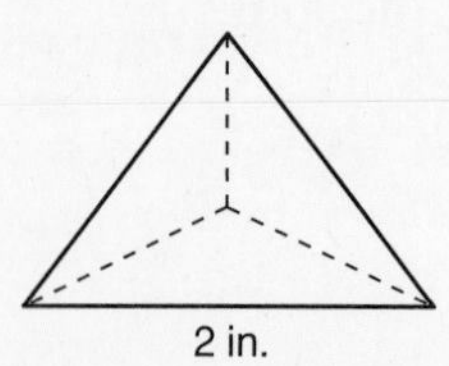

Name ______________________

Date ______________________

7. Find the surface area of this right cone to two decimal places.

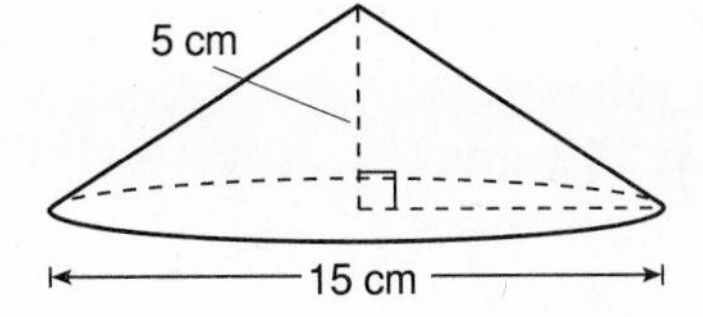

7. ______________

8. A sector of a paper circle is curled to form a drinking cup in the shape of a right cone, as shown. Explain how to determine the diameter of the base of the cup.

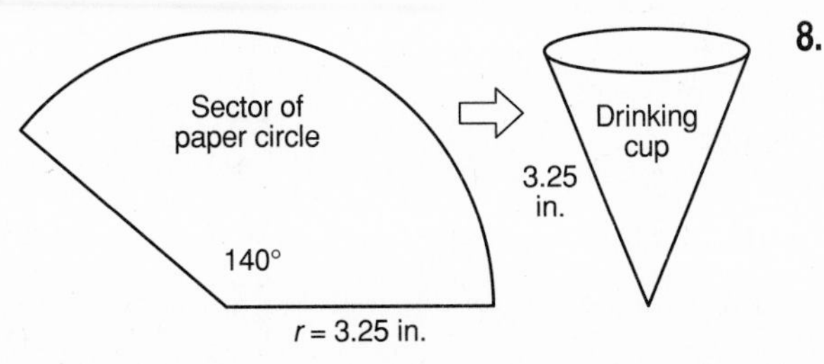

8.

9. A prism has congruent parallelograms for bases. One pair of parallel sides of the parallelogram measures 12 feet and are 5 feet apart. The altitude of the prism is 13 feet. Find the volume of the prism.

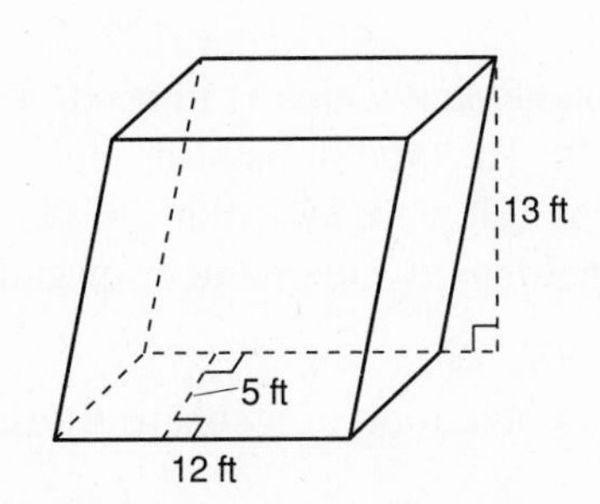

9. ______________

10. Which has a greater volume: a cube of edge length x or a cylinder with height of x and diameter of x? Explain.

10.

11. A cylindrical can is 20 cm in diameter and 16 cm in height. You want to reduce the diameter of the can to 16 cm. What must the height be if the new can still has the same volume? Explain your answer.

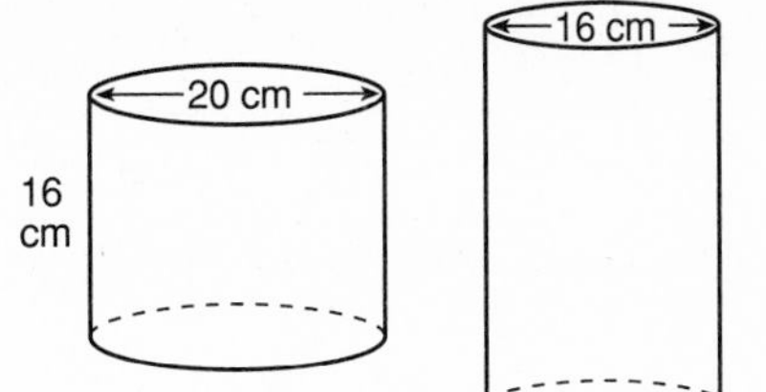

11.

12. The base of the pyramid is a parallelogram with dimensions as shown. Find the volume of the pyramid to the nearest tenth of a cubic inch.

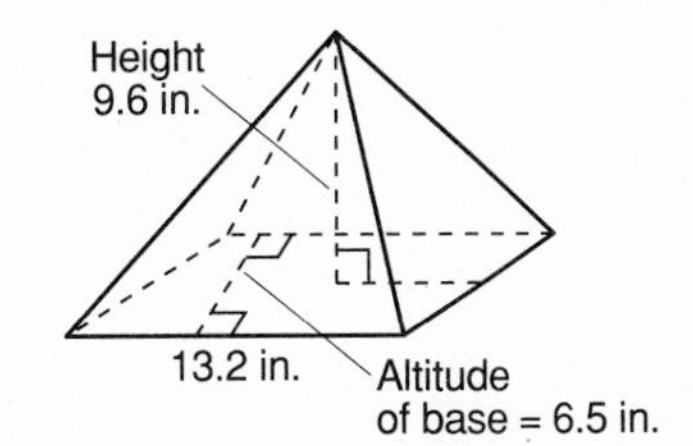

12. ______________

Name ____________________

Date ____________________

13. A by-product of an industrial operation is fine-grained yellow sulfur. One conical pile is 15 yards in diameter and 16 yards high. How many truckloads would be needed to haul it away if each truck bed had dimensions 6 ft by 12 ft by 4 ft?

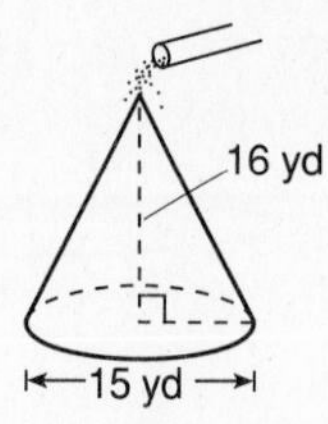

13. ____________________

14. A solid consists of a cylinder attached at one base to an off-center cone as shown. Write a formula for the volume enclosed.

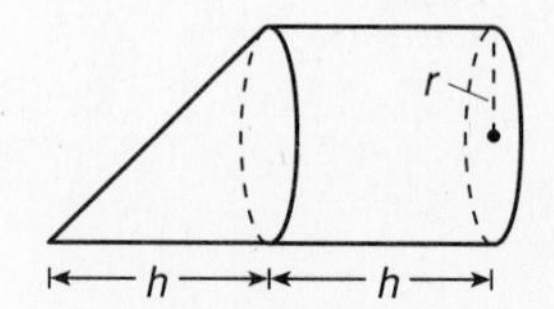

14. ____________________

15. A company has a spherical storage tank which is in need of painting. The radius of the tank is 35.4 ft. The type of paint used will cover approximately 160 ft^2 per gallon. How many gallons of paint will be needed? (Round decimal to the higher whole number of gallons.)

15. ____________________

16. Find the volume, to the nearest cubic foot, of a sphere whose surface area is 100 ft^2.

16. ____________________

17. A sphere fits snugly inside a right cylinder as shown. Find the volume lying outside the sphere but inside the cylinder to the nearest tenth of a cubic inch.

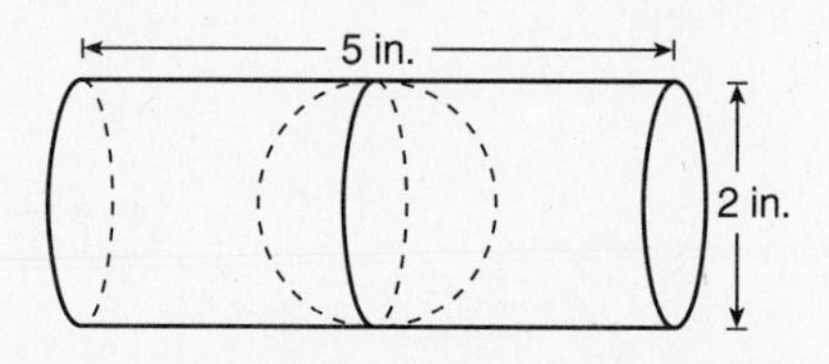

17. ____________________

18. The surface area of a sphere is 200 cm^2. If the radius were three times as large, what would the surface area be?

18. ____________________

19. Assume a person is 6 ft tall and weighs 160 lbs. If the person were 6 ft 6 in. tall but otherwise similar, what would the weight be (to the nearest pound)?

19. ____________________

Name ______________________

Date ______________________

1. Match the name of the transformation with its diagram.

a. b.

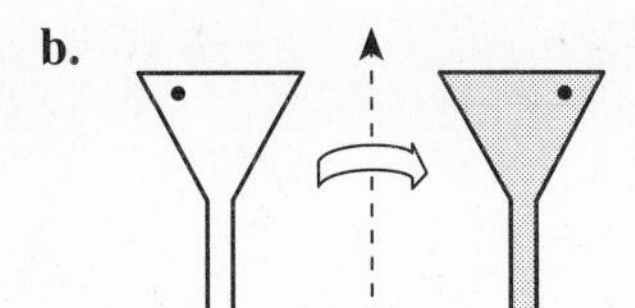

c.

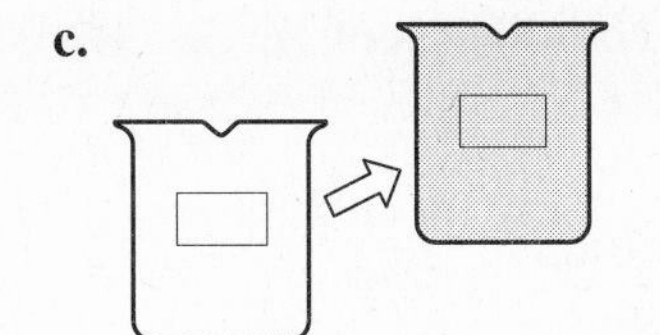

i. translation ii. rotation iii. reflection

1. a. ______
b. ______
c. ______

2. When is a transformation not an isometry?

2.

3. How many lines of symmetry does the figure have?

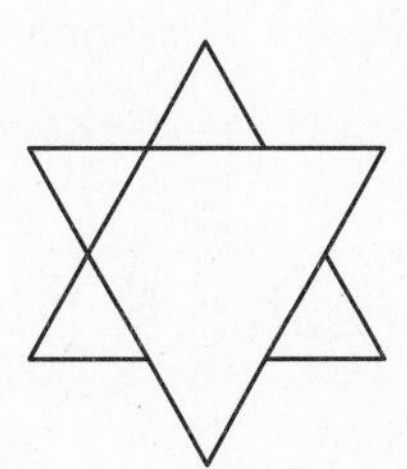

3. ______

4. Lines ℓ and m intersect at point O. A figure is reflected in ℓ followed by a reflection in m. ℓ and m form the sides of a 38° angle. The overall effect is a rotation of how many degrees about O?

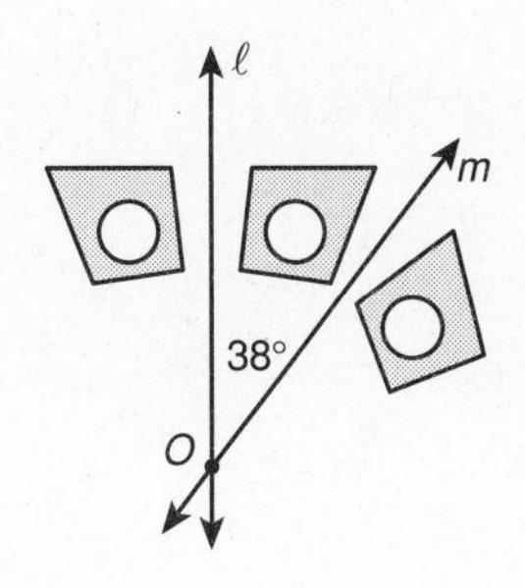

4. ______

5. Draw the reflection of $\overline{PQ}$ in the x-axis.

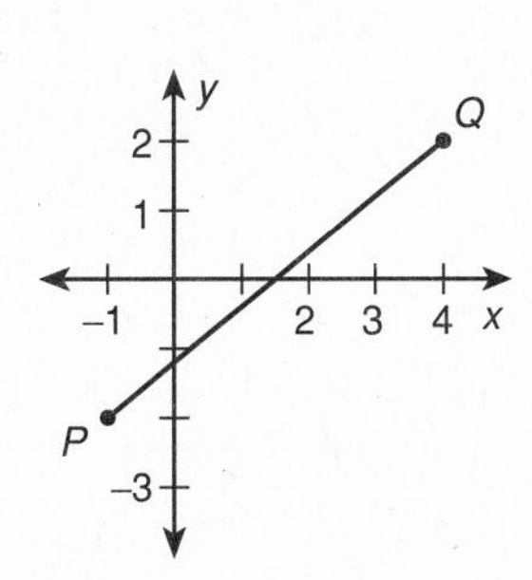

5. *Use graph at left.*

6. Draw, if possible, a hexagon with exactly two lines of symmetry.

6.

Name ____________________

Date ____________________

7. Point A is translated by the vector $\overrightarrow{u} = \langle -3, 5 \rangle$. If the image of A is $A'(7, -1)$, find the coordinates of A.

7. ____________________

8. Classify the frieze pattern.

8. 

9. A quarterback completes at least 55% of the passes he throws. If he throws 37 passes, what is the least number of completions he could have made?

9. ____________________

10. Find the geometric mean of 7 and 49.

10. ____________________

11. If $\frac{a}{b} = \frac{c}{d}$, decide whether it is true or not that $\frac{a-b}{b} = \frac{c-d}{d}$. Explain your reasoning.

11.

12. Simplify the rate. Include the unit of the product.

$$\frac{45 \text{ mi}}{1 \text{ hr}} \cdot \frac{5280 \text{ ft}}{1 \text{ mi}} \cdot \frac{1 \text{ hr}}{60 \text{ min}} \cdot \frac{1 \text{ min}}{60 \text{ sec}}$$

12. ____________________

13. Calculate the slope of the line. Does it matter which points are used? Explain.

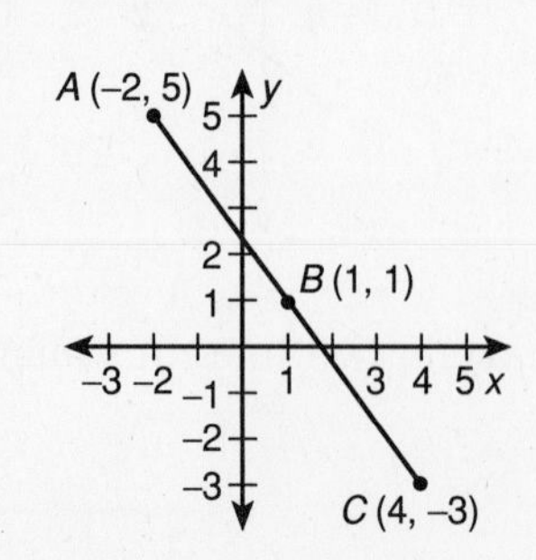

13.

14. If two hexagons are similar, what is true of their corresponding angles and corresponding sides?

14.

15. What value of x will make the two triangles similar.

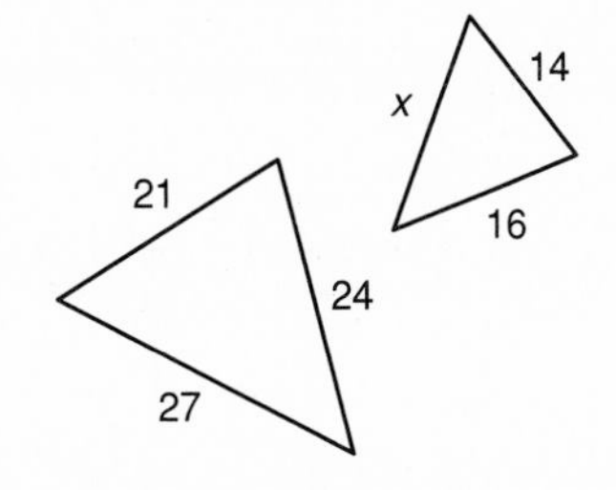

15. ______

16. Find the value of x to the nearest hundredth.

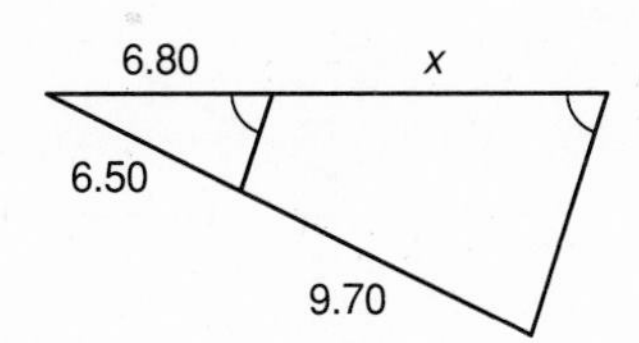

16. ______

17. Find the scale factor for the dilation shown. (Round the result to two decimal places.)

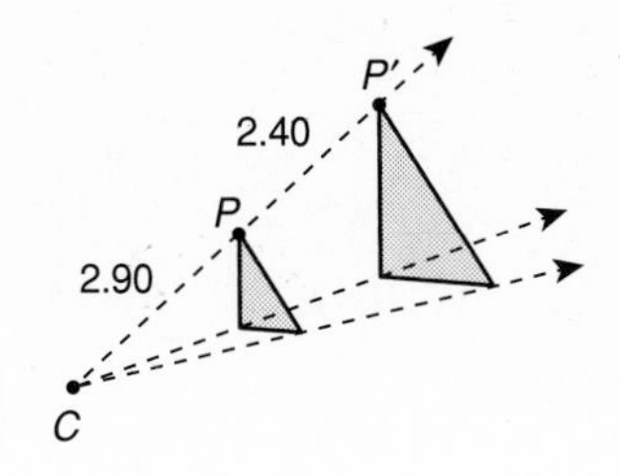

17. ______

18. **a.** BD is the geometric mean of which segment lengths?

b. If $AD = 4$ and $DC = 9$, what is the length of $\overline{BC}$?

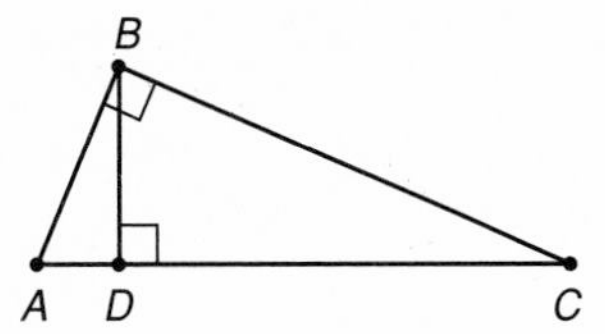

18. a. ______

b. ______

19. Choose the sets that could be the side lengths of a right triangle.

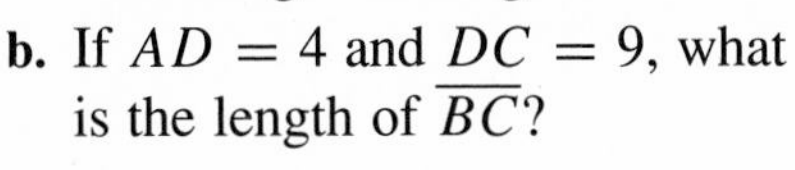

a. 1, 2, 5 **b.** 3, 4, 5 **c.** $1, \sqrt{2}, \sqrt{3}$ **d.** $5, 12, \sqrt{13}$

19. ______

Name ______________________

Date ______________________

20. A triangle has side lengths of 6, 7, and 9. Decide whether it is an acute, right, or obtuse triangle. Explain.

20.

21. If $PQRS$ is a rectangle, what value must QS have?

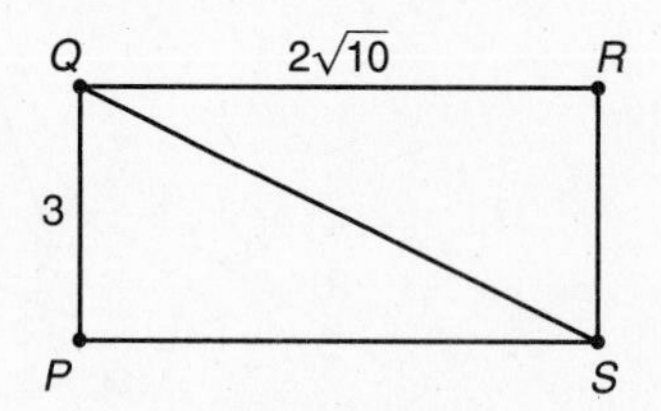

21. ______________

22. Write the trigonometric ratios for

a. $\tan A$. **b.** $\cos B$. **c.** $\sin A$.

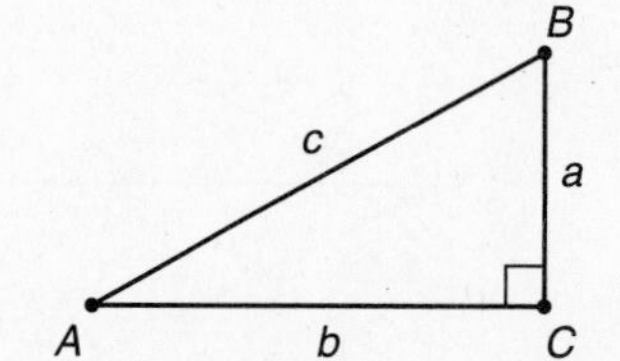

22. a. ______________

b. ______________

c. ______________

23. Use a calculator to determine which is larger: $\sin 10°$ or $\cos 10°$. Explain why this is so.

23. ______________

24. Assume $\angle B$ is an acute angle. If $\tan B = 2.719$, find the measure of $\angle B$ to two decimal places.

24. ______________

25. What is the length of the altitude of an equilateral triangle whose side length is 8?

25. ______________

26. From a point 50.0 feet from the base of a flagpole, the angle from ground level to the top of the pole is 51.3°. Find the height of the flagpole to the nearest tenth of a foot.

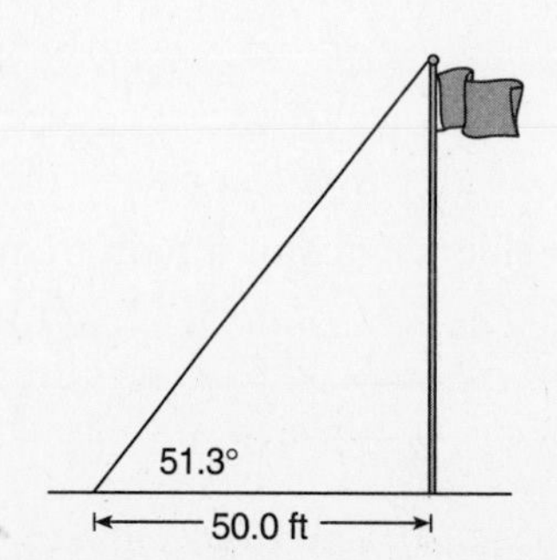

26. ______________

27. Draw a common internal tangent to $\odot A$ and $\odot B$.

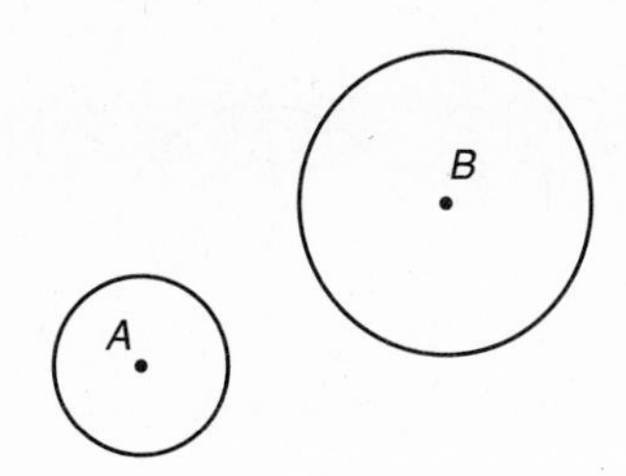

27. *Use figure at left.*

28. *Given*: $\overline{RQ}$ is tangent to $\odot P$ at Q; $PQ = 7$, and $QR = 18$.
Find PR.

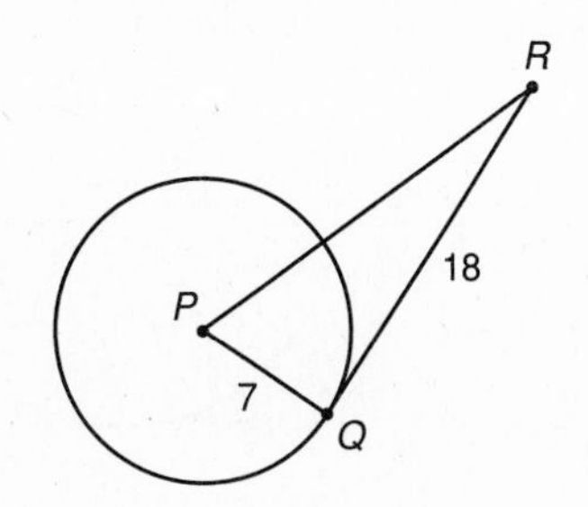

28. ____________

29. Find $m\widehat{ADB}$ in $\odot Q$, as shown.

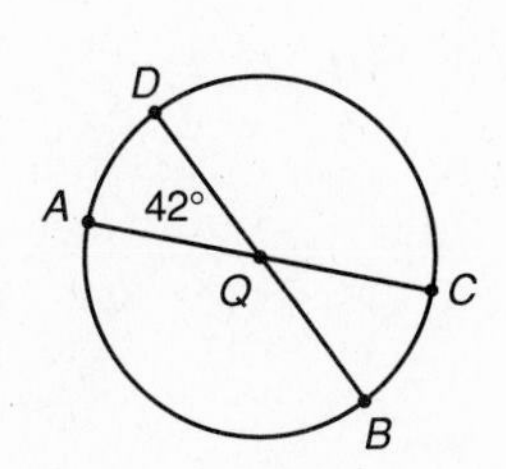

29. ____________

30. Find ST in $\odot C$, as shown.
Explain your reasoning.

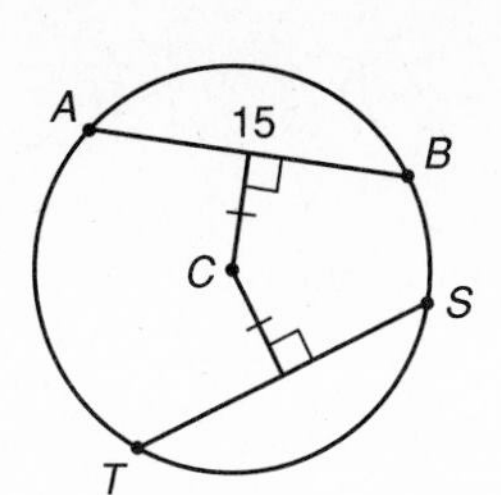

30.

31. Find $m\widehat{AD}$ and $m\angle ACD$.

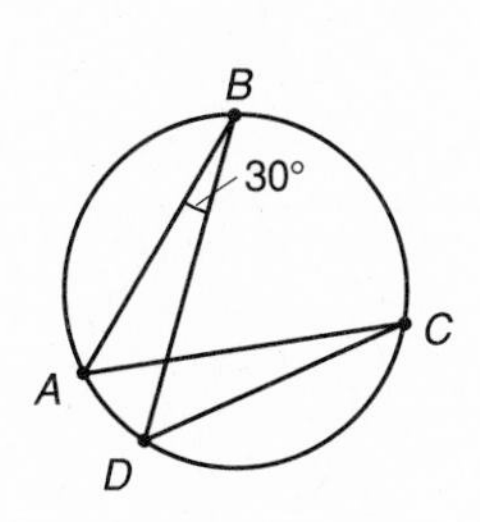

31. ____________

32. Write the standard equation of the circle whose center is $(4, -2)$ and whose radius is $3\sqrt{2}$.

32. ____________

Name ____________________

Date ____________________

33. *Given*: The line is tangent to the circle at P.

If $m\widehat{POQ} = 250°$, find the measure of $\angle 1$.

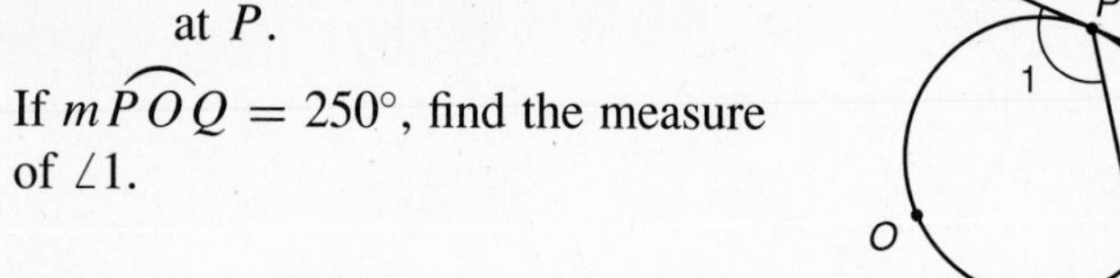

33. ____________________

34. Find the measure of $\angle 1$.

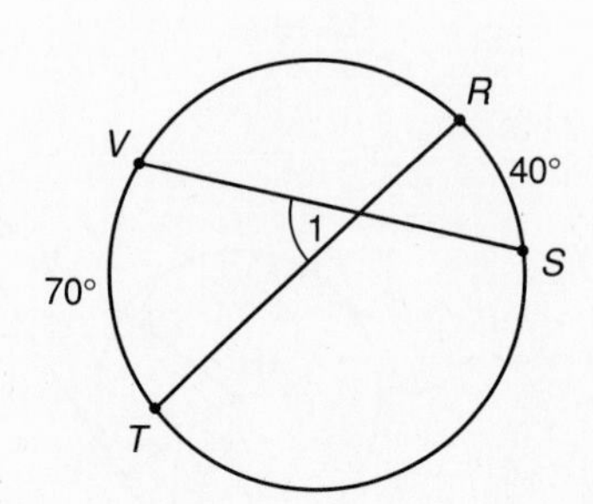

34. ____________________

35. Find the perimeter and the area of the polygon.

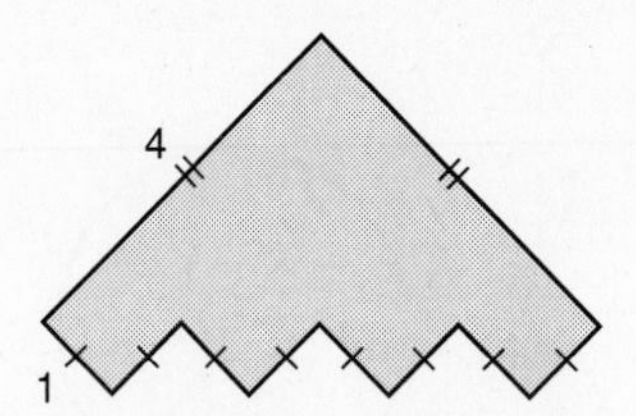

35. ____________________

36. Draw an example to show that two rectangles with the same perimeter need not have the same area.

36.

37. Find the area of $\triangle ABC$.

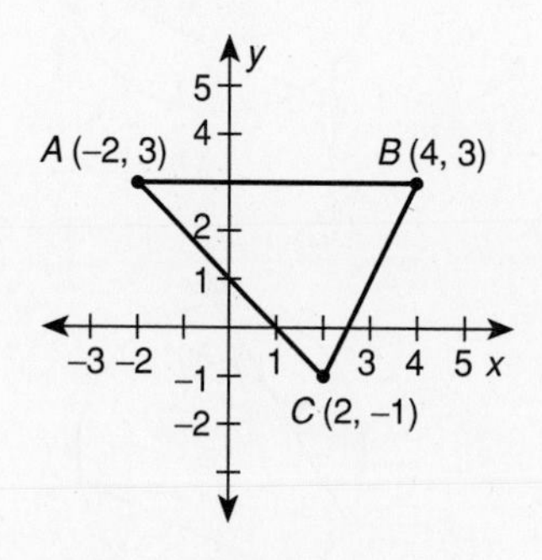

37. ____________________

38. $\odot P$ has a radius of $2\sqrt{2}$ units. Find the area of a square inscribed in the circle.

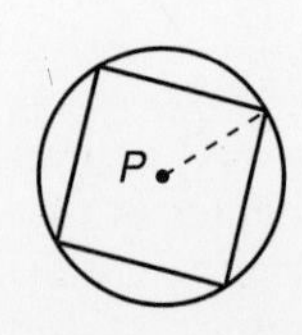

38. ____________________

Name ________________

Date ________________

39. The diagonals of quadrilateral $PQRS$ are perpendicular. Find the area of the quadrilateral.

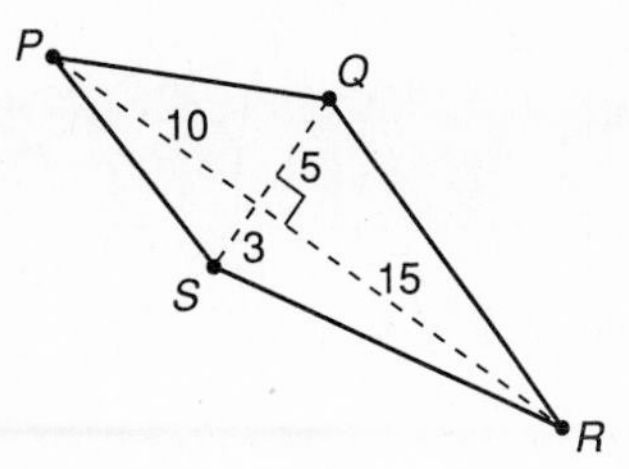

39. ________________

40. The tires of an automobile have a diameter of 22 in. If the wheels revolve 15 times, how far does the automobile move? (Round the result to the nearest tenth of a foot.)

40. ________________

41. $\triangle ABC$ and $\triangle A'B'C'$ are similar triangles with $\dfrac{BC}{B'C'} = \dfrac{2}{3}$. If the area of $\triangle A'B'C'$ is 54 square units, find the area of $\triangle ABC$.

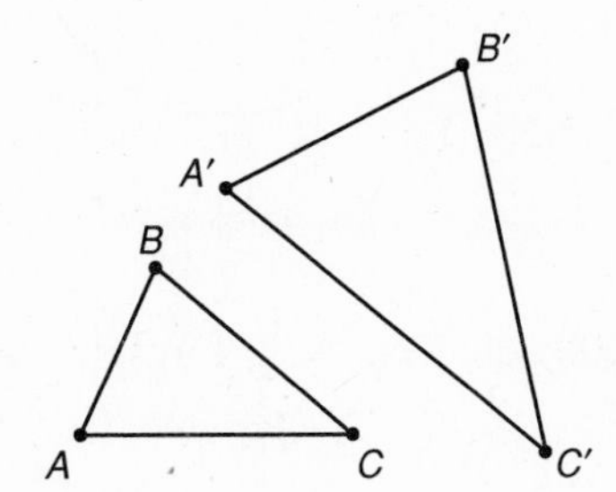

41. ________________

42. Draw an example of a polyhedron which is not convex.

42.

43. Draw a polyhedron with 8 faces. How many vertices does it have? How many edges?

43.

44. Find the surface area and volume of this prism.

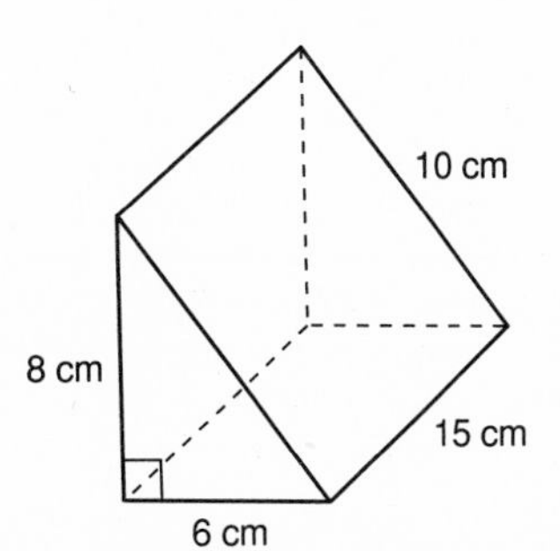

44. ________________

Name ______________________

Date ______________________

45. Find the lateral area of a right cone whose base radius is 10 inches and whose slant height is 25 inches.

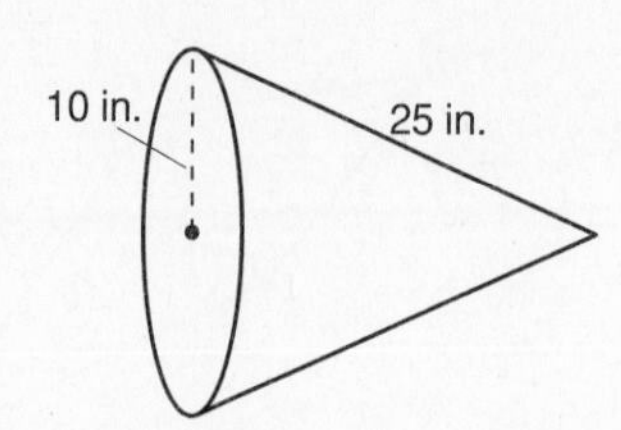

45. ______________

46. This superball has a radius of 2.3 cm. Find the volume of rubber that is needed to make one. (Round the result to the nearest tenth of a cubic centimeter.)

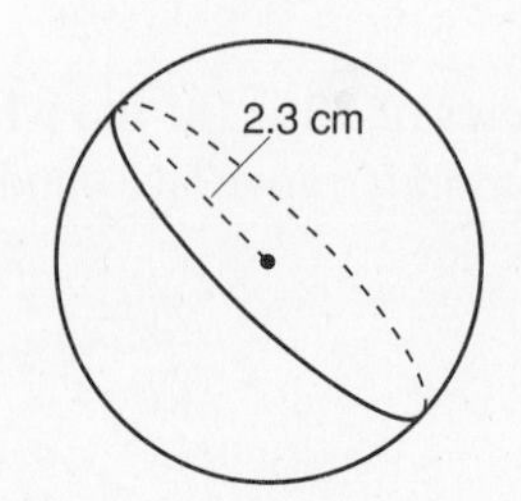

46. ______________

47. The surface area of a sphere is 800 square inches. Find the radius of the sphere to three decimal places.

47. ______________

48. A pyramid has a rectangular base, 16 yd by 9 yd, and an altitude of 10 yd.

a. Find its volume.

b. Is the volume of the shaded portion half the volume of the entire figure? Explain.

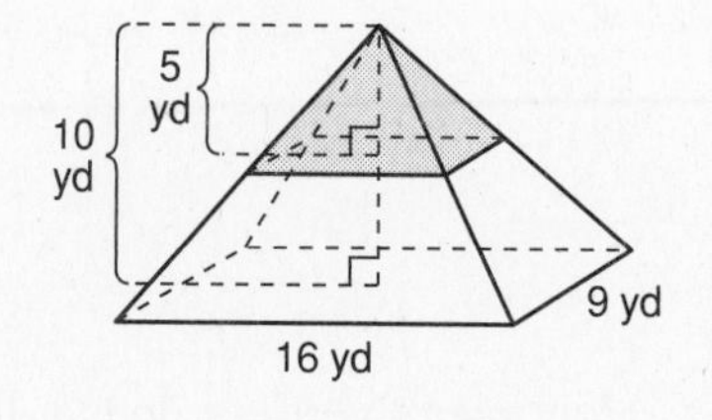

48. a. ______________

b. ______________

49. A cone has a volume of 900 cubic meters. If a area of its base is 300 square meters, then what is its height, *h*.

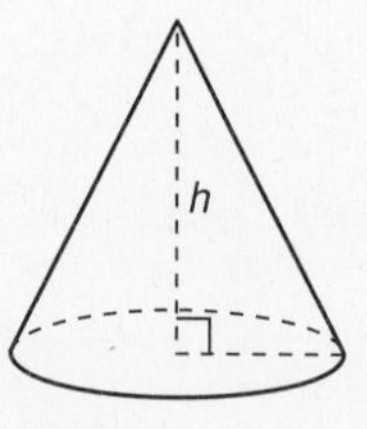

49. ______________

50. The diameter of sphere *B* is twice the diameter of sphere *A*. The surface area of sphere *A* is 452.4 in.^{2} and its volume is 904.8 in.^{3}. What is the surface area and volume of sphere *B*? (Round the result to the nearest whole number.)

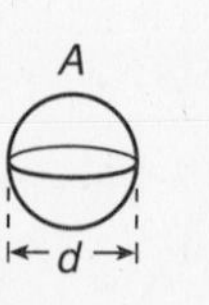

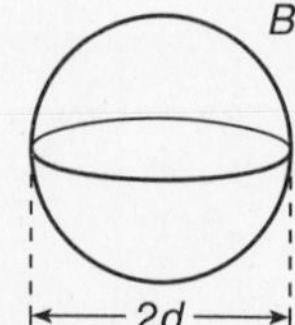

50. ______________

13.2 Short Quiz

Name ______________________

Date ______________________

1. Sketch and describe the locus of all points in a plane that are equidistant from the sides of $\triangle ABC$.

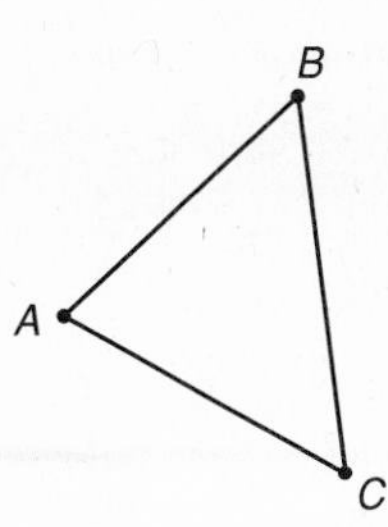

1.

2. The circle at the right has a radius of 2 units. Sketch and describe the locus of all points in a plane that are less than 1 unit from the points on the given circle.

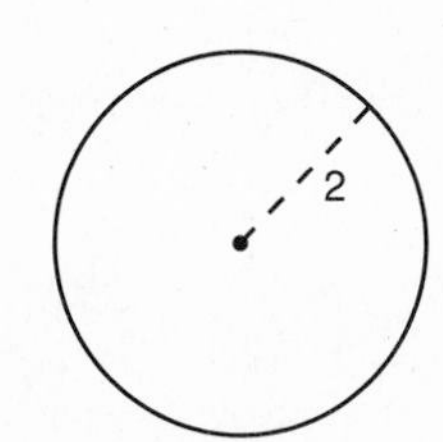

2.

3. Sketch and describe the locus of all points in a plane that are equidistant from vertices A and C of the given square.

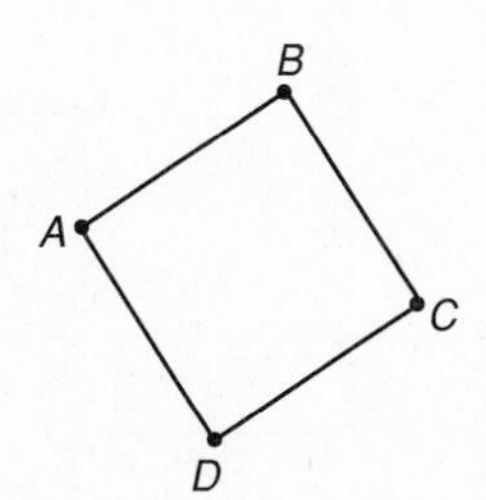

3.

4. Describe the locus of all points in space that are equidistant from two parallel lines.

4.

5. Describe the locus of all points in space that are less than 5 units from a point O.

5.

6. The space figure shown is a cylinder with congruent hemispheres at each end. Give a locus description of the figure with respect to the line segment that joins the centers, A and B, of the cylinder's bases.

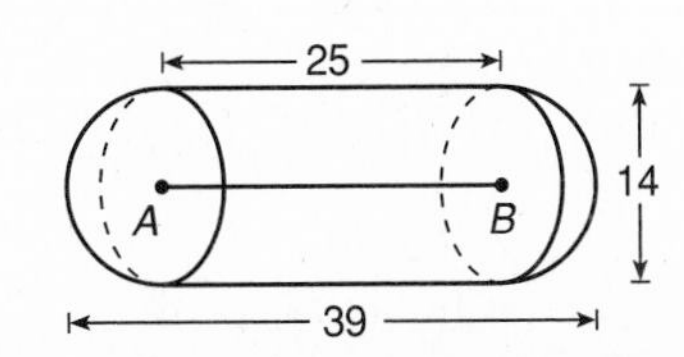

6.

Mid-Chapter 13 Test Form A

(Use after Lesson 13.3)

Name ____________________

Date ____________________

1. What is the locus of all points in a plane equidistant from the points on a circle? **1.**

2. Sketch and describe the locus of all points in a plane that are equidistant from lines ℓ and n. **2.**

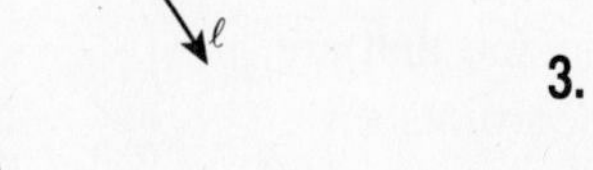

3. Sketch and describe the locus of all points in a plane that are outside the circle and 1 unit from it. **3.**

4. Sketch the locus of all points in a plane that are outside the triangle but no more than $\frac{1}{8}$ inch from it. **4.**

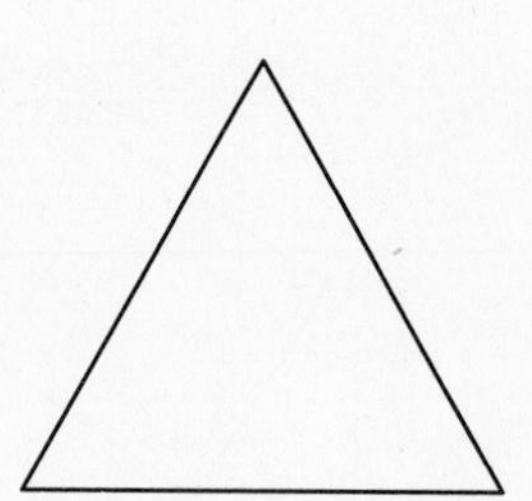

5. Describe the locus of all points in space that are equidistant from two distinct points A and B. **5.**

In Problems 6 and 7, refer to the line segment at the right.

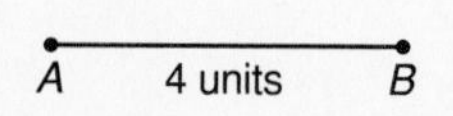

6. Sketch the locus of all points in a plane that are 1 unit from the line segment. **6.**

7. Sketch the locus of all points in space that are 1 unit from the line segment. **7.**

8. Sketch the locus of all points in a coordinate plane that are no more than 3 units from the origin and no more than 2 units from the y-axis. **8.**

Mid-Chapter 13 Test Form B

(Use after Lesson 13.3)

Name ______________________

Date ______________________

1. What is the locus of all points in a plane equidistant from the vertices of an equilateral triangle?

1.

2. Describe the locus of all points in space equidistant from the vertices of a triangle that lies in plane P.

2.

3. Sketch and describe the locus of all points in a plane that are equidistant from the two parallel lines ℓ and n.

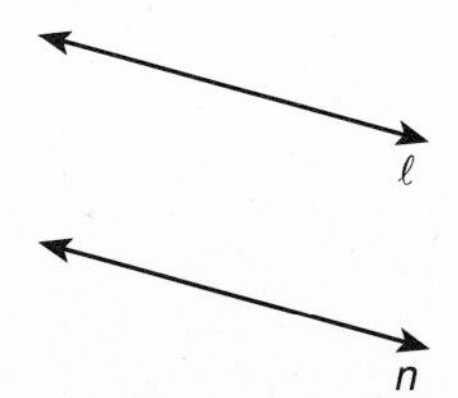

3.

4. Sketch and describe the locus of all points in the plane of a circle that are inside the circle and 1 unit from it.

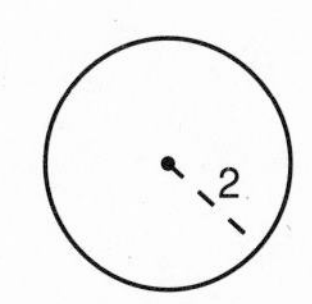

4.

5. Assume the square shown has side lengths of 1 unit. Sketch the locus of all points in a plane that are 1 unit from the square.

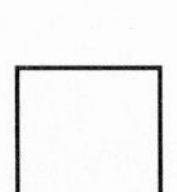

5. *Use figure at left.*

In Problems 6 and 7, refer to the line segment at the right.

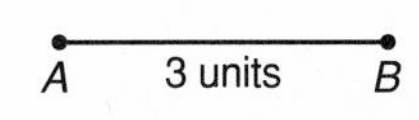

6. Sketch the locus of all points in a plane that are 1 unit from the line segment.

6.

7. Sketch the locus of all points in space that are 1 unit from the line segment.

7.

8. Sketch the locus of points in a coordinate plane that are no more than 4 units from the origin and no more than 2 units from the x-axis.

8.

13.4 Short Quiz

Name ____________________

Date ____________________

1. Sketch all possible types of intersections of two circles. How many points of intersection can there be? (Assume that the circles are not coincident.)

1.

2. Sketch the locus of all points in a coordinate plane that are 4 units from the origin and 3 units from the x-axis.

2.

3. Describe the locus of all points in a coordinate plane that are equidistant from the x- and y-axes and $3\sqrt{2}$ units from the origin.

3.

4. Find the intersection of the graphs of the system of equations.

$$\begin{cases} 2x + 5y = 3 \\ x - 2y = 6 \end{cases}$$

4. ____________________

5. Find the locus of all points in a coordinate plane that are equidistant from points $(-2, 4)$ and $(3, 1)$.

5. ____________________

6. Sketch the locus of all points that satisfy the system of inequalities.

$$\begin{cases} x + y < 4 \\ 2x + y \geq 4 \\ y \geq 0 \end{cases}$$

6.

13.6 Short Quiz

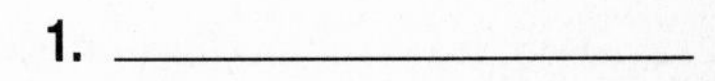

Name ______________________

Date ______________________

1. Describe the possible cross sections of a plane and a sphere? 1. ____________

2. Describe the cross section of the plane and the regular pyramid, as shown. Assume $AB = CD$. 2. ____________

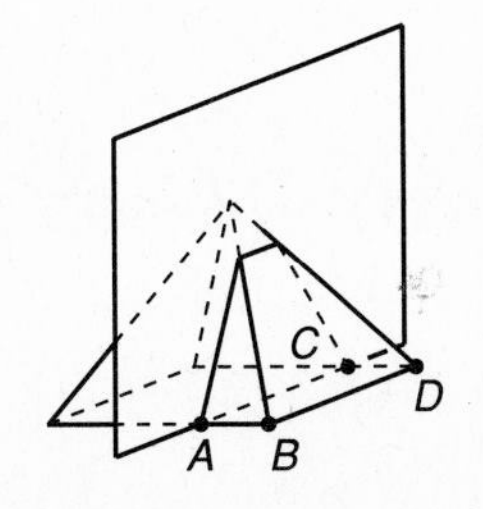

3. The intersection of a plane and a right cylinder does not include any points from either base of the cylinder. Describe the possible cross sections. 3. ____________

4. List three types of measures that are studied in geometry. 4.

5. Describe the difference between a chord and a secant line. 5.

6. Describe the difference between a postulate and a theorem. 6.

Chapter 13 Test Form A

Name ______________________

Date ______________________

1. Sketch and describe the locus of all points in a plane that are equidistant from the vertices of $\triangle ABC$.

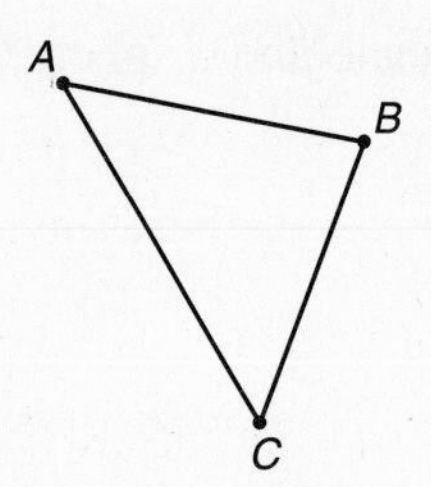

1.

2. Sketch and describe the locus of all points in a plane equidistant from two parallel lines.

2.

3. Sketch the locus of all points in a plane that are outside the given circle but no more than $\frac{1}{2}$ the radius of the circle from it.

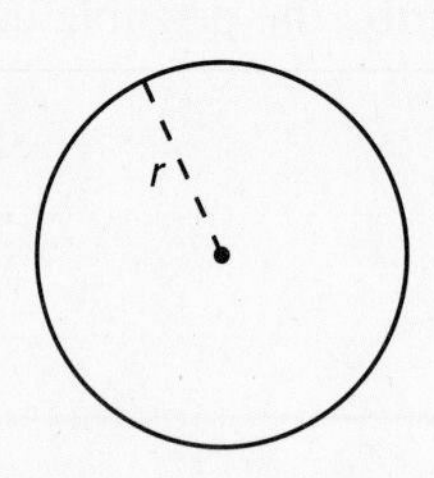

3. *Use figure at left.*

4. Describe the locus of the tip of a pendulum as it swings.

4.

5. Sketch the locus of all points in a plane exactly one-half the side length, $\frac{s}{2}$, from the square and outside the square as shown.

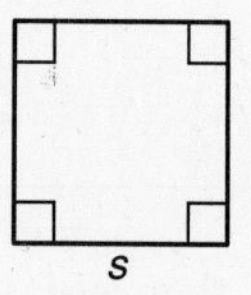

5. *Use figure at left.*

6. Sketch the locus of all points in space that are $\frac{1}{4}$ inch from a line segment AB.

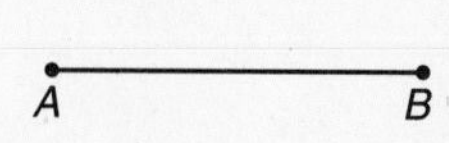

6. *Use graph at left.*

Name ____________________

Date ____________________

7. Describe the locus of all points in space that are equidistant from all of the points on a circle.

7.

8. A torus is a doughnut-shaped solid that can be described as the locus of all points in space that are a given distance from what type of plane curve?

8. ____________________

9. Give the locus description of the figure shown.

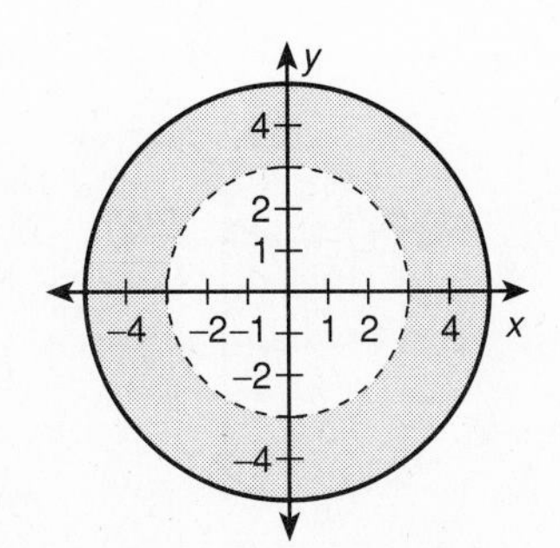

9.

10. Sketch the locus of all points in a coordinate plane with positive coordinates that are no more than 3 units from the y-axis, and whose y-coordinates are less than their x-coordinates.

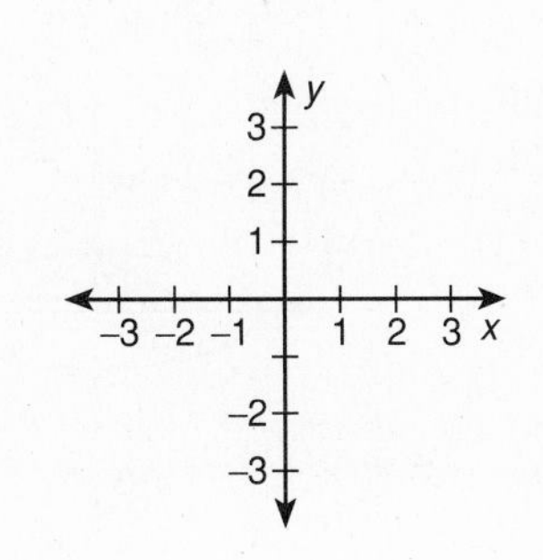

10. *Use graph at left.*

11. Find the locus of all points that lie on the graphs of both $x + 2y = 1$ and $2x - y = 12$.

11. ____________________

12. Sketch the locus of points that satisfy all three inequalities.

$$\begin{cases} 2x - y > -2 \\ x \leq 0 \\ y \geq 0 \end{cases}$$

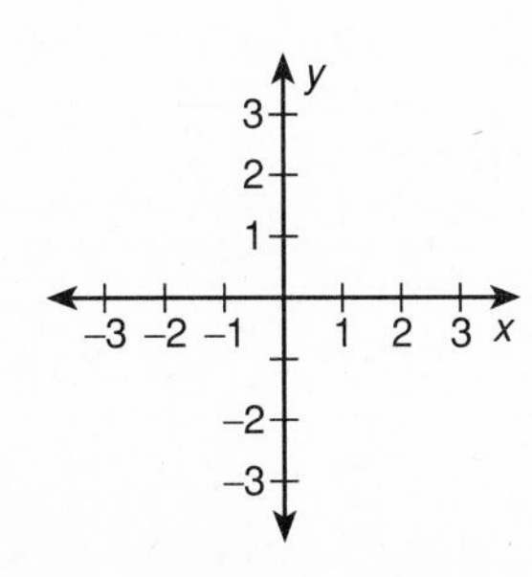

12. *Use graph at left.*

Name ______________________

Date ______________________

13. Find the locus of all points in a coordinate plane that are equidistant from points $(-3, 0)$ and $(0, 3)$.

13. ______________________

14. Describe the cross section obtained when a plane intersects a sphere of radius 4 in., through its center.

14.

15. A cylinder has a base diameter smaller than its height. Describe the cross section obtained when a plane intersects the cylinder perpendicular to its bases. Illustrate your description on the figure.

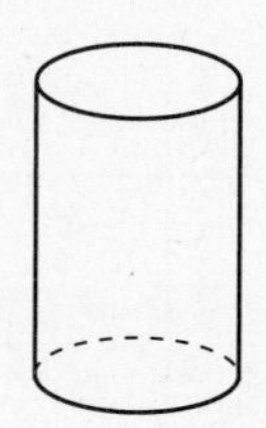

15. ______________________
Use figure at left.

16. Use the definition of an ellipse to determine the constant sum of this ellipse.

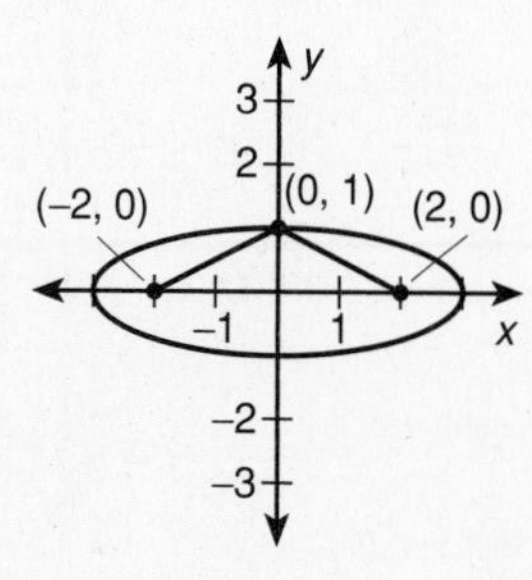

16. ______________________

17. If the diameter of a spherical ball is doubled, by what factor does its surface area increase?

17. ______________________

18. State the Pythagorean Theorem.

18.

Chapter 13 Test

Form B

(Page 1 of 3 pages)

Name ______________________

Date ______________________

1. Choose the word that best describes "The locus of all points in a plane equidistant from two fixed, distinct points."

   a. a point **b.** a line **c.** a ray **d.** a circle

   1. ________

2. The locus of the tip of the minute hand of a clock is [?].

   a. a parabola **b.** an arc **c.** a circle **d.** an ellipse

   2. ________

3. The locus of all points in a plane that are equidistant from the vertices of $\triangle ABC$ is [?].

   a. a point **b.** a circle **c.** a line segment **d.** a triangle.

   3. ________

4. The locus of all points in a plane that are one unit from a circle of radius 3 units is [?].

   a. a circle of radius 4 units
   b. two circles concentric with the given circle
   c. a circle of radius 2
   d. the center of the given circle only

   4. ________

5. The locus of points represented by the shaded area is best described by [?].

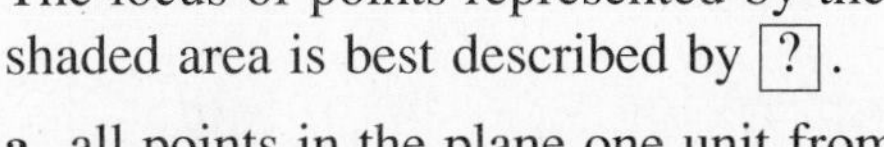

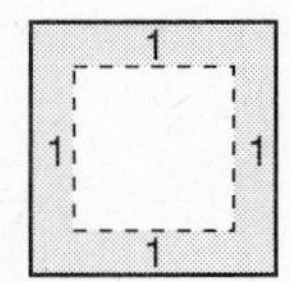

   a. all points in the plane one unit from the square
   b. all points in the square greater than one unit from the square
   c. all points in the plane no more than one unit from the square
   d. all points inside the square less than one unit from the square

   5. ________

6. The locus of all points in space that are equidistant from two distinct fixed points is [?].

   a. a line $\perp$ to the segment joining the points
   b. a circle $\perp$ to the line passing through the two points
   c. a plane through the midpoint of, and $\perp$ to, the segment joining the two points
   d. two line segments parallel to, and on opposite sides of, the segment joining the two points

   6. ________

Name ______________________

Date ______________________

7. A torus is a solid that can be described as the locus of all points in space that are a given distance from a circle. It can also be described as [?].

a. cylindrical
b. a capsule
c. doughnut shaped
d. sausage shaped

7. ______

8. All points in space that are 3 inches from a given point are [?].

a. a sphere of diameter 6 inches
b. a sphere of diameter 3 inches
c. a cube of side length 6 inches
d. a circle of diameter 6 inches

8. ______

9. Which describes the locus in a plane, as shown?

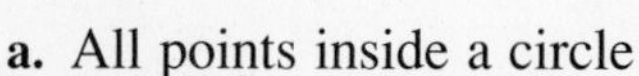

a. All points inside a circle
b. All points 2 units from point (0, 1)
c. All points equidistant from point (1, 0)
d. All points no less than 2 units from point (0, 1)

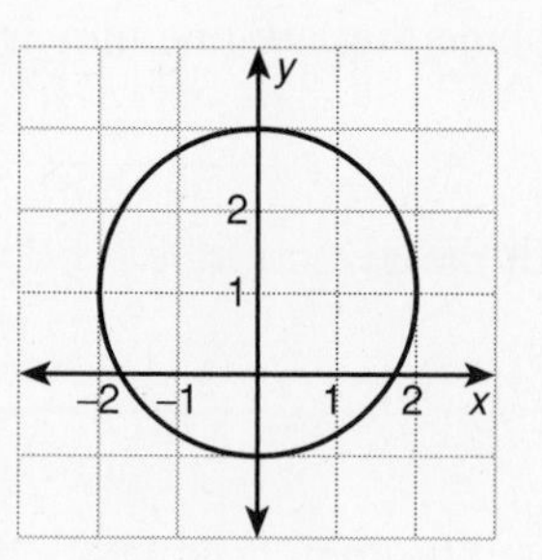

9. ______

10. The shaded area represents the locus in a plane of [?].

a. all points less than 5 units from the origin
b. all points more than 2 units from the origin but less than 5 units from the origin
c. all points more than 2 units from point (0, –2)
d. all points less than 5 units from the origin and at least 2 units from point (0, –2)

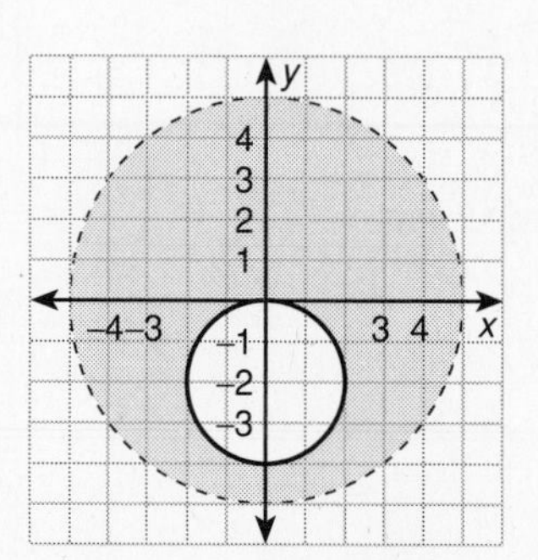

10. ______

11. The surface shown represents the locus of all points in space [?].

a. no more than 2 units from $\overline{AB}$
b. two units from $\overleftrightarrow{AB}$
c. two units from $\overline{AB}$
d. at least two units from $\overline{AB}$

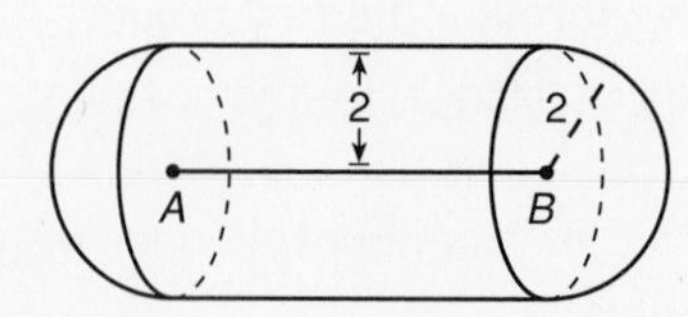

11. ______

Name ____________

Date ____________

12. The locus of all points in a plane that lie on the graphs of both $x - 2y = 5$ and $2x + 4y = 2$ is $\boxed{?}$. 12. ______

a. all points on either line
b. the point with coordinates $(-3, 1)$
c. the point with coordinates $(3, -4)$
d. the point with coordinates $(3, -1)$

13. The locus of all points in a coordinate plane that are equidistant from $(-1, 1)$ and $(1, -1)$ is $\boxed{?}$. 13. ______

a. $x^2 + y^2 = 2$ **b.** $x + y = 0$ **c.** $x^2 + y^2 = 1$ **d.** $x - y = 0$

14. Which is not a possible cross section of a plane and a cylinder? 14. ______

a. a circle **b.** a trapezoid **c.** an ellipse **d.** a rectangle

15. A regular pyramid with a square base is intersected by a plane. Which is not a possible cross section? 15. ______

a. an isosceles triangle **b.** a hexagon
c. a square **d.** a kite

16. Only one of the statements is true. Which one? 16. ______

a. A plane can intersect a cylinder in a hexagon.
b. A plane can intersect a sphere in an ellipse.
c. A plane can intersect a cone in a circle.
d. A plane can intersect a pyramid in a circle.

17. Geometry as an axiomatic system does not formally use $\boxed{?}$. 17. ______

a. postulates **b.** hunches **c.** theorems **d.** undefined terms

18. Which of the following does not represent a triangle congruence postulate or theorem? 18. ______

a. AAA **b.** SAS **c.** ASA **d.** SSS

Chapter 13 Test Form C

(Page 1 of 3 pages)

Name ______________________

Date ______________________

1. Sketch and describe the locus of all points in a plane that are equidistant from the points on a circle.

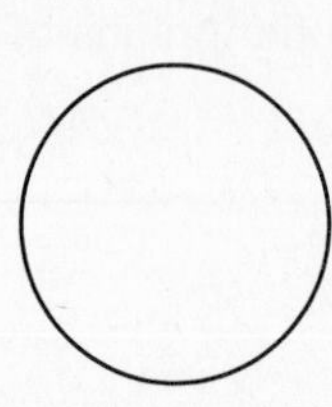

1. ______________

2. Describe the locus of all points in a plane interior to an angle and equidistant from the sides of the angle. Sketch the locus.

2.

3. A walk one yard wide is to surround a flower garden. (Assume a scale of $\frac{1}{8}$ inch = 1 yard.) Sketch the locus of all points in the plane representing the walk area on the scale drawing if every point on the outer border of the walk should be one yard from an edge of the triangular garden.

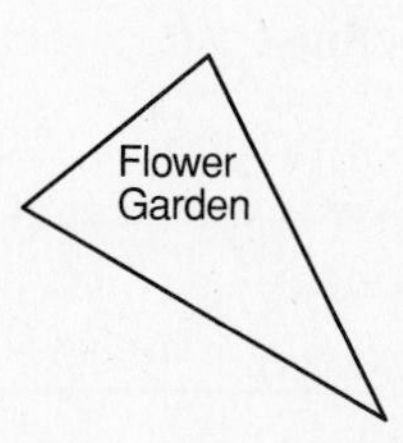

3. *Use figure at left.*

4. A javelin is thrown. Describe and sketch its locus.

4.

5. Sketch the locus of all points in a plane that are exactly a distance r from a semicircle of radius r.

5. *Use figure at left.*

6. Describe and sketch the locus in space of all points that are equidistant from the vertices of $\triangle ABC$.

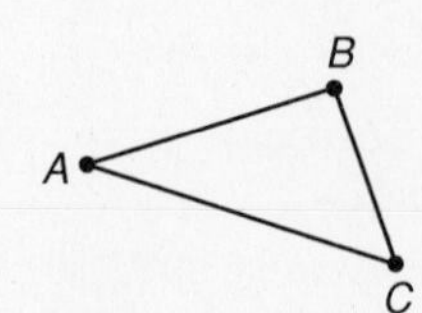

6. ______________

Name ______________________

Date ______________________

7. Describe the locus of all points in a plane that are equidistant from two parallel lines of the plane.

7.

8. A torus is a solid figure shaped like an inflated inner tube. It is the locus of all points in space at a given distance from a circle. If the torus is intersected by a plane perpendicular to the defining circle and through it center, then what is its cross section?

8.

9. What is meant by a compound locus?

9.

10. Sketch the compound locus in the coordinate plane of all points three units from the origin and at least one unit from the x-axis.

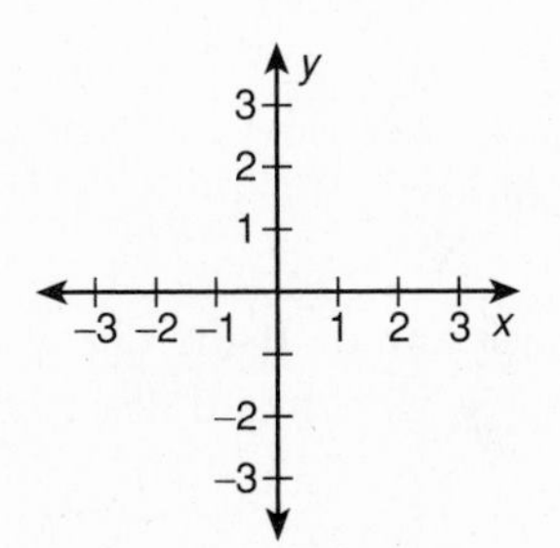

10. *Use graph at left.*

11. Find the locus of all points that lie on the graphs of both $2x + y = -1$ and $4x + 3y = 5$.

11. ______________________

12. Sketch the locus of all points that satisfy the system of inequalities.

$$\begin{cases} 2x - y > -2 \\ x + y \leq 2 \\ y + 1 > 0 \end{cases}$$

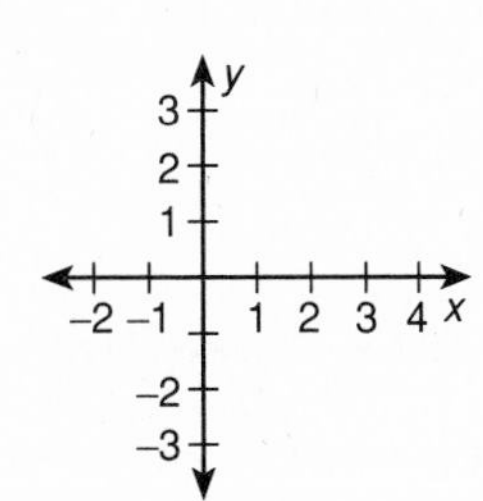

12. *Use graph at left.*

Chapter 13 Test Form C

(Page 3 of 3 pages)

Name ______________________

Date ______________________

13. Find the locus of all points in the coordinate plane that are equidistant from points $(-3, 1)$ and $(3, 5)$.

13. ______________

14. A cylindrical tank about half full of oil is tilted during an earthquake. Identify the cross section formed by the surface of the liquid as indicated in the sketch.

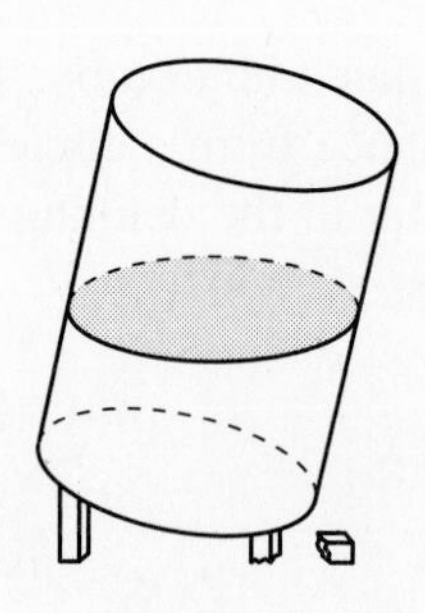

14. ______________

15. A cube has six faces. Describe the possible cross sections of a plane and a cube.

15.

16. A plane contains the tip of a convex pyramid. If the plane intersects the base of the pyramid perpendicularly, sketch and describe its cross section.

16.

17. If the diameter of a spherical ball is doubled, by how much does its volume increase?

17. ______________

18. State the converse of the Pythagorean Theorem.

18.

1.2 Short Quiz

1. a., c., f. **2.** c., d., h., j.
3. a., c., e., f., g., i.
4. (−2, 4), (4, 0)
5. b., e., h. **6.** a. and d.

1.4 Short Quiz

1. Many solutions. **2.** Similar figures
3. A pentagon has 5 sides; a quadrilateral has 4.
4. a. and c.
5. Other solutions possible.

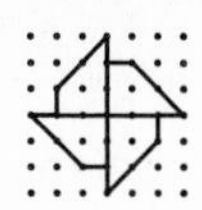

6. Rotational **7.** $\left(-\frac{1}{2}, \frac{1}{2}\right)$

Mid-Chapter Test 1-A

1.

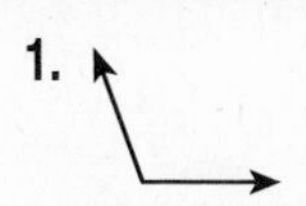

The measure of an obtuse angle is between 90° and 180°.

The measure of an acute angle is less than 90°.

2. Answers vary.

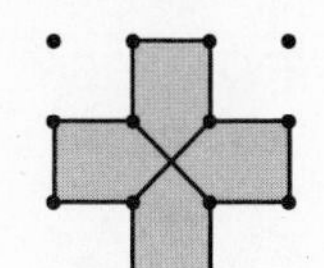

3.

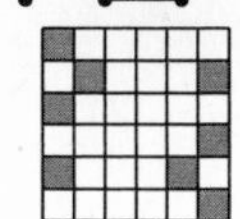

4.

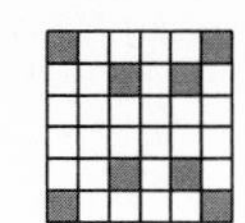

5. $\left(\frac{5}{2}, 2\right)$
6. a. and d. are congruent.
   b. and c. are similar.
   a. and d. are also similar.
7. a. and b.

Mid-Chapter Test 1-B

1.

A right angle is 90°.

A straight angle is 180°. Answers vary. For example: sides are collinear.

2. Answers vary.

3. **4.**

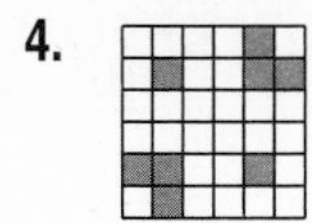

5. $\left(\frac{13}{2}, -2\right)$
6. a. and c. are congruent.
   b., a., and c. are similar.
7. a. and c.

1.6 Short Quiz

1. $-\frac{2}{3}$ **2.** −3 **3.** b.
4. Perpendicular; slopes are negative reciprocals.
5.

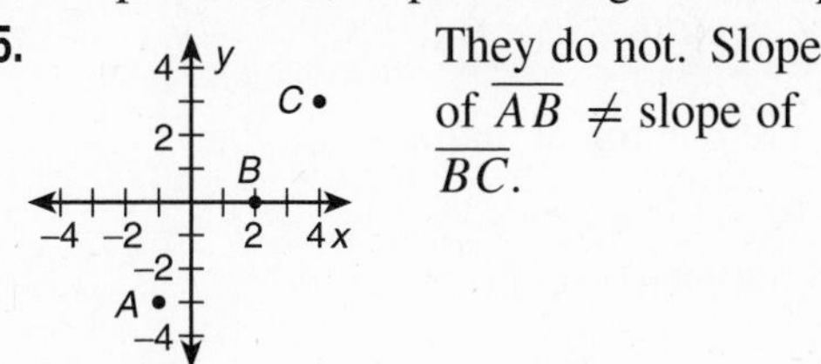

They do not. Slope of $\overline{AB} \neq$ slope of $\overline{BC}$.

6. Perimeter: $10 + 3\sqrt{2}$; area: $9\frac{1}{2}$

Chapter Test 1-A

1. b. and d. **2.** a. and c.
3. a. and b. **4.** d.
5. 5; yes, all are similar; yes, the 4 small triangles are congruent.
6. They are similar. **7.** 24 ft by 14 ft
8. Answers vary.

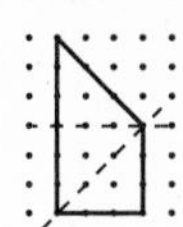

9. d. and e. **10.** Rotational symmetry
11. (−1, 1) **12.** Yes, 5 **13.** Rotational
14. Symmetry about a vertical line **15.** $-\frac{1}{2}$
16. They are parallel. (Alternate answers possible.) Slopes are equal.
17. 7
18. Check students' constructions.

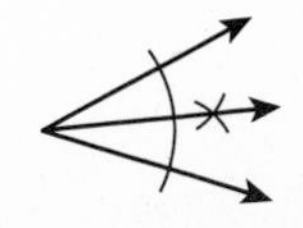

Chapter Test 1-B

1. b. **2.** c. **3.** a. **4.** d.
5. c. **6.** c. **7.** d. **8.** a.
9. b. **10.** c. **11.** a. **12.** d.
13. a. **14.** d. **15.** b. **16.** c.
17. d.

Chapter Test 1-C

1. b. and d.

2. All sides must be equal in length.

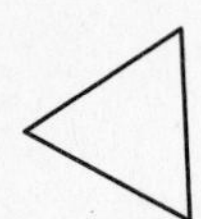

3. Point E only

4. 180°

Both sides lie along the same line in opposite directions.

5. 4

6. They are congruent and also similar.

7. 28 ft by 36 ft

8. Answers vary.

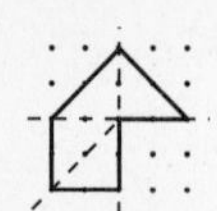

9. a. and c.

10. Rotational symmetry; eight lines of symmetry.

11. $D(4, 4)$ **12.** One line of symmetry

13. 7

14. 8 lines of symmetry; rotational symmetry.

15. Answers vary. **16.** $-\frac{1}{6}$

17. They are perpendicular. (Alternate answers possible.) Slopes are negative reciprocals.

18. 5

19.

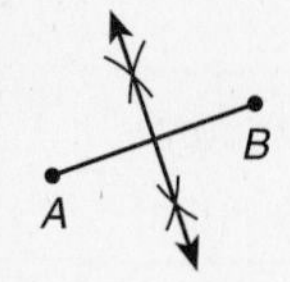

2.2 Short Quiz

1. The line through points T and R.

2. M is between L and N.

3. **a.** $\angle AOC$, $\angle AOB$, $\angle BOC$
b. $\angle AOC$

4. $LM = 8$

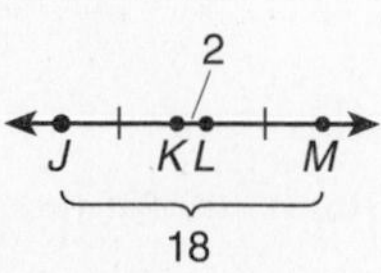

5. $m\angle BOD = 56°$

6. Answers vary.

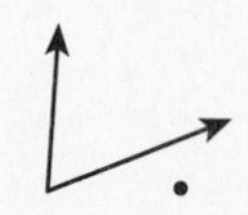

7. 9, 16, 25

Mid-Chapter Test 2-A

1. R is between S and T.

2. $m\angle VOT = 31°$

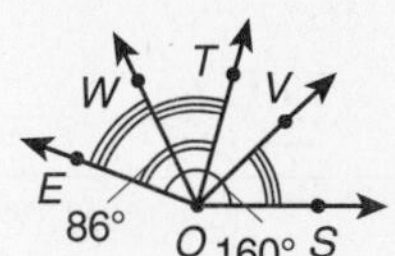

3. $5\sqrt{2}$ **4.** $\overline{RM}$

5. $\angle SQM$ **6.** Q

Mid-Chapter Test 2-B

1. T is between R and S.

2. $m\angle TOV = 36°$

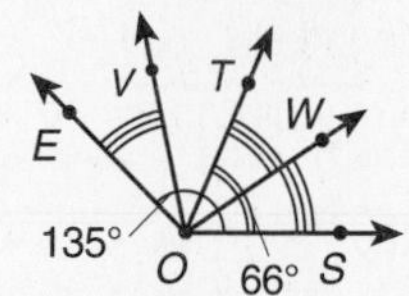

3. $5\sqrt{2}$ **4.** $\overline{SR}$

5. $\angle MTS$ **6.** T

2.4 Short Quiz

1.

2. Check that student's triangle is congruent to $\triangle ABC$.

3. $\sqrt{65}$ **4.** Two right angles

5. If a line exists, then at least two distinct points exist. True, by Postulate 6.

6. **a.** Yes
b. If x^2 is equal to 9, then x is equal to 3.
c. False. x could be -3.

2.6 Short Quiz

1. e. 2. c. 3. d.
4. a. 5. b.
6. 3. Substitution Property of Equality
   6. Division Property of Equality
7. Sketches vary.
   a. 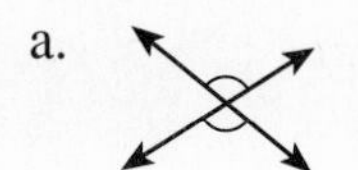 b. 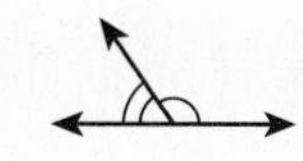
8. 75°, 60°, 43°, 25°
9. 145°, 108°, 57°, 11°

Chapter Test 2-A

1. $\overrightarrow{QP}$
2. $\angle AOY$ and $\angle YOB$ are adjacent angles.

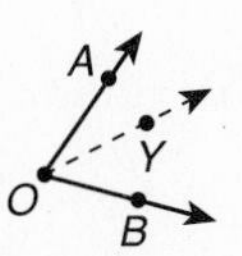

3. 7, 9, 11
4. Sketches vary.

C D A B

5. 12

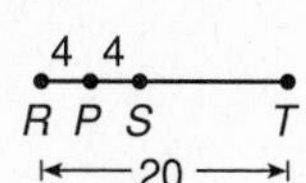

6. 128°

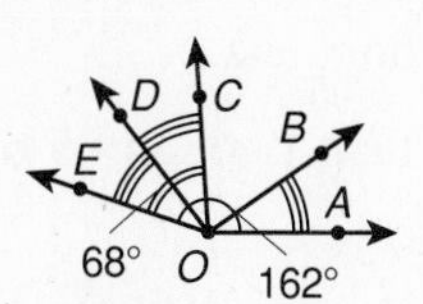

7. $\sqrt{170}$
8. $m\angle AOB = 40°$; bisector $\overrightarrow{OP}$

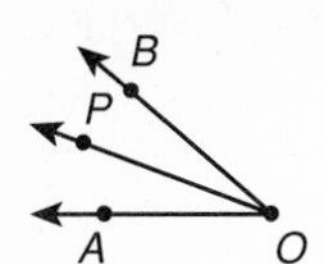

9. $\overline{CD}$
10. 38

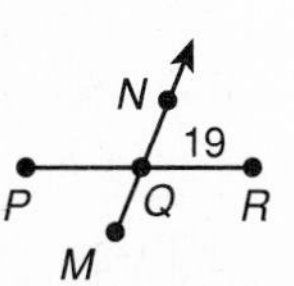

11. If two acute angles are obtained in bisecting an angle, the angle is obtuse. The converse is false. It could be acute.
12. The conclusion
13. 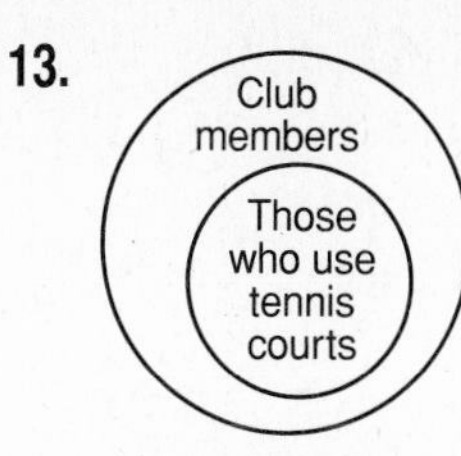

    Answers may vary, for example: If I use the tennis courts, then I am a club member.
14. Three points. They are not collinear.
15. $2(PQ) + 3(5) = 27$; $PQ = 6$
16. Answers vary. A single counterexample: Show one woggle that is not a boggle.
17. Transitive only 18. 107°
19. Two angles are supplementary if the sum of their measures is 180°.
20. 3. Linear Pair Postulate: If two angles form a linear pair, then the sum of their measures is 180°.
    5. Subtraction Property of Equality

Chapter Test 2-B

1. d.	2. b.	3. d.	4. b.
5. c.	6. b.	7. c.	8. d.
9. c.	10. a.	11. b.	12. c.
13. a.	14. d.	15. b.	16. c.
17. b.	18. a.	19. d.	20. a.

Chapter Test 2-C

1. A ray from R through S

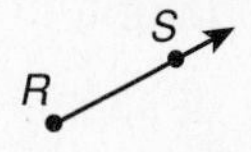

2. All of the points on $\overline{PQ}$
3. Answers may vary.

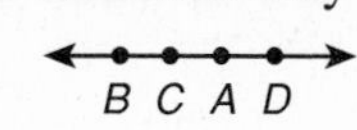

4. $m\angle TOP = 98°$
   Answers may vary.

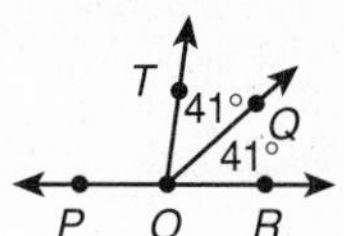

5. 27°, 250 kilometers 6. $2\sqrt{10}$
7.

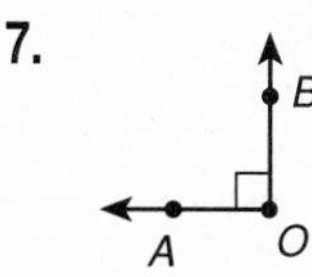

8. $\angle DAB$ 9. 34
10. Statement is true, converse is false. An acute angle bisected produces acute angles, also.
11. The hypothesis

12. 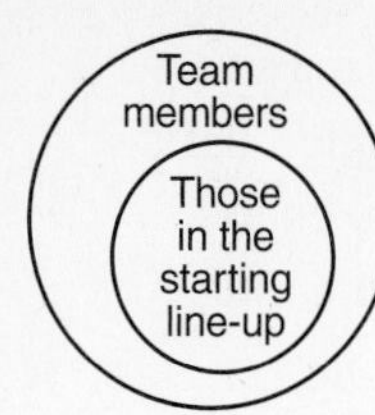

If I'm in the starting lineup, then I am a team member. Answers vary.

13. 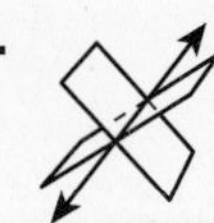

They intersect in a line (Postulate 10).

14. Observe several examples and make a conjecture from them. Examples vary.

15. A single counterexample is sufficient as it must be true in *all* cases.

16. Yes, no, yes **17.** 113°

18. Two angles are complementary if the sum of their measures is 90°.

19.

Statements	*Reasons*
1. $\angle 1$ and $\angle 2$ are vertical angles.	**1.** Given
2. $m\angle 1 = m\angle 2$	**2.** Vertical ∡ Thm.
3. $\angle 1$ and $\angle 3$ are a linear pair.	**3.** Given
4. $m\angle 1 + m\angle 3 = 180°$	**4.** Linear Pair Post.
5. $m\angle 2 + m\angle 3 = 180°$	**5.** Substitution Property
6. $\angle 2$ and $\angle 3$ are supplementary	**6.** Def. of suppl. ∡

3.2 Short Quiz

1. ℓ_1 is parallel to ℓ_2. Transitive Prop. of Parallel Lines

2. Two lines that do not lie in the same plane

3. Sketches vary.

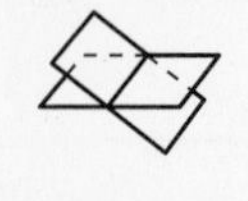

4. 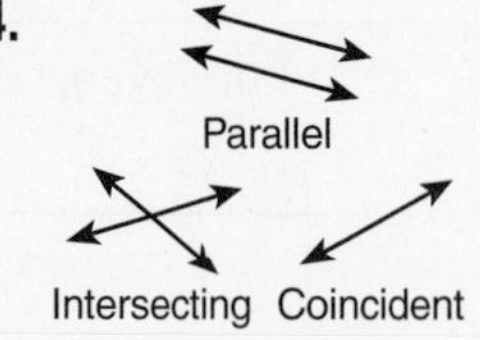

5. (2, −3); check: $2(2) - (-3) = 7$, $3(2) + (-3) = 3$

6. $y = \frac{1}{3}x - 1$

3.4 Short Quiz

1. If the property is not mine, then someone filed a lien.

2. If the property is mine, then no one filed a lien.

3. If A, B, and C are distinct points on a circle, then the slope of $\overline{AB}$ and the slope of $\overline{BC}$ are not equal.

4. *Statements*: d, h, c, e
Reasons: f, a, b, g

Mid-Chapter Test 3-A

1. Parallel lines are coplanar lines that do not intersect.

2. They are parallel.

3. They intersect but not in a right angle.

4. $y = \frac{1}{3}x + 2$

5. If I don't get a B in the course, then I didn't get an 85 or better on the final test.

6. If I get a B in the course, then I got an 85 or better on the final test.

7.

Statements	*Reasons*
1. $\angle 1$ and $\angle 2$ are supplementary.	**1.** Given
2. $\angle 2$ and $\angle 3$ are supplementary.	**2.** Given
3. $m\angle 1 + m\angle 2 = 180°$	**3.** Def. of suppl. ∡
4. $m\angle 2 + m\angle 3 = 180°$	**4.** Def. of suppl. ∡
5. $m\angle 1 = 180° - m\angle 2$	**5.** Subtraction Prop. of =
6. $m\angle 3 = 180° - m\angle 2$	**6.** Subtraction Prop. of =
7. $m\angle 1 = m\angle 3$	**7.** Transitive Prop. of =
8. $\angle 1 \cong \angle 3$	**8.** Def. of congruence

Mid-Chapter Test 3-B

1. Perpendicular lines are lines that intersect in a right angle
2. They are parallel.
3. $\ell_3 \parallel \ell_2$, but they need not all be coplanar.
4. $y = -3x - 8$
5. If I cannot get a scholarship, then my grade point average is less than 3.00.
6. If I can get a scholarship, then my grade point average is at least 3.00.
7. 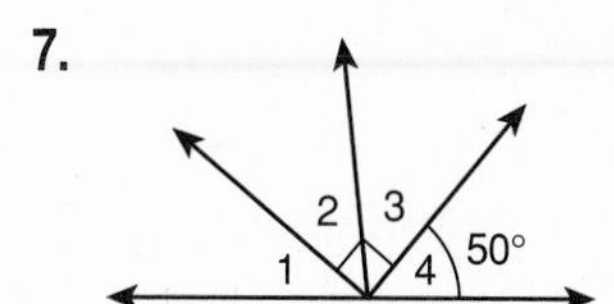

Since $\angle 2$ and $\angle 3$ are complementary by definition the sum of their measures is $90°$. By the definition of a straight angle and the Angle Addition Postulate, $m\angle 1 + m\angle 2 + m\angle 3 + m\angle 4 = 180°$. Using the Substitution Property of Equality, the previously determined $m\angle 2 + m\angle 3 = 90°$, and the given $m\angle 4 = 50°$, we have $m\angle 1 + 90° + 50° = 180°$. The Subtraction Property of Equality then gives $m\angle 1 = 40°$.

3.6 Short Quiz

1. $\angle 1$ and $\angle 8$, or $\angle 2$ and $\angle 7$
2. $\angle 3$ and $\angle 5$, or $\angle 4$ and $\angle 6$
3.  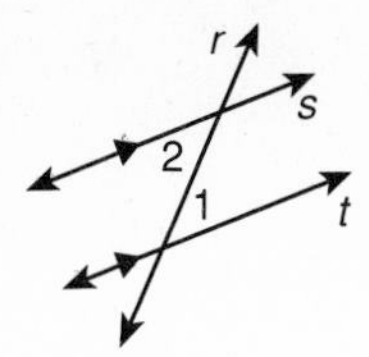

$\angle 1$ and $\angle 2$ are alternate interior angles whose measure is $45°$, as drawn.

4. $m\angle 1 = 50°$; $m\angle 2 = 130°$
Reason: $\angle 1$ and the $50°$ angle are a pair of congruent alternate exterior angles, because $\ell_1 \parallel \ell_2$. $\angle 1$ and $\angle 2$ are supplementary. Answers vary.
5. If two parallel lines are cut by a transversal, then the pairs of consecutive interior angles are supplementary. The converse is true (it is Theorem 3.7).

Chapter Test 3-A

1. 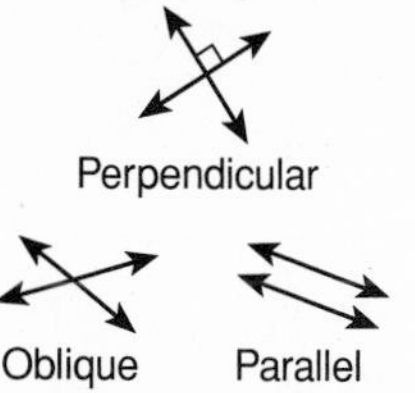

2. $\ell_2 \parallel \ell_3$, but ℓ_1, ℓ_2, and ℓ_3 need not be coplanar.
3. Answers vary. $\overleftrightarrow{AB}$ and $\overleftrightarrow{DH}$ are skew lines; $\overleftrightarrow{EH}$ and $\overleftrightarrow{GC}$ are skew lines.
4. 
5. $(10, 4)$; $2(10) - 3(4) = 8$, $10 - 2(4) = 2$
6. $y = \frac{1}{3}x$
7. Exactly one
8. If we are in the playoffs, then we won the division title.
9. If we are not in the playoffs, then we did not win the division title.
10. If A, B, and C are distinct points on a circle, then they always form a triangle.
11. Conclusion: If the three points A, B, and C are not collinear, then the slope of $\overleftrightarrow{AB}$ is the negative reciprocal of the slope of $\overleftrightarrow{BC}$.
12. 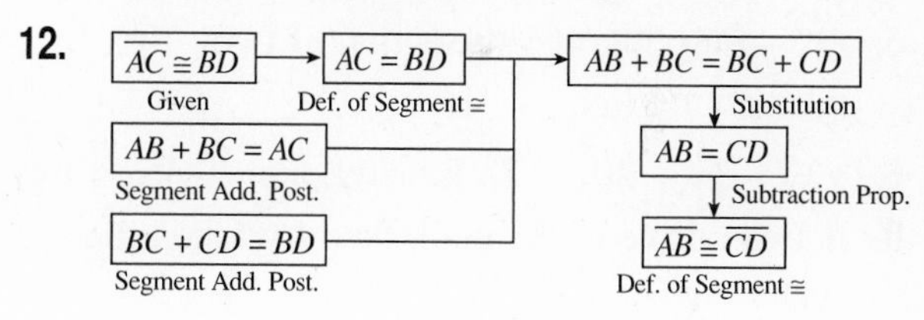

13. Sketches vary.

14. Answers vary.

If $m \parallel p$, then $\angle 1 \cong \angle 2$.

15. $\ell_1 \parallel \ell_2$
16. Answers vary.

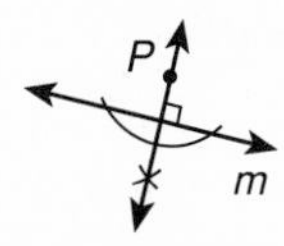

17. $\langle 6, 13\rangle$; length $\sqrt{205}$

18. a. $\langle -3, 12\rangle$ b. 17

Chapter Test 3-B

1. c.	2. a.	3. b.	4. b.
5. d.	6. d.	7. a.	8. c.
9. b.	10. c.	11. b.	12. b.
13. c.	14. c.	15. b.	16. a.
17. d.	18. a.	19. d.	20. c.

Chapter Test 3-C

1. 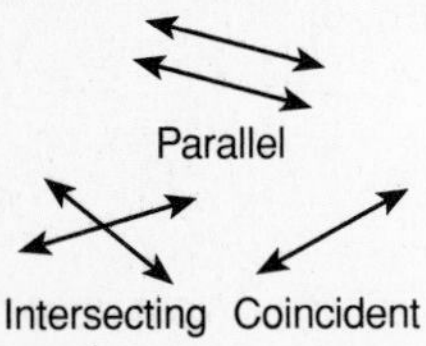

2. $\ell_1 \parallel \ell_3$

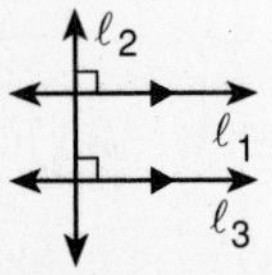

3. Skew lines are lines which do not lie in the same plane.

4. 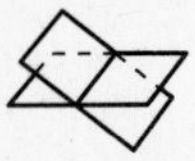

5. (3, −5); point of intersection of two intersecting lines

6. $y = -3x + 20$ 7. Exactly one

8. If I can drive to the mall, then I washed the car.

9. If I cannot drive to the mall, then I did not wash the car.

10. If $\overline{AC}$ is a diameter of a circle, then the slope of $\overleftrightarrow{AB}$ is the negative reciprocal of the slope of $\overleftrightarrow{BC}$

11. We have to sell the house.

12. 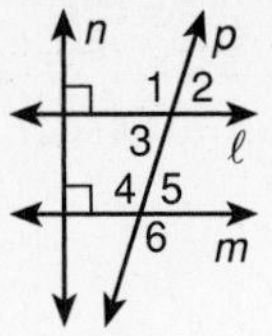

Statements	*Reasons*
1. $\ell \perp n, m \perp n$	1. Given
2. $\ell \parallel m$	2. Prop. of $\perp$ Lines
3. $\angle 1$ and $\angle 3$ are suppl.	3. Linear Pair Post.
4. $m\angle 1 + m\angle 3 = 180°$	4. Def. of suppl.
5. $\angle 1 \cong \angle 6$	5. Alt. Ext. $\angle$s Thm.
6. $m\angle 6 + m\angle 3 = 180°$	6. Substitution
7. $\angle 6$ and $\angle 3$ are suppl.	7. Def. of suppl.

13. Answers vary.

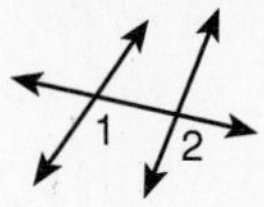

14. *Given*: $\ell \parallel m$
Prove: $\angle 1 \cong \angle 2$
Answers vary.

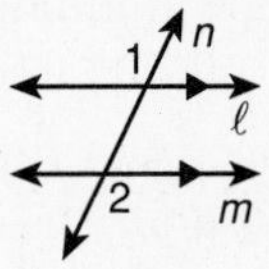

15. $m_1 \parallel m_2$

16. Answers vary.

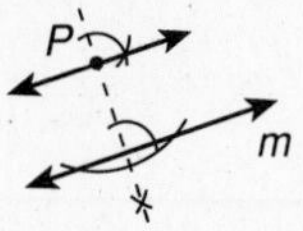

17. $\langle 11, -12\rangle$; length: $\sqrt{265}$

18. a. $\langle 7, -10\rangle$ b. 7

Cumulative Test 1–3

1. a. Rotational symmetry
b. Neither line nor rotational symmetry

2. a. Line and rotational symmetry
b. Line symmetry only

3. Answers vary; no right angles

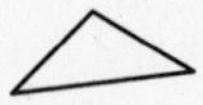

4. 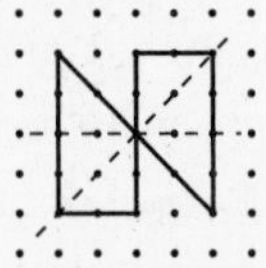

5. Similar figures have the same shape but are not necessarily the same size. Many examples possible.

6. $(2, \frac{5}{2})$ 7. 19 ft by 30 ft

8. a. and c. 9. 10 lines of symmetry

10. Answers vary.

11. They are perpendicular. Slopes are negative reciprocals.

12. ≈ 2 quarts 13. $\frac{3}{8}$

14.

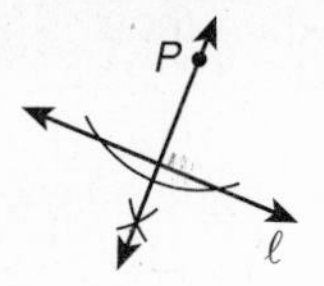

15. $\overleftrightarrow{RS}$ or $\overleftrightarrow{SR}$

16. A C B

17. Answers vary.

B A D C

18. 6, 10

19. $m\angle COD = 45°$

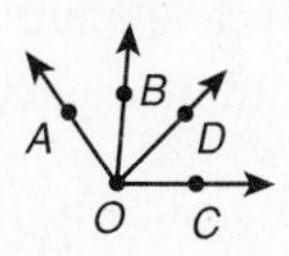

20. $6\sqrt{2}$

21. $m\angle AOB = 70°$; $\overrightarrow{OP}$ is the bisector.

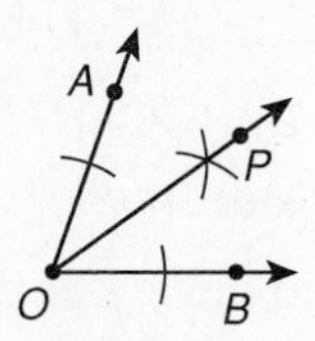

22. $QR = 13$

23. If two acute angles are obtained in bisecting an angle, then the angle was acute. The converse is false. The bisected angle could be obtuse or right.

24. $\overline{DA}$ 25. The conclusion

26.

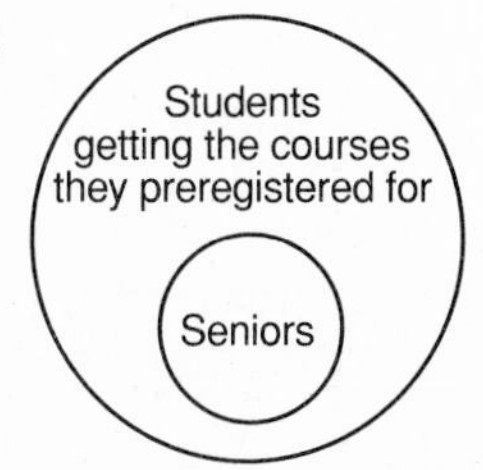

If I am a senior, then I am assured of getting the course I preregistered for.

27. Exactly one plane

28. $2(AB) + 3(10) = 54$; $AB = 12$

29. $m\angle 3 = 107°$

30. A single counterexample where A is true and B is not true.

31. Symmetric only

32. They are both adjacent and supplementary.

33. 1. Given
2. Def. of supplementary angles
3. Given
4. Substitution Property of =
5. Subtraction Property of =

34. $\ell_1 \parallel \ell_3$

35. Skew lines are two lines that do not lie in the same plane.

36. Answers vary.

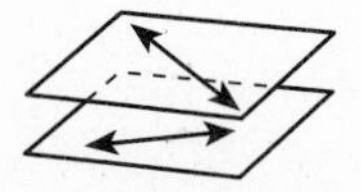

37. Answers vary.

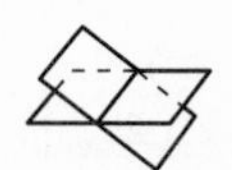

38. Answers vary, for example: $\overleftrightarrow{AD}$ and $\overleftrightarrow{DH}$; $\overleftrightarrow{DC}$ and $\overleftrightarrow{DH}$.

39. $(5, -6)$; $6(5)+5(-6) = 0$; $5(5)+4(-6) = 1$

40. $y = -4x - 6$ 41. $y = \frac{1}{4}x + 11$

42. If I can buy the car, then I can raise the down payment.

43. If I cannot buy the car, then I cannot raise the down payment.

44.

Statements	*Reasons*
1. $\angle AOD \cong \angle COB$	1. Given
2. $m\angle AOD = m\angle COB$	2. Def. of $\cong$
3. $m\angle AOD = m\angle AOC + m\angle COD$	3. Angle Add. Post.
4. $m\angle COB = m\angle COD + m\angle DOB$	4. Angle Add. Post.
5. $m\angle AOC + m\angle COD = m\angle COD + m\angle DOB$	5. Substitution Prop. of =
6. $m\angle AOC = m\angle DOB$	6. Subtr. Prop. of =
7. $\angle AOC \cong \angle DOB$	7. Def. of $\cong$

45. If I get a 92 on the final test, then I can finish college.

46. 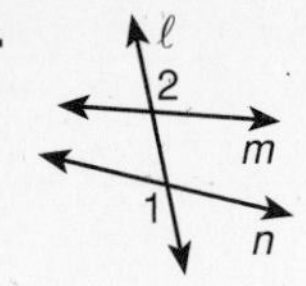

Answers vary; $\angle 1$ and $\angle 2$ are alternate exterior angles.

47. They are parallel.

48. $\overrightarrow{u} = \langle -6, -13 \rangle$; length of $\overrightarrow{u} = \sqrt{205}$

49.
$$\overrightarrow{u_1} = \langle 1, -6 \rangle$$
$$\overrightarrow{u_2} = \langle 10, -1 \rangle$$
$$\overrightarrow{u_1} + \overrightarrow{u_2} = \langle 11, -7 \rangle$$
$$\overrightarrow{u_1} \cdot \overrightarrow{u_2} = 16$$

4.2 Short Quiz

1. Isosceles
2. 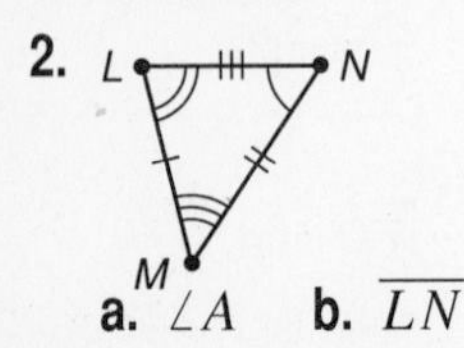

   a. $\angle A$ **b.** $\overline{LN}$
3. **a.** No. **b.** Yes.
   A scalene triangle is one in which no sides have the same length. An isosceles triangle is one in which at least 2 sides are congruent; therefore, it can have 3 sides congruent, which is also equilateral.
4. $m\angle 3 = 70°$
5. **a.** 35° **b.** 35° **c.** 55° **d.** 28°

4.4 Short Quiz

1. No. SSA is not enough by itself.
2. Yes; SSS is enough for congruence.
3. No. Sides are not necessarily congruent.
   No. Sides are not necessarily congruent.
4. ASA Congruence Postulate
5. SSA and AAA are not sufficient. Diagrams vary.

Mid-Chapter Test 4-A

1. $\overline{SR}$
2. If two angles have the same measure, then they are congruent.
3. No. Any two angles of a triangle must total less than 180°.
4. Sketches vary.
5. **a.** $x = 8$; **b.** No.
6. No. Postulates are taken to be true without proof in an axiomatic system.
7. ASA Congruence Postulate

Mid-Chapter Test 4-B

1. $\overline{LJ}$
2. If two angles are congruent, then their measures are equal.
3. No. By definition, an acute triangle has only acute angles.
4. Sketches vary.
5. **a.** $x = 5$; **b.** Yes; $AB = AC = BC$.
6. Every triangle is congruent to itself (Reflexive Property of Congruence).
7. SSS Congruence Postulate

4.6 Short Quiz

1. SSS Congruence Postulate
2. $\overline{LD} \cong \overline{ED}$; if two angles of a triangle are congruent, then the sides opposite them are congruent.
3. Equilateral or equiangular triangle
4. $\overline{RE}$
5. The Hypotenuse-Leg Congruence Theorem

Chapter Test 4-A

1. 67°
2. 2 or 3; a scalene triangle has to have 2 or 3 acute angles because if it has only 1 acute angle, the sum of the angle measures of the triangle would be greater than 180°.
3. **a.** $\overline{LN}$ **b.** $\angle UVW$ 4. Transitive
5. 89°
6. **a.** $x = 8$ **b.** No
7. $m\angle A = 103°$; $m\angle B = 77°$; $m\angle C = 46°$
8. **a.–d.** Sketches vary.
9. $x = 82°$
10. It can have two acute angles if it is a right isosceles or an obtuse isosceles triangle; otherwise it will have 3 acute angles.
11. SSS Congruence Postulate
12. Answers vary; for example: vertical angles.
13. **a.** Yes, $\triangle QRT \cong \triangle SRT$ by SAS, and $\overline{QR} \cong \overline{SR}$ by CPCTC **b.** Isosceles

14. 127°

15. Yes. Corresponding sides and angles are preserved.

16. $LM = 290$; ASA Congruence Postulate

17. $\angle ABD \cong \angle CDB$; Alternate interior angles are congruent.

18.

Statements	Reasons
1. $\overline{BC} \cong \overline{BA}$, $\overline{CD} \cong \overline{AD}$	1. Given
2. $\overline{BD} \cong \overline{BD}$	2. Reflexive Prop. of $\cong$
3. $\triangle ABD \cong \triangle CBD$	3. SSS Congruence Post.

19. 1; diagrams vary.

■ Chapter Test 4-B

1. d.	2. b.	3. c.	4. b.
5. a.	6. a.	7. c.	8. b.
9. c.	10. a.	11. d.	12. c.
13. d.	14. c.	15. c.	16. b.
17. b.	18. d.		

■ Chapter Test 4-C

1. $\overline{QR}$ 2. Sometimes

3. Yes. Theorems must be proven using postulates and/or previously proven theorems.

4. 131° 5. $x = 5$; yes.

6. Equilateral, since $\overline{AB} \cong \overline{CA} \cong \overline{BC}$

7. No. In order for $\triangle DEF \cong \triangle FED$, $\triangle DEF$ must be isosceles or equilateral.

8. Right isosceles

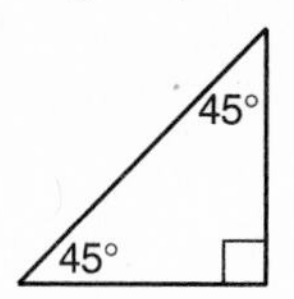

9. 72°, 52°, 56°, $(x = 36°)$

10. 1; a scalene triangle cannot have 2 or 3 right angles because the sum of the measures of the angles in a triangle is 180°.

11. SAS or SSS congruence Postulates

12. Base angles of isosceles triangles are equal; answers vary.

13. a. No. There is no information about the measures of the other angles.
    b. Obtuse. If $\angle F$ is less than 45°, $\angle FEH$ is greater than 45°. Since the 2 triangles are congruent by SAS, $\angle GEH$ is greater than 45°. Therefore, $\angle FEG$ is greater than 90°, which is obtuse.

14. 79°

15. Isosceles, at least, since $\triangle ABC \cong \triangle DEF$, $\overline{AB} \cong \overline{DE}$, $\overline{BC} \cong \overline{EF}$. Given that $\overline{AB} \cong \overline{EF}$, and $\overline{EF} \cong \overline{BC}$, then $\overline{AB} \cong \overline{BC}$ by the Transitive Property of Congruence. Therefore, $\triangle ABC$ is isosceles.

16.

The length AB could be determined, as shown. Point C is located at a distance from A and B. $\overline{CB}$ is extended to $E(\overline{BC} \cong \overline{CE})$, and $\overline{CA}$ is extended to $D(\overline{AC} \cong \overline{CD})$. $\angle ACB$ and $\angle DCE$ are vertical angles, therefore, $\angle ACB \cong \angle DCE$. The two triangles are congruent by SAS Congruence Postulate. Therefore, $\overline{DE} \cong \overline{AB}$.

17.

Statements	Reasons
1. $\overline{AB} \cong \overline{DE}$	1. Given
2. $\angle B \cong \angle E$	2. Given
3. $\angle ACB \cong \angle DCE$	3. Vertical ∠s Thm.
4. $\triangle ABC \cong \triangle DEC$	4. AAS Congruence Thm.

18.

Statements	Reasons
1. $\overline{AB} \cong \overline{ED}$	1. Given
2. $\angle 1 \cong \angle 2$	2. Given
3. $\angle ACB \cong \angle ECD$	3. Vertical ∠s Thm.
4. $\triangle ABC \cong \triangle EDC$	4. AAS Congruence Thm.

Proofs may vary.

19. AAA and SSA

5.2 Short Quiz

1. a. $\angle BCD$ b. $\cong$
2. a. $\angle CBO$ b. $\cong$
3. $(5, 5)$ 4. a. 5 b. 3
5. A vertex of the triangle and the midpoint of its opposite side

Mid-Chapter Test 5-A

1. Centroid 2. $DB = 6$
3. $(\frac{3}{2}, \frac{7}{2}), (\frac{7}{2}, \frac{9}{2}), (3, 2)$
4. $(-2, 1), (4, 3), (2, -3)$
5.

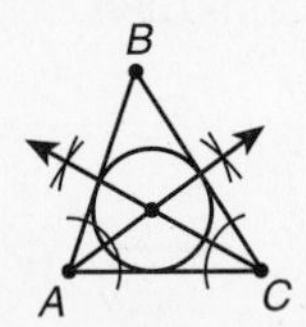

Mid-Chapter Test 5-B

1. Incenter 2. $m\angle OBC = 65°$
3. a. A line segment joining the midpoints of two sides of the triangle.
   b. Three
   c. Parallel to the third side; or half the length of the third side.
4. $(0, 3), (6, 1), (-2, -3)$
5.

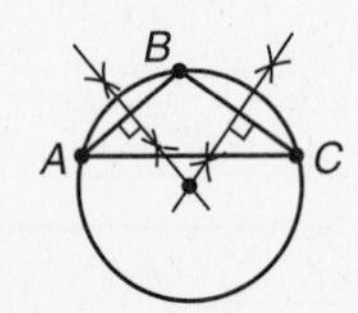

5.4 Short Quiz

1. Midpoints of sides 2. 7
3. 26 4. $3 < x < 13$
5. Shortest $\overline{AB}$, longest $\overline{AC}$
6. Smallest $\angle L$, largest $\angle M$

5.6 Short Quiz

1. $=$ 2. $=$ 3. $<$ 4. $>$
5. $30°$

Chapter Test 5-A

1. Incenter 2. Vertices
3. $LO = 4$, $MN = 6$; HL $\cong$ Thm., CPCTC.
4. $(0, 1), (2, -2), (4, 0)$
5. A vertex with the midpoint of the opposite side
6. 3
7.

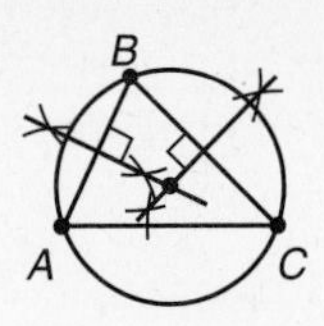

8. $(-\frac{1}{2}, -\frac{3}{2})$ 9. $LN = 7.5$ 10. 7
11. $55°$ 12. Centroid
13. $\overline{BF}$ 14. $\overleftrightarrow{GF}$
15. $\angle BCA$ 16. $3 < x < 19$
17. $>$ 18. $x > \frac{3}{4}$ 19. A right triangle
20.

Statements	*Reasons*
1. $\overrightarrow{AE}$ bisects $\angle A$	1. Given
2. $\angle BAE \cong \angle CAE$	2. Def. of $\angle$ bisector
3. $\triangle ABC$ is equilateral	3. Given
4. $\overline{AB} \cong \overline{AC}$	4. Def. of Equil. $\triangle$
5. $\overline{AE} \cong \overline{AE}$	5. Reflex. Prop. of $\cong$
6. $\triangle AEB \cong \triangle AEC$	6. SAS Congruence Pos
7. $\angle AEB \cong \angle AEC$	7. CPCTC
8. $AE \perp BC$	8. If two lines intersect to form adjacent $\cong$ $\angle$s the lines are $\perp$.

Chapter Test 5-B

1. b. 2. c. 3. a. 4. d.
5. c. 6. a. 7. d. 8. c.
9. b. 10. a. 11. b. 12. c.
13. b. 14. b. 15. a. 16. d.
17. a. 18. c.

Chapter Test 5-C

1. Circumcenter 2. Sides
3. $RS = 8$, $UT = 5$
4. $(1, -4), (9, 4), (-5, 2)$
5. A segment joining the midpoints of two sides
6. 3
7.

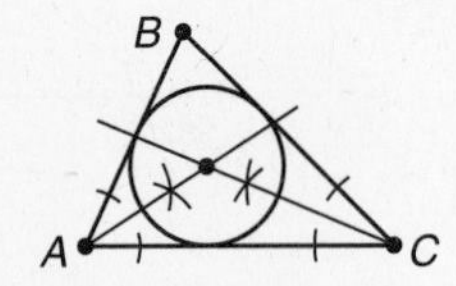

8. $(-1, -3)$
9. $\overline{AB} \parallel \overline{DF}$ and $AB = \frac{1}{2}DF$
10. 28 11. $131°$
12. Orthocenter 13. $\overleftrightarrow{BE}$
14. $\overline{BD}$ 15. $\angle ACD$
16. $4 < x < 24$ 17. $m\angle BAE > m\angle DAE$
18. $x > \frac{1}{2}$ 19. Equilateral

20.

Statements	Reasons
1. $\overline{AB} \cong \overline{AC}$	1. Given
2. $\overline{AE} \cong \overline{AE}$	2. Reflexive Prop. of $\cong$
3. $\angle AEC$ and $\angle AEB$ are rt. $\angle$s.	3. Def. of altitude
4. $\triangle AEB \cong \triangle AEC$	4. Hypotenuse-Leg Congruence Thm.
5. $\angle BAE \cong \angle CAE$	5. CPCTC

Therefore, $\overline{AE}$ bisects $\angle BAC$.

6.2 Short Quiz

1. Sketches vary.

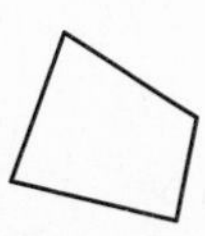

2. Sketches vary.

3. Sketches vary.

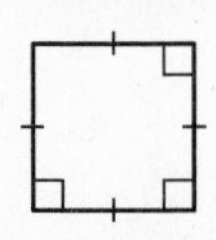

A regular polygon is a polygon that is equilateral and equiangular.

4. 2   5. 108°   6. 60°

Mid-Chapter Test 6-A

1. Sketches vary.

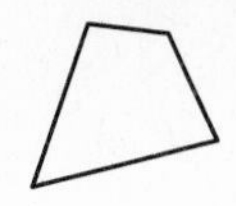

If a polygon is convex, then no line which contains a side of the polygon also contains a point interior to the polygon.

2. Sketches vary.

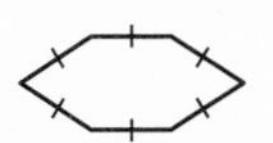

3. 5   4. 720°   5. 60°
6. a. 10;   b. 100°

7.

Statements	Reasons
1. $\overline{PQ} \parallel \overline{SR}$, $\overline{PS} \parallel \overline{QR}$	1. Given
2. Draw diagonal $\overline{SQ}$.	2. Two points determine a line.
3. $\angle PQS \cong \angle QSR$	3. Alt. Interior $\angle$s Thm.
4. $\angle SQR \cong \angle QSP$	4. Alt. Interior $\angle$s Thm.
5. $\overline{SQ} \cong \overline{SQ}$	5. Reflexive Prop. of $\cong$
6. $\triangle SQP \cong \triangle QSR$	6. ASA Congruence Post.
7. $PQ = SR$	7. CPCTC
8. $PS = RQ$	8. CPCTC

Answers vary.

Mid-Chapter Test 6-B

1. Sketches vary.

If a polygon is not convex, then a line that contains a side of the polygon also contains points interior to the polygon.

2.

Sketches vary.

3. 14   4. 540°   5. 113°
6. a. 6;   b. 70°

7.

Statements	Reasons
1. $ABCD$ is a ▱.	1. Given
2. Draw diagonal $\overline{AC}$.	2. Two points determine a line.
3. $\overline{AB} \cong \overline{DC}$, $\overline{BC} \cong \overline{AD}$	3. If a quad. is a ▱, then its opposite sides are $\cong$.
4. $\overline{AC} \cong \overline{AC}$	4. Reflexive Prop. of $\cong$
5. $\triangle BCA \cong \triangle DAC$	5. SSS Congruence Post.
6. $\angle BCA \cong \angle DAC$	6. CPCTC

6.4 Short Quiz

1. 120°   2. $RS = 15$
3. 60°   4. $QR = 23$

5. Since $AD = BC = 10$ and $AB = DC = \sqrt{73}$, $ABCD$ is a parallelogram.

6. Sketches vary.

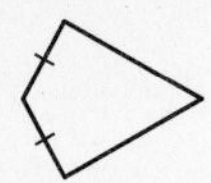

7. Sketches vary.

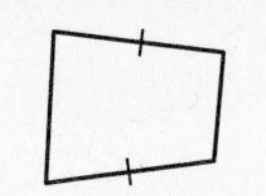

6.6 Short Quiz

1. Check students' constructions.

2. A., C., D. **3.** B., E.
4. A., B., C., D., E.
5. a. Isosceles trapezoid; b. $m\angle B = 130°$; c. measure of midsegment = 9.

Chapter Test 6-A

1.

Sketches vary.

2. 5
3. The polygon is equilateral and equiangular.

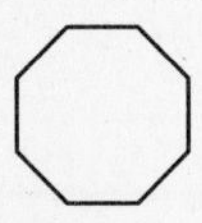

Sketches vary. The sketch shown is a regular octagon.

4. Not all sides are line segments.
5. 135° **6.** 720° **7.** 72°
8. 112° **9.** 19 **10.** 68° **11.** 34
12.

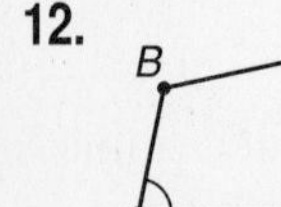

In this sketch, $\overline{AB} \parallel \overline{CD}$ but $AB \neq CD$. Sketches vary.

13.

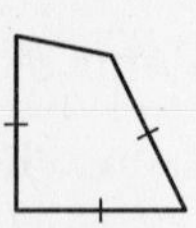

Sketches vary.

14. Since $AB = CD = \sqrt{53}$ and $BC = AD = 8$, $ABCD$ is a parallelogram.
15. a. True.
    b. If a parallelogram is a rhombus, then it is a square.
    c. False.
16. Rectangle or square
17. Drawings vary. Midsegment = $1\frac{1}{4}$ in.

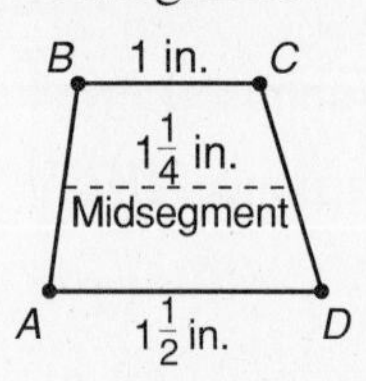

18. An isosceles trapezoid
19. Drawings vary.

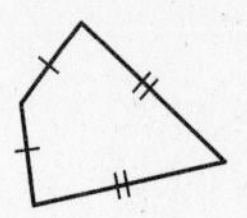

20. One diagonal bisects the other, and they are perpendicular.

Chapter Test 6-B

1. b. **2.** c. **3.** a. **4.** b.
5. d. **6.** a. **7.** c. **8.** c.
9. b. **10.** a. **11.** c. **12.** d.
13. c. **14.** c. **15.** d. **16.** b.
17. d. **18.** b. **19.** c.

Chapter Test 6-C

1.

Sketches vary.

2. 9

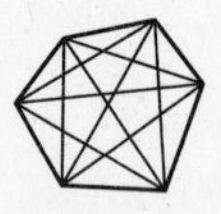

3. The polygon is equiangular and equilateral. Drawings vary, for example, this is a regular hexagon.

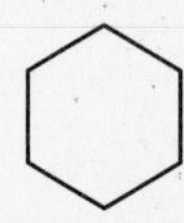

4. Answers vary, for example, not every side intersects exactly two other sides.
5. 108° **6.** 1080° **7.** 8 **8.** 46°
9. 24 **10.** 134° **11.** 30

12. Drawings vary.

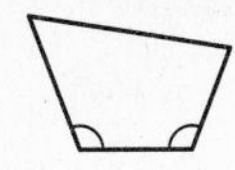

13. Drawings vary.

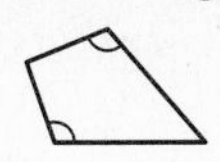

14. If $ABCD$ is a parallelogram, then $AB = DC$. Since $AB = \sqrt{37}$ and $DC = 2\sqrt{10}$, $ABCD$ is *not* a parallelogram.

15.

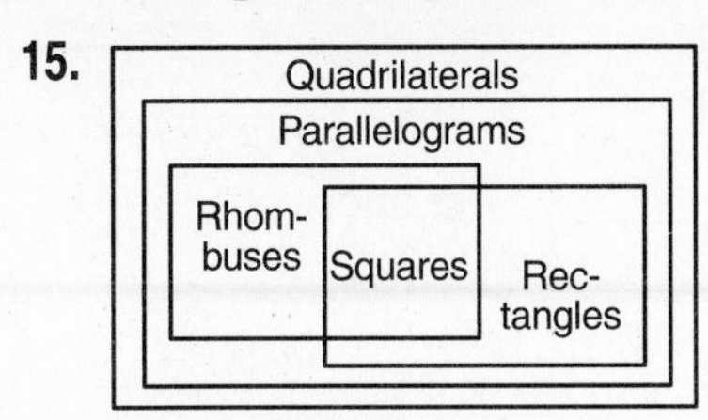

Diagrams vary.

16. A rhombus or square

17. Sketches vary. In this trapezoid, $\overline{BC} \parallel \overline{AD}$.

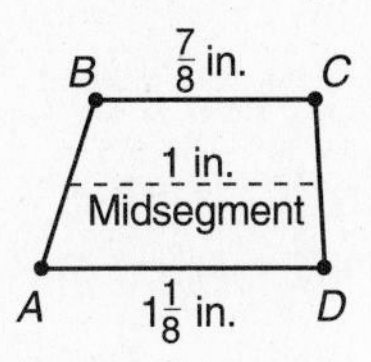

18. Not possible

19.

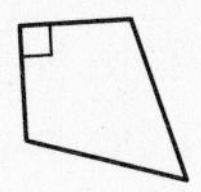

20. Answers vary.

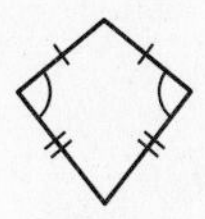

By Theorem 6.25, exactly one pair of opposite angles are congruent. So a kite is neither a parallelogram nor a trapezoid.

Cumulative Test 1–6

1. Sketches vary.

2. It has rotational symmetry and three lines of symmetry.

3.

4. They are similar. **5.** $(-5, 1)$ **6.** $\frac{1}{7}$

7. Check students' constructions.

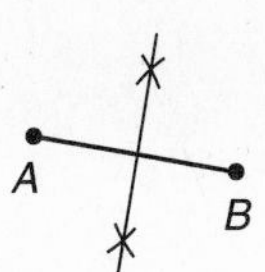

8. The lines are perpendicular because their slopes are negative reciprocals of each other.

9. The ray from P through point N.

P N

10. Q is between S and T.

11. $BE = 8$ **12.** $m\angle AOD = 123°$ **13.** 74

14. **a.** It is a rhombus.
    b. Yes. A square is a parallelogram that is both a rhombus and a rectangle. Both a square and a rhombus have 4 congruent sides.
    c. If the quadrilateral is a rhombus, then it is a square.
    d. No. A rhombus need not be equiangular.

15. If he does not vote for my bill, then I do not vote for his.

16. 69°

17. Two angles are supplementary if the sum of their measures is 180°.

18. Since parallelism of lines is transitive, $\ell_2 \parallel \ell_3$, but ℓ_3 need not be in the same plane as ℓ_1 and ℓ_2.

19. Skew lines

20. $(-4,\ 5)$; $2(-4)+3(5) = 7,\ (-4)+2(5) = 6$

21. $y = -\frac{2}{3}x + 6$

22. Sketches vary.

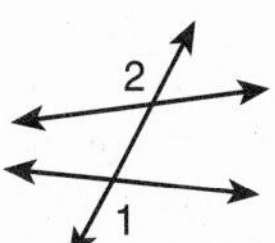

23. $m_1 \parallel m_2$

24.

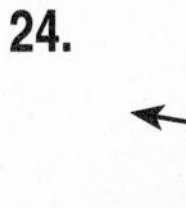

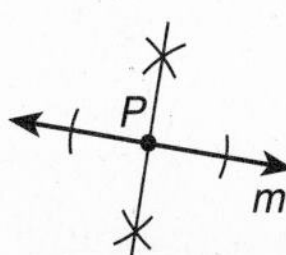

25. $\sqrt{53}$ **26.** $\langle 3, -4\rangle$ **27.** $2 - 21 = -19$

28. $\vec{u} + \vec{v} = \langle 490, 400 \rangle$;
New speed $= \sqrt{490^2 + 400^2} \approx 632.5$ miles per hour

29. Yes, because of the Transitive Property of Congruence.

30. $53°$

31. $m\angle A = 123°$, $m\angle B = 57°$, $m\angle C = 48°$

32. It may or may not have an obtuse angle so it can have 2 or 3 acute angles.

33. Yes. Corresponding sides and angles are preserved.

34. $\angle ACB \cong \angle ECD$ by vertical angles. So $\triangle ABC \cong \triangle EDC$ by ASA. Thus, $AB = 99$ by CPCTC.

35.

Statements	*Reasons*
1. $\overline{AB} \cong \overline{CD}$	**1.** Given
2. $\overline{BD} \cong \overline{BD}$	**2.** Reflexive Prop. of $\cong$
3. $\overline{AB} \parallel \overline{CD}$	**3.** Given
4. $\angle ABD \cong \angle CDB$	**4.** Two lines $\parallel \Rightarrow$ alt. int. $\measuredangle$ are $\cong$
5. $\triangle ABD \cong \triangle CDB$	**5.** SAS $\cong$ Post.

36. (4, 3)

37. The centroid

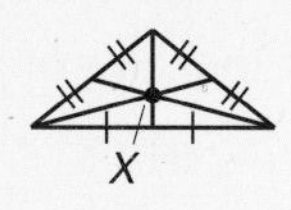

38. A segment that connects the midpoints of two of the sides of a triangle. For example,

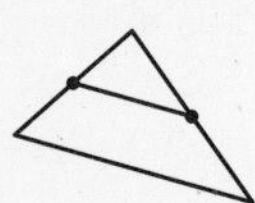

39. Check students' construction.

40. 17 **41.** $m\angle GDE = 70°$

42. a. $\overline{BD}$ altitude; $\overline{BE}$, angle bisector; $\overline{BF}$, median; $\overline{GF}$, perpendicular bisector.
b. $\overline{BE}$

43. $\angle B$; it is opposite the largest side.

44. A right triangle

45.

Statements	*Reasons*
1. Draw $\overline{BD}$.	**1.** Two points determine a line.
2. $\overline{BC} \cong \overline{DA}$; $\overline{AB} \cong \overline{CD}$	**2.** Given
3. $\overline{BD} \cong \overline{BD}$	**3.** Reflexive Prop. of $\cong$
4. $\triangle BCD \cong \triangle DAB$	**4.** SSS
5. $\angle ADB \cong \angle CBD$ $\angle CDB \cong \angle ABD$	**5.** CPCTC
6. $\overline{BC} \parallel \overline{DA}$, $\overline{AB} \parallel \overline{CD}$	**6.** Alt. int. $\measuredangle \cong$ $\Rightarrow$ lines are $\parallel$.
7. $ABCD$ is a $\square$.	**7.** Def. of $\square$

46. $8 < x < 20$ **47.** 14 **48.** $900°$

49. Using distance formula, $BC = DA = \sqrt{130}$, $AB = CD = \sqrt{101}$; opposite sides are equal in length. It *is* a parallelogram. (Slopes may be used instead.)

50. A square

51. a. 20
b. $AB = CD$ and/or $\angle DAB \cong \angle ADC$

7.2 Short Quiz

1. a. Translation **b.** Rotation
c. Reflection **d.** Non-rigid transformation

2. a. Same: shape and size; Changed: orientation
b. Same: shape; Changed: size and orientation

3. Side lengths are not preserved

4. 8

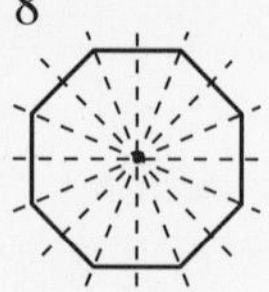

5.

Mid-Chapter Test 7-A

1. Every image is congruent to its preimage.

2. a. and d.

3. 3

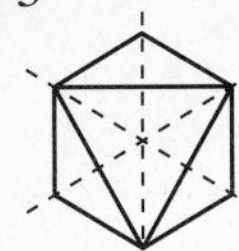

4.

5. $A'(-1, 1)$, $B'(1, 4)$, $C'(5, 0)$
6. 90°

7.

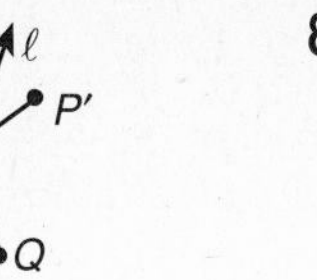

8. 2

9. Drawings vary. One possible answer: A kite.

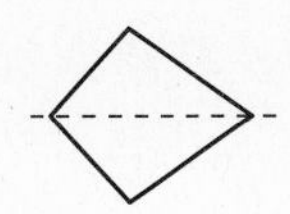

10. $P'(4, 5)$, $Q'(0, 5)$, $R'(0, 0)$
11. A 160° clockwise rotation of P about O.
12. Yes, 180°
13. $2d$, or twice the distance between ℓ and m.
14. $\vec{v} = \langle 5, -4 \rangle$ 15. $(13, -8)$
16. $A'(3, 2)$
17. For many compositions, different results are obtained when the order is altered. Each case must be examined individually.
18. TR

Mid-Chapter Test 7-B

1. The transformation preserves lengths.
2. b. and c.
3. 8

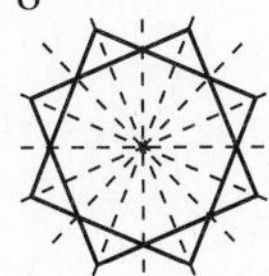

4.

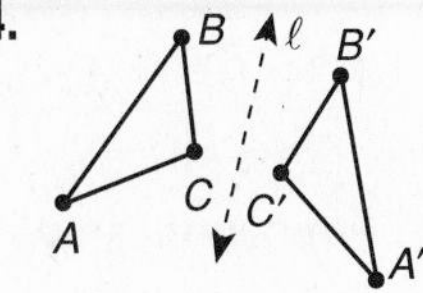

5. $A'(-2, 1)$, $B'(2, 4)$, $C'(3, -2)$
6. 60°

7.4 Short Quiz

1. Choose 3 from H, I, N, O, X, Z (S sometimes, depends on style.)
2. 96°
3. Four lines of symmetry; 90° and 180° rotational symmetries.
4. $A'(1, 0)$, $B'(4, 2)$, $C'(6, -3)$
5. 7 units

7.6 Short Quiz

1. A glide reflection is a transformation that consists of a translation by a vector followed by a reflection in a line that is parallel to the vector.
2. $P'(-1, 1)$; $Q'(-4, 3)$; $R'(-4, 1)$
3. Yes, it matters.
4. $(3, -3)$ 5. TV

Chapter Test 7-A

1. Every image is congruent to its preimage.
2. Translation 3. Reflection
4. No. It is not a rigid transformation and is not an isometry.
5. $A'(2, -6)$; $B(8, -4)$ 6. $(-x, y)$

Chapter Test 7-B

1. c.	2. a.	3. c.	4. c.
5. d.	6. a.	7. b.	8. c.
9. c.	10. d.	11. b.	12. a.
13. c.	14. d.	15. d.	16. a.
17. b.	18. a.		

Chapter Test 7-C

1. When it preserves lengths
2. Reflection 3. Rotation
4. It is not a rigid transformation and is not an isometry.
5. $A'(-3, 10)$; $B'(1, 2)$ 6. (y, x)
7. ℓ is the perpendicular bisector of $\overline{PP'}$

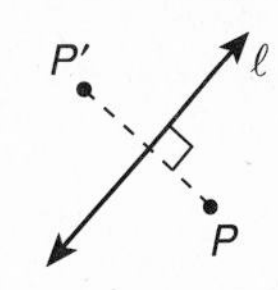

8. 6

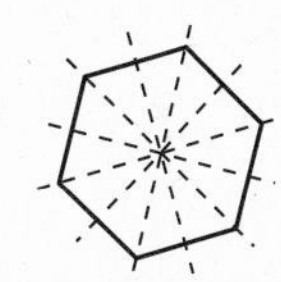

9. Find the reflection of A in ℓ (call it A'). Draw $\overline{A'B}$ and place the tap-in where $A'B$ intersects ℓ. This is the shortest distance from B to ℓ to A.

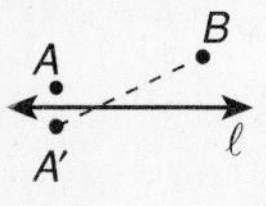

10. $P'(0, -2)$; $Q'(-1, 5)$

11. 68°

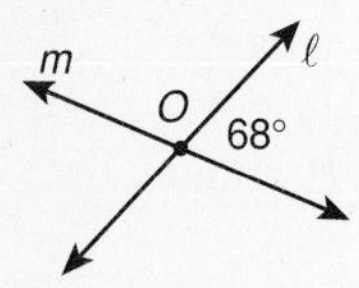

12. Yes, 180°

13. 8

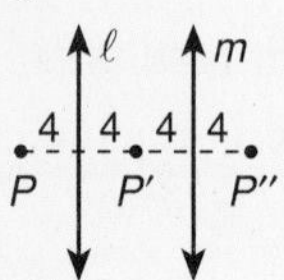

14. $\vec{v} = \langle -6, -2 \rangle$ 15. $(-2, -4)$

16. $A'(2, 1)$; $B'(6, -4)$

17.

Statements	*Reasons*
1. $\triangle ABC \Rightarrow \triangle A'B'C'$	1. Given
2. $AB = A'B'$,	2. Def. of isometry
2. $BC = B'C'$,	
2. $AC = A'C'$	
3. $\triangle ABC \cong \triangle A'B'C'$	3. SSS $\cong$ Post.

18. Since each step preserves length, the overall composition preserves length and is an isometry.

19. TR

8.2 Short Quiz

1. $\frac{14}{11}$ 2. $d = \frac{6}{5}$

3. $\dfrac{3 \text{ ft}}{4 \text{ ft}} = \dfrac{36 \text{ in.}}{48 \text{ in.}} = \dfrac{3}{4}$ 4. $\frac{60}{69} = \frac{20}{23}$

5.

$\frac{a}{b} = \frac{c}{d}$	Given
$\frac{b}{a} = \frac{d}{c}$	Reciprocal Property
$\frac{b}{a} + 1 = \frac{d}{c} + 1$	Addition Property of =
$\frac{b+a}{a} = \frac{d+c}{c}$ or	Rewrite as fractions.
$\frac{a+b}{a} = \frac{c+d}{c}$	Commutative Property of +

It is true.

6. 42 ft 7. 9

Mid-Chapter Test 8-A

1. $\dfrac{180 \text{ min}}{90 \text{ min}} = \dfrac{3 \text{ hr}}{1\frac{1}{2} \text{ hr}} = \dfrac{2}{1}$

2. The product of the extremes must also increase because the two products must remain equal in order for it to be a proportion.

3. $b = \frac{4}{3}$

4. $y = -8$. Both ratios represent the slope of the line and must be equal.

5. $3\sqrt{10}$ 6. \$42.90

7. Yes; since $\frac{4}{8} = \frac{5}{10}$ corresponding sides of the rectangles are proportional.

8. Sketches vary; for example,

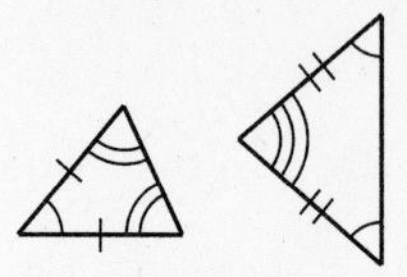

Mid-Chapter Test 8-B

1. $\dfrac{32 \text{ ounces}}{40 \text{ ounces}} = \dfrac{2 \text{ pounds}}{2\frac{1}{2} \text{ pounds}} = \dfrac{4}{5}$

2. The product of the means must also increase because the two products must remain equal in order for it to be a proportion.

3. $b = \frac{3}{2}$

4. $y = 7$. Both ratios represent the slope of the line and must be equal.

5. $4\sqrt{5}$ 6. \$154.83

7. No; since $\frac{16}{11} \neq \frac{20}{14}$, corresponding sides of the rectangles are not proportional.

8. Sketches vary; for example,

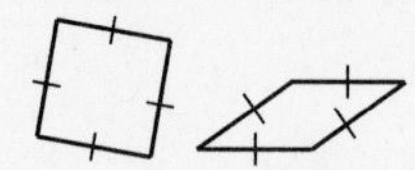

8.4 Short Quiz

1. $x = 36$

2. Two polygons are similar if their corresponding angles are congruent and the lengths of corresponding sides are proportional.

3. Yes; b. and d.

4. No; since $\frac{8}{10} \neq \frac{11}{14}$, corresponding sides are not proportional.

5. Pentagon A is similar to pentagon C. This is an example of the Transitive Property of Similarity.

6. Slope $= \frac{18-6}{20-0} = 0.6$; no; properties of similar triangles can be used to verify that slope ratios for any two points on a line are equal.

7. Since all equilateral triangles have angle measures of 60°, all equilateral triangles are equiangular. Thus, by the AA Similarity Postulate, all equilateral triangles are similar.

8.6 Short Quiz

1. They are similar since $\frac{60}{50} = \frac{108}{90} = \frac{156}{130} = 1.2$, corresponding sides are proportional.
2. Yes; a. and c. **3.** $LM = 18$; $LN = 15$
4. SSS Similarity Theorem
5. AA Similarity Postulate
6. $VS = 13.5$ **7.** $BPQ \cong \angle A$

Chapter Test 8-A

1. $\dfrac{420 \text{ sec}}{400 \text{ sec}} = \dfrac{7 \text{ min}}{6\frac{2}{3} \text{ min}} = \dfrac{21}{20}$
2. The product of the extremes must also increase because the two products must remain equal in order for it to be a proportion.
3. $x = \frac{15}{14}$ **4.** 158 **5.** 12
6. It is true. Both proportions have the same cross product: $ad = bc$.
7. 5 ft
8. Hexagon A is similar to hexagon C; Transitive Property of Similarity.
9. 55 feet **10.** Slope $= \dfrac{4-(-4)}{3-(-1)} = 2$. No.
11. Sketches vary; for example,

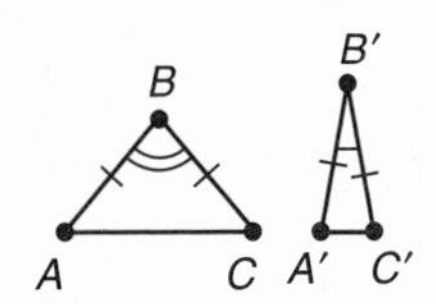

12. Yes. Since all regular hexagons are both equilateral and equiangular, they can be shown to be similar.
13. a. **14.** 60
15. SAS Similarity Theorem
16. AA Similarity Postulate

17. Since $\triangle PST \sim \triangle PQR$, $\angle PST \cong \angle Q$ and $\angle PTS \cong \angle R$. Since the pairs of angles are corresponding angles, then $\overline{ST} \parallel \overline{QR}$ by the Corresponding Angle Converse Postulate.
18. $x = 7.6$ **19.** 4

Chapter Test 8-B

1. c.	2. b.	3. b.	4. c.
5. b.	6. a.	7. c.	8. a.
9. d.	10. b.	11. c.	12. d.
13. d.	14. a.	15. d.	16. b.
17. c.	18. c.	19. a.	20. b.

Chapter Test 8-C

1. $\dfrac{108 \text{ in.}}{48 \text{ in.}} = \dfrac{9}{4}$ or $\dfrac{3 \text{ yd}}{1\frac{1}{3} \text{ yd}} = \dfrac{9}{4}$
2. 6 **3.** $x = \frac{7}{2}$ **4.** 141
5. Yes; when $a = b$ the arithmetic mean is the same as the geometric mean.
6. $\frac{a}{b} = \frac{c}{d}$ Given

   $\frac{a}{b} - 1 = \frac{c}{d} - 1$ Subtraction Prop. of =

   $\frac{a}{b} - \frac{b}{b} = \frac{c}{d} - \frac{d}{d}$ Substitution Prop. of =

   $\frac{a-b}{b} = \frac{c-d}{d}$ Algebra
7. $-\frac{3}{2}$; no, the slope remains the same.
8. $\triangle ABC \sim \triangle A''B''C''$; $x \approx 6.6$; $y \approx 43.7$
9. Yes, x and y are each the length of a side of a given triangle.
10. $\frac{56}{5}$ in. $= 11\frac{1}{5}$ in.
11. Sketches vary; for example,

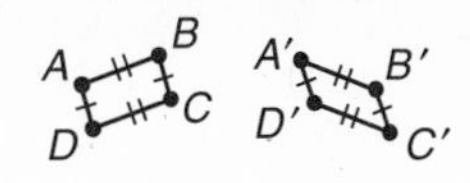

12. No. The measures of the base angles could vary.
13. c. **14.** $x = 60$
15. $\overline{BF} \parallel \overline{AE}$ and $\overline{AC} \parallel \overline{EF}$, so $ABFE$ is a ▱; since $ABFE$ is a ▱, $\angle A \cong \angle F$. $\angle AEC \cong \angle FDE$ because they are alternate interior angles and $\overline{BF} \parallel \overline{AE}$. Then $\triangle ACE \sim \triangle FED$ by the AA Similarity Postulate.
16. $\overline{UV} \parallel \overline{TS}$ **17.** $x \approx 4.7$ **18.** $\frac{34}{15} \approx 2.27$

9.2 Short Quiz

1. $\triangle WSR$ and $\triangle WRT$ **2.** WR
3. RW **4.** 20 **5.** $\sqrt{12} = 2\sqrt{3}$
6. $x = 4\sqrt{5}$, $y = 8\sqrt{5}$

Mid-Chapter Test 9-A

1. $4\sqrt{6}$ **2.** $4\sqrt{5}$
3. $\sqrt{10}$ **4.** a. and c.
5. The arithmetic mean is greater than the geometric mean. The arithmetic mean of 4 and 40 is 22 while the geometric mean of 4 and 40 is $4\sqrt{10} \approx 12.6$.
6. Since $6^2 + 7^2 > 9^2$, it is acute.
7. $8\sqrt{2}$ **8.** $2\sqrt{5}$ and $4\sqrt{5}$

Mid-Chapter Test 9-B

1. $4\sqrt{10}$ **2.** $8\sqrt{5}$
3. $\sqrt{8} = 2\sqrt{2}$ **4.** b. and d.
5. The arithmetic mean is greater than the geometric mean. The arithmetic mean of 6 and 54 is 30 while the geometric mean of 6 and 54 is 18.
6. Since $7^2 + 8^2 < 11^2$, it is an obtuse triangle.
7. $4\sqrt{2}$ **8.** $3\sqrt{13}$

9.4 Short Quiz

1. $5\sqrt{2}$ **2.** $6\sqrt{3}$ **3.** b. and d.
4. Since $9^2 + 13^2 < 16^2$, it is an obtuse triangle.
5. $x = 45°$, $y = 6\sqrt{2}$
6. $x = 20$, $y = 10\sqrt{3}$

9-6 Short Quiz

1. **a.** iii **b.** i **c.** ii
2. $\frac{5}{12}$ **3.** $\frac{8}{17}$ **4.** 18.00°
5. $x \approx 24$ feet **6.** $m\angle A = 17.82°$

Chapter Test 9-A

1. $\triangle MON$ or $\triangle LMN$ **2.** ML
3. LO and ON **4.** $3\sqrt{2}$
5. The arithmetic mean is greater than the geometric mean. The arithmetic mean of 4 and 12 is 8 while the geometric mean of 4 and 12 is $4\sqrt{3} \approx 6.9$.
6. b. and d. **7.** $90\sqrt{2}$ ft ≈ 127.3 ft **8.** b.
9. Since $6^2 + 9^2 < 11^2$, it is an obtuse triangle.
10. $\sqrt{65}$ **11.** $\sqrt{56}$ or $2\sqrt{14}$ **12.** 14
13. 12 **14.** 6 **15. a.** $\frac{a}{c}$ **b.** $\frac{b}{a}$ **c.** $\frac{b}{c}$
16. 2 **17.** 0.7547 **18.** 83.72°
19. No. You must know a side because a similar triangle can have the same angles but have different side lengths.
20. $x \approx 155$ ft

Chapter Test 9-B

1. c. **2.** a. **3.** b. **4.** a.
5. b. **6.** c. **7.** d. **8.** d.
9. c. **10.** d. **11.** a. **12.** b.
13. d. **14.** c. **15.** c. **16.** a.
17. b. **18.** a. **19.** a. **20.** b.

Chapter Test 9-C

1. $\triangle SRT$ or $\triangle WRS$ **2.** TS
3. $\overline{RW}$ and $\overline{RT}$ **4.** $\frac{25}{3}$
5. Arithmetic mean **6.** b. and c.
7. $90\sqrt{2}$ ft ≈ 127.3 ft **8.** b.
9. Since $7^2 + 9^2 > 11^2$, it is an acute triangle.
10. $0 < PR < 10$ and $PR \neq \sqrt{50}$ or $5\sqrt{2}$
11. $d = \dfrac{\sqrt{3}}{16} \approx 0.108$ **12.** $11\sqrt{2}$
13. $4\sqrt{3}$ **14.** $\dfrac{2}{\sqrt{3}}$ or $\dfrac{2\sqrt{3}}{3}$
15. **a.** $\frac{a}{c}$ **b.** $\frac{a}{b}$ **c.** $\frac{b}{c}$ **16.** $\frac{4}{7}$
17. 0.956, 0.799, .545, 0.225; decreases, the ratio of the adjacent side to the hypotenuse becomes smaller.
18. 3.1821 **19.** Another side length, or an acute angle
20. $\dfrac{100 + x}{50 + 10} = \tan 66°$, $x \approx 35$ ft

Cumulative Test 7–9

1. The transformation preserves lengths (or equivalent answer).
2. It is *not* rigid, because it does not preserve lengths.
3. Reflection **4.** Rotation **5.** $(x, -y)$
6. $C'(0, 1)$, $D'(5, -6)$
7. V, ℓ, V′, W′, W
8. 8 **9.** 2

10. An isosceles triangle

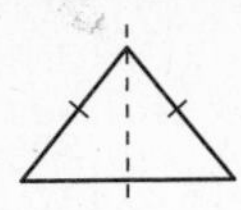

11. $R'(2, -1)$, $S'(-2, 4)$
12. Yes; 90°, 180°
13. A 60° clockwise rotation of Q about O
14. A translation of distance $2d$.
15. $\vec{v} = \langle 3, 2 \rangle$ 16. (1, −4)
17. (5, 3) 18. (−3, 1) 19. TRVG
20. $\frac{108 \text{ in.}}{180 \text{ in.}} = \frac{3}{5}$ or $\frac{3 \text{ yd}}{5 \text{ yd}} = \frac{3}{5}$
21. $x = \frac{19}{2}$ 22. 450
23. True; $\frac{a+b}{b} = \frac{a}{b} + 1$, $\frac{c+d}{d} = \frac{c}{d} + 1$, $\frac{a}{b} + 1 = \frac{c}{d} + 1$ if $\frac{a}{b} = \frac{c}{d}$
24. $6\sqrt{5}$ 25. 646 ft
26. Slope $= \frac{1}{2}$; no, any two distinct points can be used.
27. No; but their corresponding angles must be congruent and the ratios of their corresponding sides must be equal.
28. b.
29. Sketches vary. For example,

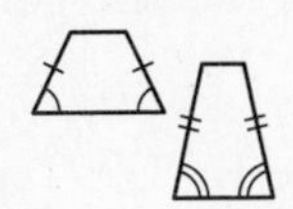

30. $x = 32$ 31. SSS Similarity Theorem
32. $\overline{GH} \parallel \overline{DF}$
33. $x \approx 4.2$ 34. a. and c. 35. 2.3
36. $\triangle UVW$ or $\triangle UYV$ 37. UY and YW
38. UW 39. $UV = 6$ 40. 32
41. $2\sqrt{6}$ or $\sqrt{74}$

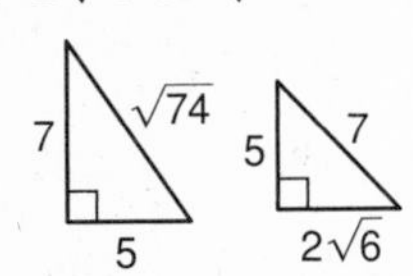

42. 19.0 in.
43. 1.618. The smaller rectangle appears to be a golden rectangle also.
44. a. obtuse b. right c. acute
45. 12 46. $3\sqrt{5}$
47. No, rectangles with a diagonal of 10 cm can have different sizes and shapes.
48. a. $\frac{p}{r}$ b. $\frac{p}{q}$ c. $\frac{p}{r}$
49. $\frac{3}{\sqrt{34}} = \frac{3\sqrt{34}}{34}$ 50. 1.3270
51. 13.89° 52. 0.943 53. 1580 ft

10.2 Short Quiz

1. $\overline{AC}$ 2. E 3. $\overline{AP}$ or $\overline{BP}$
4. D 5. ℓ, or $\overleftrightarrow{AD}$ 6. m, or $\overleftrightarrow{AC}$
7. $\overleftrightarrow{BR}$ 8. $\overline{CR}$
9. $\overleftrightarrow{AE}$ 10. $\angle CRO$

Mid-Chapter Test 10-A

1. 6.37 inches 2. $2\sqrt{21}$
3. 13 4. $\sqrt{97} - 4 \approx 5.85$
5. a. 125° b. 100° c. 60°
6. $m\widehat{AB} \approx 119°, m\widehat{BC} \approx 133°, m\widehat{CD} \approx 29°, m\widehat{DA} \approx 79°$

Mid-Chapter Test 10-B

1. 9.39 inches 2. $6\sqrt{2}$
3. $6\sqrt{2}$ 4. $5\sqrt{10} - 5 \approx 10.8$
5. a. 185° b. 45° c. 120°
6. $m\widehat{AB} \approx 76°, m\widehat{BC} \approx 112°, m\widehat{CD} = 54°$ $m\widehat{DA} \approx 119°$

10.4 Short Quiz

1. $m\widehat{AC} = 134°$
2. L.A. 288; Bus. 342; Sci. 189; Other 81.
3. $\angle AOB$ 4. $\widehat{AB}$, $\widehat{AC}$, or $\widehat{BC}$
5. $\widehat{ACB}$, $\widehat{CBA}$, or $\widehat{BAC}$
6. $\overline{RL} \cong \overline{LS}$ 7. 14

10.6 Short Quiz

1. 90° 2. $m\angle ACD = 50°$ 3. 66°
4. 180°; If a quadrilateral is inscribed in a circle, then its opposite angles must be supplementary.
5. 65° 6. $m\angle x = 44°$

Chapter Test 10-A

1. A chord of a circle is a segment whose endpoints are on the circle.
2. Two possible answers.

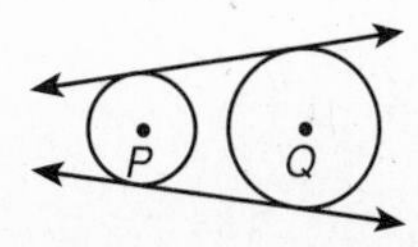

3. A secant of a circle is a line that intersects a circle twice. Sketches vary. For example,

4. $\angle QAO$ 5. $\sqrt{425} = 5\sqrt{17} \approx 20.6$
6. Each side of the polygon must be tangent to the circle.
7. 238°
8. Type O: 470 units, Type A: 260 units, Type B: 200 units, Type AB: 70 units
9. 140°
10. $\overline{AB} \cong \overline{CD}$. $\angle AOB \cong \angle COD$ by the Vertical Angles Theorem, so $\widehat{AB} \cong \widehat{CD}$; then $\overline{AB} \cong \overline{CD}$ because congruent arcs have congruent chords (Theorem 10.5).
11. $RS = 7$. In a circle, two chords that are equidistant from the center are congruent. (Theorem 10.8)
12. Yes, $RT = TS$. A diameter perpendicular to a chord bisects the chord and its arc. (Theorem 10.6)
13. (Alternate arguments are possible.) They are congruent if $\overline{AD} \cong \overline{CB}$ but similar otherwise. $\angle DAB \cong \angle DCB$ since they both intercept $\widehat{DB}$, and $\angle ADC \cong \angle CBA$ since they both intercept $\widehat{AC}$. Therefore, $\triangle ADP \sim \triangle CBP$ by the AA Similarity Postulate. If $\overline{AD} \cong \overline{CB}$, $\triangle ADP \cong \triangle CBP$ by the ASA Congruence Postulate.
14. $m\widehat{BC} = 80°$, $m\angle BDC = 40°$
15. $m\angle C = 100°$, $m\angle D = 70°$
16. 70° 17. 68° 18. 32°
19. $(x+3)^2 + (y-5)^2 = \frac{25}{4}$
20.

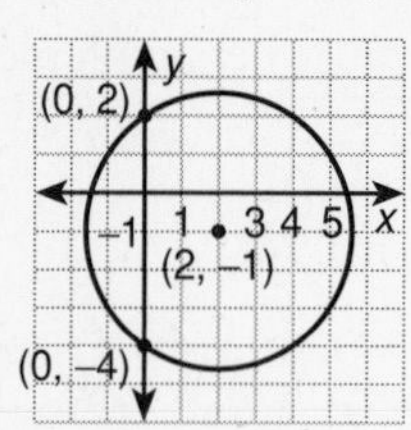

Chapter Test 10-B

1. c. 2. b. 3. d. 4. d.
5. a. 6. b. 7. d. 8. c.
9. a. 10. d. 11. b. 12. c.
13. a. 14. b. 15. c. 16. a.
17. b. 18. d. 19. c. 20. d.

Chapter Test 10-C

1. A diameter is a chord that passes through the center of a circle.
2. Two possible answers.

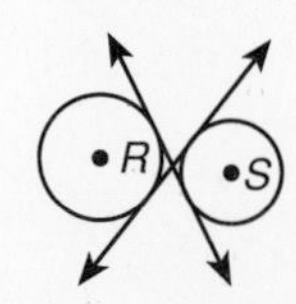

3. A tangent is a line that intersects a circle at exactly one point.

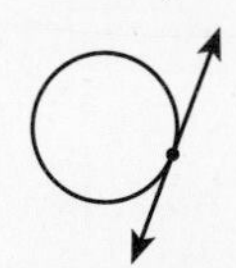

4. $\angle QAO$ and $\angle QCO$
5. $\sqrt{448} = 8\sqrt{7} =\approx 21.2$
6. A circle that is circumscribed about a polygon must pass through each vertex of the polygon.
7. $(x - 1^2) + (y + 1)^2 = 13$
8. Type O: 169°, Type A: 94°, Type B: 72°, Type AB: 25°
9. $m\widehat{CEB} = 215°$, $m\widehat{EA} = 50°$
10. $\overline{PT} \cong \overline{RV}$. $\angle PQT \cong \angle RQV$ by the Vertical Angles Theorem, so $\widehat{PT} \cong \widehat{RV}$; then $\overline{PT} \cong \overline{RV}$ because congruent arcs have congruent chords (Theorem 10.5).
11. (Alternate arguments are possible.) Since $\overline{AB}$ and $\overline{DC}$ are perpendicular to the same line, $\overline{AB} \parallel \overline{CD}$ by the Property of Perpendicular Lines. $AB = CD$ because they are chords equidistant from the center. Since one pair of opposite sides of $ABCD$ are both congruent and parallel, the quadrilateral is a parallelogram.
12. (Alternate arguments are possible.) $\widehat{PQ} \cong \widehat{RS}$ since $\overline{PQ} \cong \overline{RS}$ (Theorem 10.5). $\widehat{PQ} + \widehat{QR} \cong \widehat{QR} + \widehat{RS}$ (Addition Property of Equality). $\widehat{PR} \cong \widehat{QS}$ (Arc Addition Postulate). Therefore, $\overline{PR} \cong \overline{QS}$ (Theorem 10.5)

13. $m\widehat{ABD} = 288°$, $m\angle ACD = 36°$

14. $m\angle B = 102°$, $m\angle C = 77°$

15. $121°$ 16. $\frac{1}{2}(36° + x) = 65°$, $x = 94°$

17. $56°$

18. 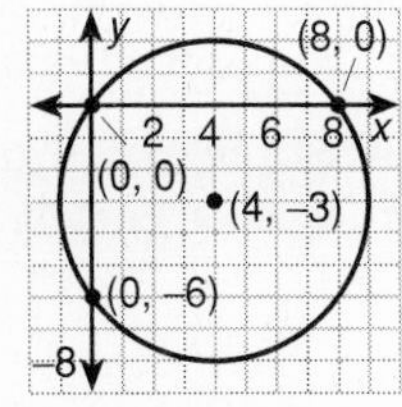

11.2 Short Quiz

1. 60 cm 2. 150 cm^2

3. Perimeter: 12.0 cm, Area: 7.56 cm^2

4. 140 ft^2 5. 17.5 units^2 6. 160 units^2

Mid-Chapter Test 11-A

1. Perimeter: $8\sqrt{13}$ units $\approx$ 28.8 units, Area: 48 units^2

2. Perimeter: 52 in., Area: 26 in.^{2}

3. 60 units^2

4. $14\sqrt{5}$ units^2 $\approx$ 31.3 units^2

5. 416 units^2 6. $x = 40$ units

Mid-Chapter Test 11-B

1. Perimeter: 24 units, Area: 28 units^2

2. Perimeter: 104 in., Area: 104 in.^{2}

3. 72 units^2 4. $2\sqrt{182}$ units^2 $\approx$ 27.0 units^2

5. 540 units^2 6. $x = 8$ units

11.4 Short Quiz

1. 100 units^2 2. $x = 45$ units

3. $50\sqrt{21}$ units^2 $\approx$ 229 units^2

4. $\sqrt{3}$ units^2 $\approx$ 1.7 units^2

5. $6\sqrt{3}$ units^2 $\approx$ 10.4 units^2

6. $\frac{1}{\tan 22.5} \approx 2.41$ units

11.6 Short Quiz

1. $\frac{28}{\pi}$ in. $\approx$ 8.91 in. 2. $\frac{13\pi}{9}$ cm $\approx$ 4.54 cm

3. $\frac{110}{3}\pi$ ft $\approx$ 115 ft 4. $\approx$ 25.97 in.^{2}

5. 45.57 cm^2 6. $\approx$ 0.61 units^2

Chapter Test 11-A

1. $(12 + 4\sqrt{2})$ units $\approx$ 17.7 units

2. 21 units^2

3. Answers vary. For example, a 1×6 rectangle and a 2×3 rectangle both have areas of 6, but perimeters of 14 and 10 respectively.

4. Yes, the two squares have equal side lengths $\left(\frac{P}{4}\right)$ and equal angle measures (90°).

5. 0.1 6. $5\sqrt{5}$ units

7. $25\sqrt{5}$ units^2 8. $160\sqrt{5}$ units^2

9. 12.5 units^2 10. 32.5 units^2

11. 38.5 units^2 12. 20 units^2

13. $3\sqrt{3}$ units^2 14. 36 units^2

15. 18.9 ft 16. 2.62 cm

17. 57.6 ft 18. $\approx$ 322 units^2

19. $$\cos AOC = \tfrac{2}{7}$$
$$m\angle AOC \approx 73.398°$$
$$m\angle AOB = 2m\angle AOC \approx 2(73.398°) \approx 146.796°$$
$$m\angle AOB = m\widehat{AB}$$
$$\text{Arc length of } \widehat{AB} = (2\pi \cdot 7)\left(\tfrac{146.796°}{360°}\right) \approx 17.93$$

20. 125 units^2

Chapter Test 11-B

1. b. 2. c. 3. a. 4. b.
5. d. 6. c. 7. b. 8. c.
9. d. 10. a. 11. b. 12. b.
13. c. 14. d. 15. b. 16. d.
17. a. 18. c. 19. a. 20. b.

Chapter Test 11-C

1. $16 + 4\sqrt{2}$ 2. 29 units^2

3. 20 4. 40 5. 800

6. $\frac{9}{2}\sqrt{319}$ units^2 7. $32\sqrt{319}$ units^2

8. $4\sqrt{53}$ 9. 224 units^2

10. 135 units^2 11. 15.5 units^2

12. Explanations will vary. The apothem, the radius, and half the side of the regular polygon form a right triangle in which the apothem, a, is a leg and the radius, r, is the hypotenuse. So $a < r$.

13. 9.50 ft 14. $9\sqrt{3}$ units^2 15. 14.45 units

16. 30 in. 17. 24.34 units^2

18. The area decreases. For example, if the number of sides of a regular hexagon is doubled, then the area of the polygon increases by six times the amount shown. Since the area of the polygon has increased, the area of the segments between the polygon and the circle must have decreased.

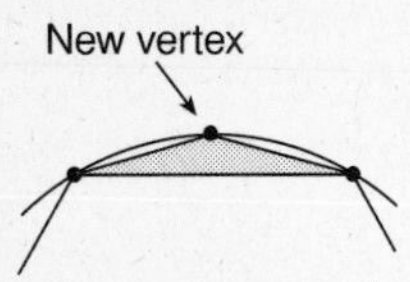

19. 18.4 units^2

12.2 Short Quiz

1. Yes. Every face is a polygon.
2. Sketches vary.

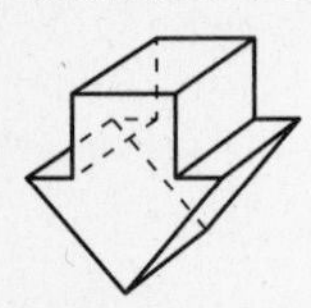

3. Sketches vary.

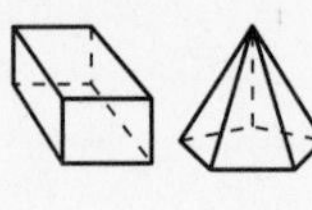

4. Sketches vary.

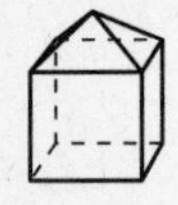

5. Not possible.
6. Sketches vary.

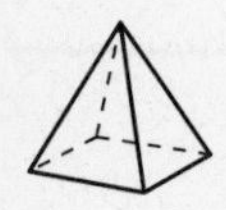

7. 7
8. 88π in.$^2 \approx 276.5$ in.^{2}

Mid-Chapter Test 12-A

1. B; Not all of the faces are polygons.
2. D; It appears to be the only one with congruent faces.
3. A; A line segment can be drawn between 2 points on different faces of the polyhedron's surface that does not lie entirely inside the polyhedron.
4. Sketches vary.

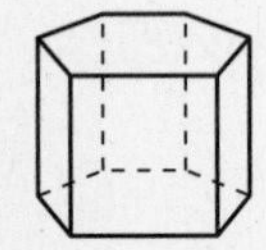

5. 120π units$^2 \approx 377$ units^2
6. $(105 + 40\sqrt{2})$ units$^2 \approx 161.6$ units^2
7. 80π units$^2 \approx 251$ units^2

Mid-Chapter Test 12-B

1. C; It appears to be the only one with congruent faces.
2. D; Not all of the faces are polygons.
3. B; A line segment can be drawn between 2 points on different faces of the polyhedron's surface that does not lie entirely inside the polyhedron.
4. Sketches vary.

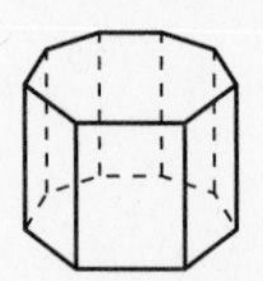

5. 140π units$^2 \approx 440$ units^2
6. 3 in., 6 in., 9 in.
7. 184π units$^2 \approx 578$ units^2

12-4 Short Quiz

1. 481 cm^2
2. 33π units$^2 \approx 103.7$ units^2
3. 259π units$^2 \approx 813.7$ units^2
4. Sketches vary. For example,

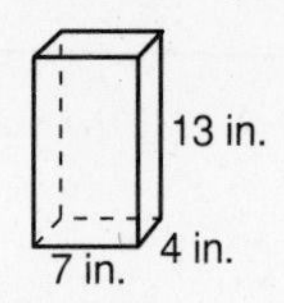

5. 540π cm$^3 \approx 1696$ cm^3
6. 18 ft

12-6 Short Quiz

1. 1575π ft$^3 \approx 4948$ ft^3
2. Volume $= 9$ yd^3

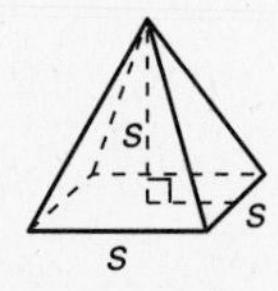

3. 192.4 in.^{3}
4. 974.3 in.^{3}
5. 475.3 in.^{2}
6. 11.070 ft

Chapter Test 12-A

1. Not possible
2. Sketches vary.

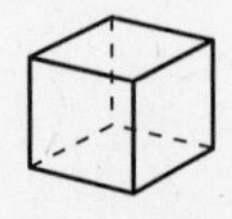

3. Sketches vary.

4. $E = 11$; 11, because $F + V = E + 2$ and $6 + 7 = 11 + 2$.

5. 54 in.^{2} 6. 251.33 cm^2
7. 95 ft^2 8. 388.77 cm^2
9. Increases 10. 825 in.^{3}
11. 595.8 ft^3
12. Area: 6π ft^2, Volume: 2π ft^3
13. 69.8 m^3 14. 1303.9 ft^3
15. 66 yd^3 16. 3761.0 in.^{2}
17. 12.6 ft^3 18. 11.2 yd^2
19. 33.1%

Chapter Test 12-B

1. d.	2. b.	3. a.	4. c.
5. a.	6. b.	7. c.	8. d.
9. b.	10. c.	11. d.	12. a.
13. c.	14. b.	15. d.	16. b.
17. a.	18. d.		

Chapter Test 12-C

1. Sketches vary.

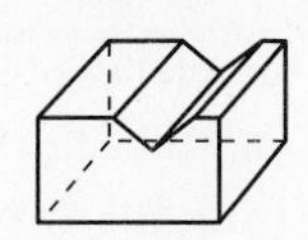

2. Not all of its vertices are formed by the same numbers of faces. Two are three-face vertices, and three are four-face vertices.
3. 14, because $F + V = E + 2$ and $9 + 14 = 21 + 2$.
4. $(60 + 8\sqrt{3})$ cm$^2 \approx 73.9$ cm^2
5. 24π in.$^2 \approx 75.4$ in.^{2}
6. $4\sqrt{3}$ in.$^2 \approx 6.93$ in.^{2}
7. 389.10 cm^2
8. Determine the length of the arc of the sector. This arc length is the circumference of the base. Then divide by π to find the diameter.
9. 780 ft^3
10. The cube; the cube has volume of x^3 while the cylinder has volume of $\frac{\pi}{4}x^3 \approx 0.79x^3$.
11. 25 cm; Volume of cans $= 1600\pi$; so $1600\pi = 64\pi h$ and $h = 25$.
12. 274.6 in.^{3} 13. 89
14. $V = \frac{4}{3}\pi r^2 h$ 15. 99 gallons
16. 94 ft^3 17. 11.5 in.^{3}
18. 1800 cm^2 19. 203 lb

Cumulative Test 7–12

1. a. ii, b. iii, c. i
2. Answers vary. For example, when it fails to preserve distances.
3. 1 4. 76°
5.
6. Sketches vary.

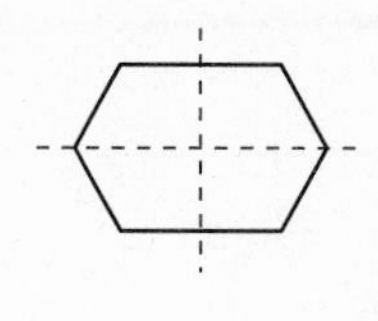

7. (10, −6) 8. TG
9. 21 10. $7\sqrt{7}$
11. True; explanation vary. For example,

$\frac{a}{b} = \frac{c}{d}$	Given
$\frac{a}{b} - 1 = \frac{c}{d} - 1$	Subtraction Prop. of =
$\frac{a}{b} - \frac{b}{b} = \frac{c}{d} - \frac{d}{d}$	Subst. Prop. of =
$\frac{a-b}{b} = \frac{c-d}{d}$	Algebra

12. 66 ft/sec
13. $\frac{5-(-3)}{-2-(4)} = \frac{8}{-6} = -\frac{4}{3}$; no; Properties of similar triangles can be used to verify that slope ratios for any two points on a line are equal.
14. Corresponding angles are congruent and corresponding sides are proportional.
15. $x = 18$ 16. 10.15 17. 1.83
18. a. AD and DC b. $BC = 3\sqrt{13}$
19. b. and c.
20. Since $6^2 + 7^2 > 9^2$, it is an acute triangle.
21. 7 22. a. $\frac{a}{b}$ b. $\frac{a}{c}$ c. $\frac{a}{c}$
23. cos 10°. $\text{sin of an } \angle = \frac{\text{side opposite the } \angle}{\text{hypotenuse}}$

$$\text{cos of an } \angle = \frac{\text{side adjacent to the } \angle}{\text{hypotenuse}}$$

Since the side adjacent to the 10°∠ is opposite an 80°∠, it is longer than the side opposite the 10°∠. So, the numerator of the cosine of the ∠ is larger than the numerator of the sine of the ∠.

24. 69.81° 25. $4\sqrt{3}$
26. 62.4 ft
27. Two possible answers.

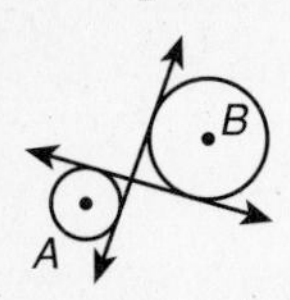

28. $\sqrt{373} \approx 19.3$ 29. 222°
30. $ST = 15$. In a circle, two chords equidistant from the center are congruent.
31. $m\widehat{AD} = 60°$, $m\angle ACD = 30°$
32. $(x-4)^2 + (y+2)^2 = 18$
33. 125° 34. 55°
35. Perimeter: 16 units, Area: 10 square units
36. Many possible answers. For example, these rectangles have the same perimeter, 20.

37. 12 units^2 38. 16 units^2
39. 100 units^2 40. 86.4 ft
41. 24 units^2
42. Sketches vary.

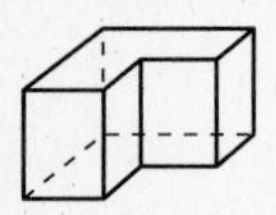

43. Sketches vary. Vertices and edges may vary also.

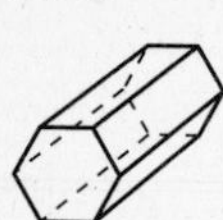

12 vertices, 18 edges

44. Surface area: 408 cm^2, Volume: 360 cm^3
45. 250π in.$^2 \approx 785.4$ in.^{2}
46. 51.0 cm^3 47. 7.979 in.
48. **a.** 480 yd^3;
  b. No, because the ratio of the volume of the upper pyramid to the volume of the entire pyramid is

$$\frac{\frac{1}{3}\left(8 \times \frac{9}{2}\right)5}{\frac{1}{3}(16 \times 9)10} = \frac{1}{8}.$$

49. 9 m
50. Surface area: 1810 in.^{2}, Volume: 7238 in.^{3}

13.2 Short Quiz

1. 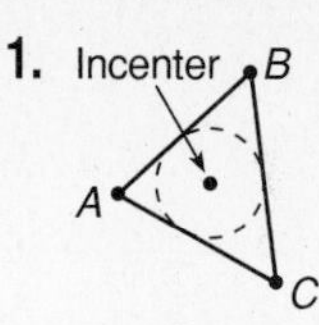

A single point; the center of the inscribed circle of the triangle.

2. 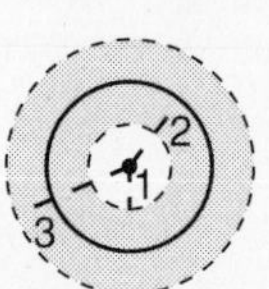

All points between circles that are concentric with the given circle and that have 1- and 3-unit radii.

3. 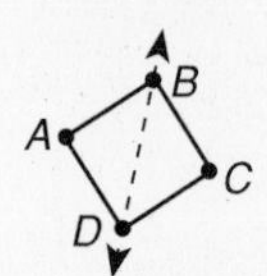

A line that contains diagonal $\overline{BD}$ of the square.

4. A plane that is perpendicular to the plane containing the two parallel lines and through the line in that plane equidistant from the two lines.
5. All points *inside* a sphere, centered at O, whose radius is 5.
6. The locus of all points in space 7 units from $\overline{AB}$.

Mid-Chapter Test 13-A

1. A single point, the center of the circle.
2. Two perpendicular lines that are the bisectors of the angles formed by ℓ and n.

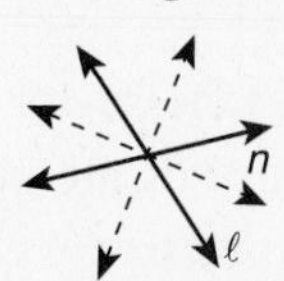

3. 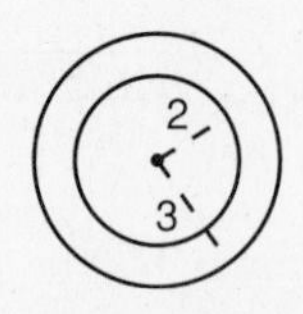

A concentric circle whose radius is 3 units.

4. A shaded triangular region, as shown.

5. A plane that is perpendicular to $\overline{AB}$ through the midpoint of $\overline{AB}$

6. 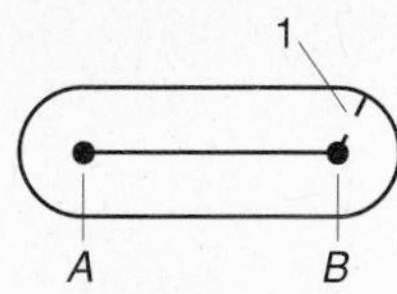

7. 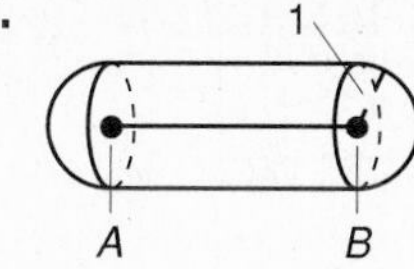

8. 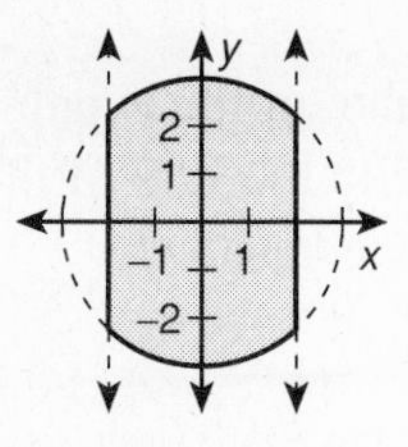

A shaded region, as shown

Mid-Chapter Test 13-B

1. A single point; the center of the circumscribed circle of the triangle

2. A line perpendicular to P containing the center of the circumscribed circle of the triangle

3. 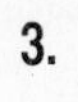 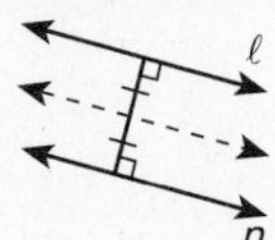

A line parallel to, and midway between, ℓ and n

4. 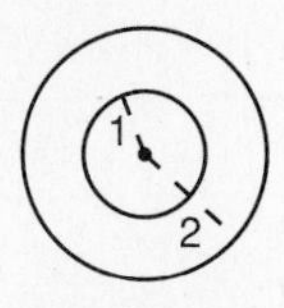

A concentric circle whose radius is 1 unit.

5. 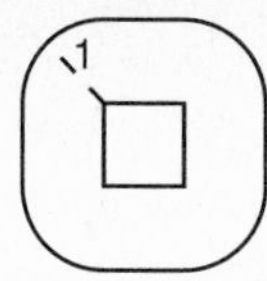

6. 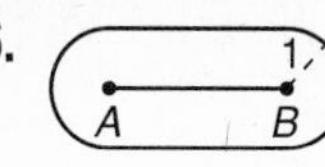

7. 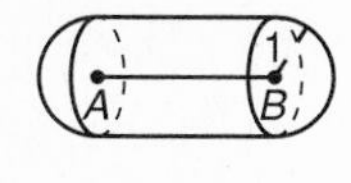

8. 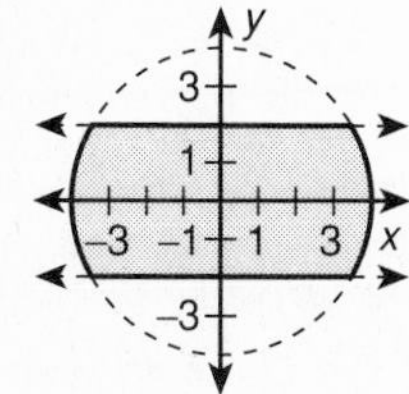

A shaded region, as shown

13.4 Short Quiz

1. 0 - 2

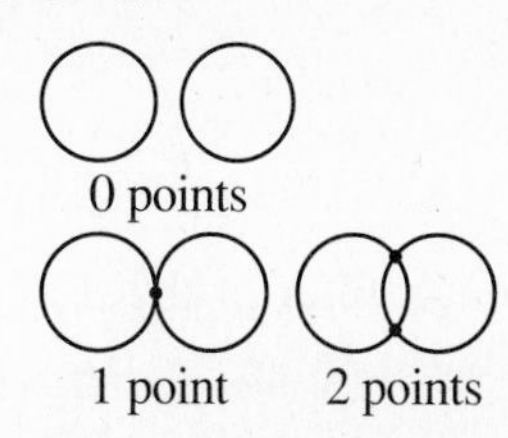

2. 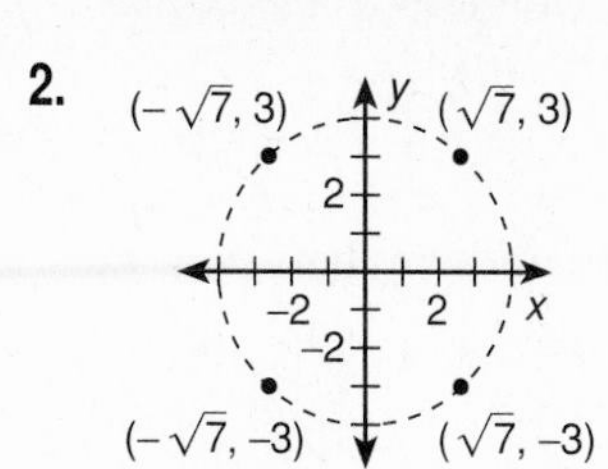

3. Four points:
(3, 3), (3, −3), (−3, 3), (−3, −3)

4. 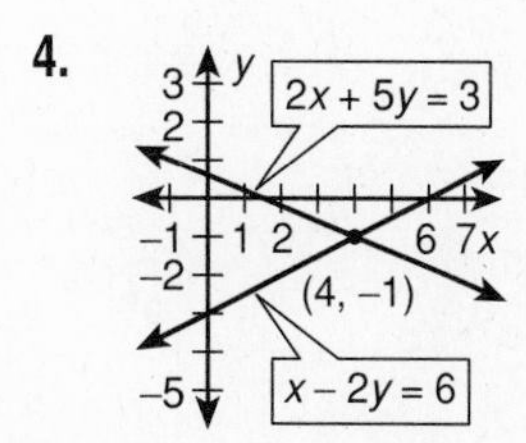

5. $5x - 3y = -5$

6. 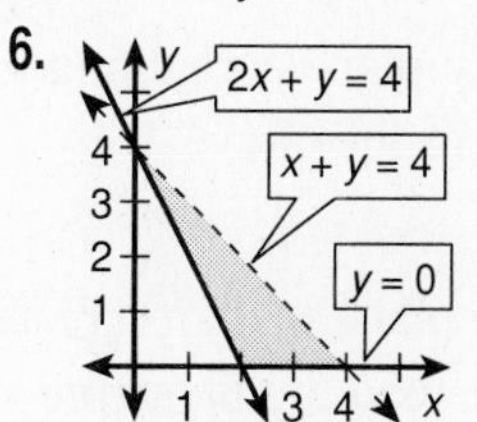

A shaded region, as shown

13.6 Short Quiz

1. A point, a circle
2. An isosceles trapezoid
3. A circle, an ellipse
4. Responses include: length, area, volume, angle measure.
5. A chord is a line segment with endpoints on a circle while a secant line is a line through two points on a circle.
6. A postulate is a statement that is assumed to be true without proof. A theorem is a statement that must be proved to be true.

Chapter Test 13-A

1.

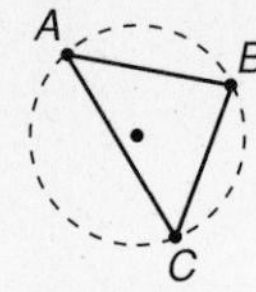

A single point; the circumcenter of the triangle.

2.

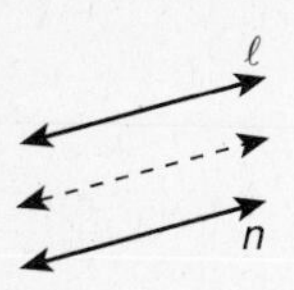

A line parallel to, and midway between, the given parallel lines.

3.

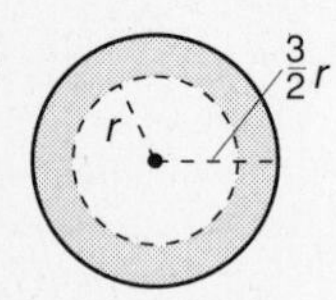

A shaded region, as shown.

4. The arc of a circle.

5.

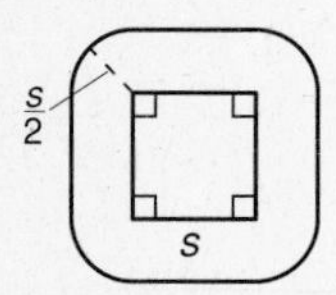

6.

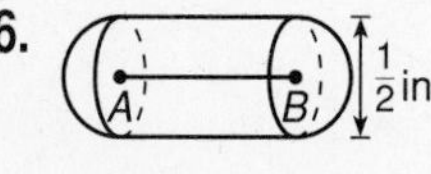

7. A line perpendicular to the plane containing the circle, through the center of the circle.

8. A circle.

9. All points in a coordinate plane that are greater than 3 units, but not more than 5 units, from the origin.

10.

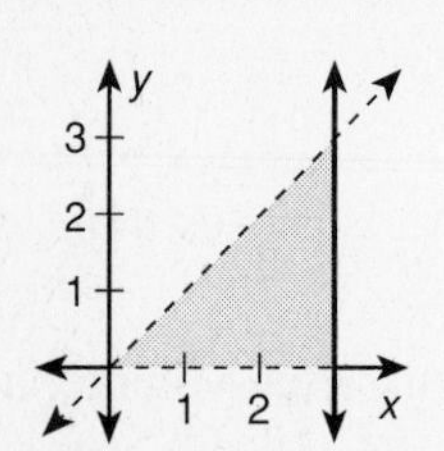

11. One point, $(5, -2)$

12.

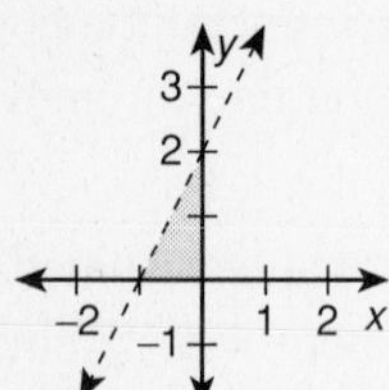

13. $x + y = 0$

14. A great circle of radius 4 in.

15. A rectangle.

16. $2\sqrt{5}$ **17.** 4

18. In a right triangle, the square of the length of the hypotenuse is equal to the sum of the squares of the lengths of the legs.

Chapter Test 13-B

1. b.	**2.** c.	**3.** a.	**4.** b.
5. d.	**6.** c.	**7.** c.	**8.** a.
9. b.	**10.** d.	**11.** c.	**12.** d.
13. d.	**14.** b.	**15.** b.	**16.** c.
17. b.	**18.** a.		

Chapter Test 13-C

1.

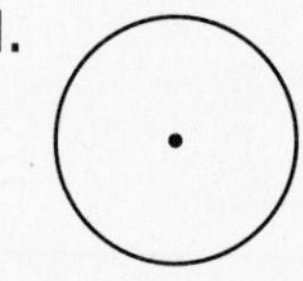

A single point; the center of the circle.

2. The points on the bisector of the angle.

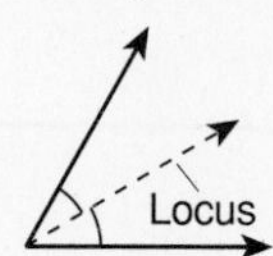

3.

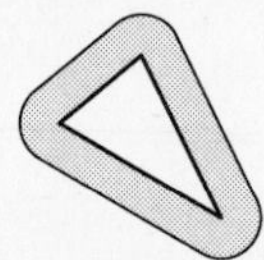

A shaded region, as shown.

4.

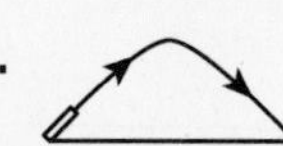

A parabola.

5.

6. A line perpendicular to the plane containing $\triangle ABC$ through its circumcenter.

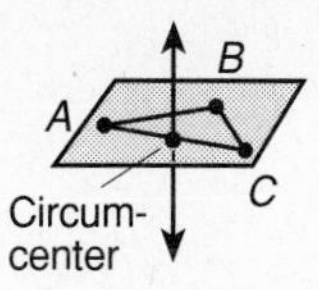

7. A line parallel to, and midway between, the given lines.
8. Two congruent circles whose centers are the endpoints of the diameter of the defining circle.
9. The intersection of loci given by two or more conditions.
10.

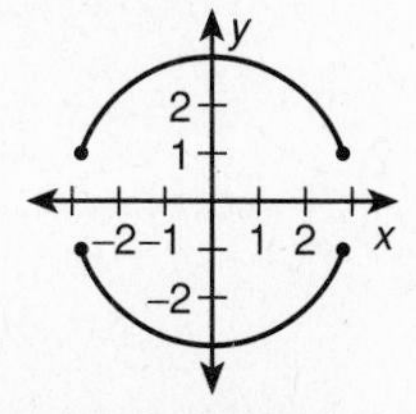

11. One point, $(-4, 7)$
12.

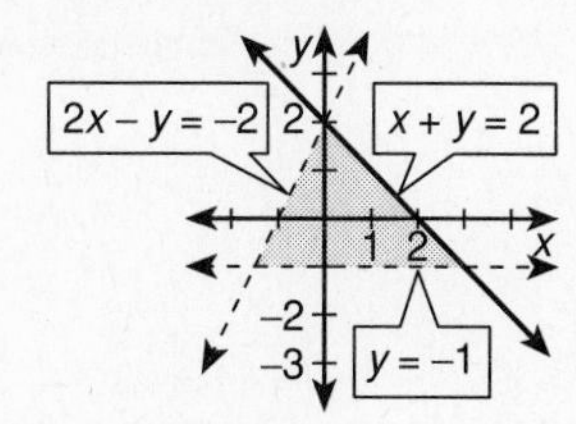

13. $3x + 2y = 6$
14. An ellipse
15. A point, a line, triangles, quadrilaterals, pentagons, hexagons
16. A triangle

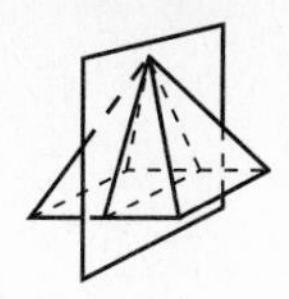

17. It is eight times greater.
18. If the square of the length of the longest side of a triangle is equal to the sum of the squares of the lengths of the two shorter sides, then the triangle is a right triangle.